Leprosy

*Bioarchaeological Interpretations of the Human Past:
Local, Regional, and Global Perspectives*

LEPROSY

Past and Present

CHARLOTTE A. ROBERTS

Foreword by Clark Spencer Larsen

University of Florida Press
Gainesville

Half of the proceeds from this book will be donated to LEPRA.

Copyright 2020 by Charlotte A. Roberts
All rights reserved
Published in the United States of America

25 24 23 22 21 20   6 5 4 3 2 1

Library of Congress Cataloging-in-Publication Data
Names: Roberts, Charlotte A., author. | Larsen, Clark Spencer, author of
    foreword.
Title: Leprosy : past and present / Charlotte A. Roberts ; foreword by
    Clark Spencer Larsen.
Other titles: Bioarchaeological interpretations of the human past.
Description: Gainesville : University of Florida Press, 2020. | Series:
    Bioarchaeological interpretations of the human past: local, regional,
    and global perspectives | Includes bibliographical references and index.
Identifiers: LCCN 2020016826 (print) | LCCN 2020016827 (ebook) | ISBN
    9781683401841 (hardback) | ISBN 9781683402251 (pdf)
Subjects: LCSH: Leprosy—History. | Leprosy—Treatment—History. |
    Leprosy—Treatment.
Classification: LCC RC154.4 .R59 2020 (print) | LCC RC154.4 (ebook) | DDC
    616.99/8—dc23
LC record available at https://lccn.loc.gov/2020016826
LC ebook record available at https://lccn.loc.gov/2020016827

University of Florida Press
2046 NE Waldo Road
Suite 2100
Gainesville, FL 32609
http://upress.ufl.edu

Dedicated to all people who have experienced leprosy in the past, all those who are living with leprosy today, and everyone in the future who may be diagnosed with this infectious disease.

I hope I have helped you all in some way by attempting to dispel some of the myths surrounding leprosy.

And thank you mum for just being you.

The often-negative experiences people with leprosy have, have had, and will have inspired me to write this book.

Contents

List of Figures ix

List of Tables xiii

Foreword xv

Preface xix

Acknowledgments xxiii

Introduction 1

1. The Biology of Leprosy Bacteria and How They Are Transmitted to Humans 20
2. How Leprosy Affects the Human Body 64
3. Past and Present Diagnosis, Treatment, and Prognosis 89
4. The Bioarchaeology of Leprosy 127
5. The Bioarchaeological Evidence of Leprosy 191
6. Reconstructing the Origin, Evolution, and History of Leprosy 281

Conclusions: A Future for Leprosy; Clinical and Bioarchaeological Perspectives 303

Appendix 1. Full List of Acknowledgments 311

Appendix 2. Questionnaire about Leprosy 319

Appendix 3. Skeletons from Archaeological Sites in the British Isles with Leprous Bone Changes 323

Appendix 4. Archaeological Hospitals, Including Leprosaria, Where Skeletons with or without Leprosy Have Been Excavated 329

Appendix 5. Useful Websites and Organizations 331

Notes 333

References 335

Index 409

Figures

I.1. Ripon "leper" chapel   5
I.2. Global map of total new "cases" detected   12
1.1. Gerhard Armauer Hansen   24
1.2. The origin and spread of leprosy using DNA data   29
1.3. Droplets being expelled from the mouth   33
1.4. A nine-banded armadillo   59
2.1. Ridley and Jopling's leprosy classification system   67
2.2. Skin lesions in advanced untreated lepromatous leprosy   72
2.3. Main nerve trunks affected in leprosy   73
2.4. Enlarged nerve next to the ear of a person with leprosy   74
2.5. Enlarged nerve in the ankle area of a person with leprosy   74
2.6. Sensory nerves affected by leprosy in the upper limbs   76
2.7. Sensory nerves affected by leprosy in the lower limbs   77
2.8. Damage caused by repeated injury to an anesthetic fingertip   78
2.9. How an ulcer develops in an anesthetic foot   78
2.10. Radiograph of hand bones showing loss of bones   79
2.11. Ulceration of the foot in a person with leprosy   80
2.12. Radiograph of the foot of a person with leprosy   81
2.13. Radiograph of the hand of a person with leprosy   82
2.14. Flexion deformity of the fingers of a person with leprosy   83
2.15. Radiograph of the foot of a person with leprosy showing flexion deformity   83
2.16. Facial palsy as a result of leprosy   84
2.17. The effects of lagophthalmos caused by leprosy   84

3.1. Destruction of the nasal bones as a result of leprosy   92
3.2. Shoes being fitted for a person with leprosy   101
3.3. An example of modern shoes used to protect the feet in leprosy   101
3.4. Image of person with leprosy by Johan Ludvig Losting   109
3.5. Leprosy being diagnosed by a committee   111
3.6. Cauterization points used for treating people with leprosy   119
3.7. Evidence of amputation of the foot   119
4.1. Skeleton diagram showing the regions most affected in leprosy   130
4.2. Loss of the anterior nasal spine and remodeling of the nasal aperture   134
4.3. Absorption of the alveolar process of the maxilla   136
4.4. Porosity of the nasal surface of the palate   137
4.5. Perforation of the oral surface of the palate   138
4.6. Porosity and destructive lesions of the oral surface of the palate   139
4.7. Teeth showing leprogenic odontodysplasia   144
4.8. Cup-and-peg deformity of foot bones   147
4.9. Cup-and-peg deformity and palmar grooves   156
4.10. Hand phalanges showing palmar grooves   157
4.11. Figure showing dorsal tarsal "bars"   157
4.12. Bone formation on dorsal surfaces of the tarsals   157
4.13. Diaphyseal remodeling of foot phalanges   161
4.14. Sharp-edged remodeling of a metatarsal   161
4.15. Enlarged nutrient foramina in hand phalanges   161
4.16. Bone changes on the tibia suggesting an overlying ulcer   165
4.17. Periosteal new bone formation on the lower leg bones   165
4.18. Periosteal new bone formation on the forearm bones   165
4.19. Imprint of blood vessels in new bone formation   165
4.20. Developmental defect of a first foot phalanx   168
4.21. Osteochondritis dissecans in a tibia   168
4.22. Porosity in the eye sockets   169
4.23. Destructive lesions of bones of the middle ear   172
4.24. Frontal bone groove   174
4.25. Lepromas on the head of a person with leprosy   174
4.26. Active bone formation   186

4.27. Healed bone formation 186
5.1. Evidence of leprosy around the world 198
5.2. European evidence of leprosy 199
5.3. Distribution of British and Irish archaeological sites with leprosy 200
5.4. Upper jaw with evidence of leprosy, Gloucester, England 203
5.5. Skeleton with leprosy, Beckford, England 204
5.6. Plan of the grave of the person with leprosy in Figure 5.5 205
5.7A. Mycolic acid profile of *M. leprae,* Winchester, England; skeleton 2 208
5.7B. Mycolic acid profile of *M. leprae,* Kiskundorozsma, Hungary; skeleton KD 517 209
5.8. Site plan of the burials at late medieval Blackfriars, England 211
5.9. Reconstruction of the late medieval site of Brough St Giles, England 212
5.10. Plaster cast of Robert the Bruce's skull 214
5.11. Notifications of leprosy by ethnicity in England and Wales, 1950–1997 215
5.12. Møller-Christensen's line drawings of the alveolar process of the maxilla and oral surface of the palate 219
5.13. Møller-Christensen's line drawings of the anterior nasal spine 220
5.14. Møller-Christensen's line drawings of the lower leg bones 221
5.15. Locations of Danish archaeological sites with evidence of leprosy 225
5.16. Hungary: Plan of the Szarvas cemetery in Hungary showing location of the burial of a person with leprosy 231
5.17. Bedroom in the Leprosy Museum, Bergen, Norway 236
5.18. Frequency of leprosy in Norway, 1857–1925 238
5.19. Relative incidence rate of leprosy in Norway, 1851–1920 239
5.20. Average incidence rates of leprosy in Norway, 1851–1920 239
5.21. Average relative air humidity in Norway in July, 1874–1913 240
5.22. Alveolar process of the maxilla of the skull of a "Comox Indian" from Vancouver Island, Canada 270
5.23. Nasal surface of the palate of a "Comox Indian" from Vancouver Island 270

5.24. Oral surface of the palate of a "Comox Indian" from Vancouver Island   270
6.1. Graph of founding dates for leprosy hospitals in Britain   295
6.2. Cemetery plan of the distribution of burials of people with leprosy, St Mary Magdalene, Chichester, England   301

Tables

I.1. Registered prevalence and number of new "cases" detected during 2018 by World Health Organization region   12
2.1. Types of leprosy   68
2.2. Skin lesions in different types of leprosy   70
2.3. WHO classification of leprosy-related impairments   73
2.4. Causes and effects of injuries   78
2.5. Effects of *M. leprae* on the motor nerves   82
3.1. Eight of the most convincing signs documented for all forms of leprosy by twelve authors dating from the first–second centuries AD to the sixteenth–seventeenth centuries AD   108
4.1. Studies of leprosy using ancient biomolecular analysis   181
5.1. Bioarchaeological evidence for leprosy in Hungary   229
5.2. Archaeological sites where skeletons have been buried "differently"   278
5.3. Some scenarios for leprosarium and non-leprosarium burial contexts where skeletons with and without leprosy have been found   279
6.1. Ancient DNA analysis of archaeological human remains revealing strain data   289
6.2. Skeletons with both leprosy and tuberculosis   298

Foreword

There is no doubt that references to leprosy in the media and in published literature bring to mind a trove of horrors and body disfigurations that accrue for the person having the disease, both now and in the past. The reference might also raise the specter of an infectious disease that is highly contagious and results in transformations of especially the face, hands, and feet. Historically, and more often than not, people with leprosy have been described as being relegated to a life of isolation. This superb volume by Charlotte Roberts succinctly presents the record of the bioarchaeology (including paleopathology), history, and epidemiology of this long-known disease that successfully dispels the myths and mythology of leprosy. Roberts points out that the myths are clearly wrong and are based on fiction and incorrect perceptions and not facts. Roberts counts ten myths that are the most egregious, underscoring the remarkable degree of misunderstanding about the nature of the disease by all but a small number of authorities in medicine, bioarchaeology, and history. In this important book, Roberts addresses the science of the disease drawn from modern research on the epidemiology and pathology, anthropology, bioarchaeology and history of leprosy and key facts that together dispel these myths and develop and provide a broader understanding of the disease via new perspectives.

The book builds on over thirty years of Roberts's research informed by new science and new discoveries and the research of others who have investigated the bioarchaeology of leprosy. I have followed Roberts's work on leprosy, especially beginning with a conference and subsequent monograph that she coedited with Mary E. Lewis and Keith Manchester (Roberts et al., 2002). The volume was a watershed, particularly in the study of the paleopathology of leprosy; it focused on the archaeological, historical, paleopathological, and clinical approaches to the disease by informed authorities from the fields of bioarchaeology, palaeopathology, and the allied sciences. The edited volume was a milestone in ongoing discussions involving multiple disciplines. That volume had a relatively low profile; it was unknown to most except those of us who may have dug deep in tracking down an informed discussion of the disease.

The current book stands apart from other treatments by offering a succinct discussion of the range of clinical, historical, archaeological, and anthropological findings on leprosy. In this regard, the book was no doubt a difficult one to write, not only with respect to conveying to the reader just how fragmented the knowledge base and presentation of the disease and the literature in general are in the twenty-first century but also with respect to broadening the context and perspective from the country-by-country treatment so characteristic of the medical history of leprosy. This book pulls together the various and complex lines of evidence drawn from the range of disciplines having an interest in the disease. Roberts evaluates the record and tells the reader what the "package" of leprosy entails for those who are infected now and for those who were infected in the past, thereby removing the specter that leprosy has built up over hundreds of years.

The book starts the reader off with a word that is often used when people with the disease are described. Right up front, Roberts points out the term often used for those infected—"leper"—is a misnomer. Although it is used widely in the media as a derogatory term and by some people in history, bioarchaeology, and other sciences, it incorrectly characterizes the new identity that has been developing for people with leprosy in recent years. At long last, we have a book that lays out the context and complexities of leprosy, the nature of the infection, the lived experience of the people and communities affected today, its origin in the Old World, its geographic distribution past and present, how the disease affects the body and how it is recognized in the skeleton, and the fascinating story of its study by bioarchaeologists and anthropologists. A large degree of credit for the bioarchaeological part of this compelling book goes to Danish physician and paleopathologist extraordinaire, Vilhelm Møller-Christensen (1961), who provided the first systematic investigation of the paleopathology of leprosy and its context with his contextualized study of skeletal remains from medieval cemeteries in Denmark. His work should always be the starting point for studies on the bioarchaeology of leprosy. Finally, Roberts gives comprehensive treatment of the range of circumstances that accompany the evolution of the pathogen and its impact and the role of other infectious diseases in its history, including the Black Death and tuberculosis in the medieval era.

Admittedly, some of my own impressions of leprosy were incorrect, likely because my first exposure to the topic was in primary school and no one was around to tell me otherwise. Clearly, this is one of the most misunderstood diseases, especially by the public and the general readership of the history of disease. Roberts's book goes a long way toward reshaping the understanding of leprosy based on science rather than myth.

Clark Spencer Larsen
Series Editor

REFERENCES CITED

Møller-Christensen, Vilhelm. 1961. *Bone Changes in Leprosy.* Copenhagen: Munksgaard.

Roberts, Charlotte A., Mary E. Lewis, and K. Manchester, eds. 2002. *The Past and Present of Leprosy: Archaeological, Historical, Palaeopathological and Clinical Approaches.* BAR International Series 1054. Oxford: Archaeopress.

Preface

This book has taken some time to write and this is the fourth iteration of it. It was started in late 2006 with a two-year Leverhulme Trust Fellowship. During that time, Dr. Tina Jakob took the lead in teaching palaeopathology in the MSc program at Durham University, for which I am grateful. A year's research leave was also awarded by Durham University in 2011–2012 and for a term in 2016, and Dr. Becky Gowland then took the reins of the MSc, again acknowledged.

I am sure that many will wonder why it has taken so long to complete this volume. Some know the reasons, but others can be less forgiving in their opinions. It is fitting that I have at last written about leprosy, a disease that had no cure at one time and that was described as the "living death." My mum was diagnosed with leukemia in 2005, a blood cancer that for her had no cure and one that is usually not the cause of death; this diagnosis came shortly before her husband died. Living on her own for the first time, she needed constant support, which I willingly gave. After her diagnosis, she was treated for clinical depression until she died in May 2012. In 2011 she was also diagnosed with dementia, specifically Alzheimer's disease, another disease that has as yet no cure and that I would argue is the living death. It is also stigmatized and people become isolated, as my mum found. This association between Alzheimer's and isolation is becoming more recognized as more people are diagnosed. A person with dementia does not fit with how we expect a person to be. Thus started a mind-numbingly tortuous journey for both of us, me for the first time learning about this devastating disease that our health and social care services are struggling to deal with and she becoming more and more incapacitated as a result of a disease she had always feared. I had to learn fast, but I wish I had had the chance to read *Where Memories Go: Why Dementia Changes Everything* (2014), by Sally Magnuson, before my mum was diagnosed. What I now know is that you cannot care effectively for a person with dementia while having a full-time job, even when carers are employed.

While I was mum's sole carer and also a full-time wholly dedicated and extremely hardworking academic, it was incredibly hard to keep my mind on tasks in hand. Some say I should have taken time out, but allowable time is very short for circumstances such as this, and continuing to work (being distracted) meant I was not constantly worrying about mum. In the years since her (thankfully peaceful) death, which was a blessing, I have picked myself up and started to focus back on this book, but eventually I needed bereavement counselling. Nevertheless, the stresses and demands of being an academic have meant that it has not always been possible to place the book at the top of my list of things to do; major service responsibilities in academia, such as being a member of a subpanel for the UK Research Excellence Framework in 2014 (aka Research Assessment), have meant that virtually another year was taken out of my life and my research in 2017–2018. Thus, I apologize to readers who were impatient and did not understand, but these are the main reasons why this book is late in arrival. However, I do appreciate the many colleagues who have been very supportive and understanding of my situation.

Why did I write this book? I started working on leprosy in 1983 after I was appointed as a research assistant to Keith Manchester's project at the University of Bradford on the antiquity of leprosy and tuberculosis that was funded by the Science and Engineering Research Council and later by the Wellcome Trust. I owe my initial interest in the topic to Keith. The first task for the project was to learn about the bone changes of leprosy, and I was fortunate to travel with Keith to Copenhagen and meet Vilhelm Møller-Christensen to learn about the bone changes in skeletons from the medieval leprosarium of Naestved. Since that time I have worked on and off on the bioarchaeology of leprosy and of course have taught many students about it. As a trained nurse, I was fascinated by this "social" disease and have made an effort to visit leprosy hospitals in Nepal (Anandaban) and India (Bandorawalla) and on islands where people with leprosy have been "incarcerated" in the past (e.g., Spinalonga, Crete, and Molokai, Hawaii, USA).

I have benefited from many discussions with colleagues in different disciplines, and I always wanted to write a book on leprosy. I started writing about the disease when I coedited the proceedings of the 1999 Leprosy Congress at the University of Bradford, *The Past and Present of Leprosy* (2002), but the pinnacle of my career in leprosy was attending the International

Leprosy Congress in late 2013 in Brussels, Belgium. What an inspiration it was to see the dedication of people working for and with people with leprosy today. People presented 265 papers and 500 posters over three days, and eight hundred fifty delegates attended. I learned so much from that conference. I was indebted to delegates there for talking to somebody not on the front line of managing people with leprosy today, and I am especially grateful to David Scollard (Louisiana, USA), who took time out to consider burning questions I still had even after all these years. Having worked on leprosy for so long, I also came to realize that people did not have much understanding of it, and I wanted to change that. I was able to attend the same conference in Beijing in September 2016.

So here it is, and any errors are all mine. I should state that I have mainly relied on English-language sources and a limited number of non-English sources that have been translated by helpful students and colleagues. I apologize in advance if I have missed key information. I have also used websites sparingly and have focused on those that I believe are more reliable than others. While I have tried to keep abreast of newly excavated skeletons with leprosy, it has not been easy because so much of that information is in gray literature and studies are published in such a wide variety of places. I do not profess to understand fully all of the material included here because I am primarily a bioarchaeologist, but it was necessary to include it all to fulfill the ultimate aim of this book: to dispel the myths of leprosy. I feel privileged to have been able to work on this disease and to experience people with leprosy in Asia. I hope that dispelling the myths may help people with leprosy in the future in some small way.

Finally, and most importantly, I am as always indebted to my partner, Stewart Gardner, who constantly and unconditionally supports me in my work and who helped unfailingly with my mum (and more recently with my aunt's failing health). Many academics receive support from their partners that enables to them to actually do their job, and without Stewart I would not be anywhere near the position I am in academia today.

Acknowledgments

Writing a book from multidisciplinary and holistic perspectives requires lots of help from many people. I am most grateful to the people and organizations listed in Appendix 1. Everyone has been so generous with their time. However, I would like to thank the following people, organizations, and dogs:

First, if the Leverhulme Trust had not awarded me a Senior Research Fellowship to write this book, it would never have got written. Second, I am most grateful to Keith Manchester (University of Bradford, UK) for highlighting this interesting disease to me back in the 1980s. Third, the University of Florida Press should be applauded for being patient but also incredibly helpful and encouraging as this book has gone through several versions. The two people who should be most thanked are Kate Babbitt, for helping in developing the manuscript through three versions to what you see now, and Meredith Morris Babb, who had unending faith in me when times looked bleak. I must not forget the very helpful reviewers too. I thank Durham University for research leave during 2011–2012 and for a term in 2016. I also thank my PhD and MSc Palaeopathology students over the years, who have been kind to leave me alone when I have needed space to work but who have also inspired me greatly at times!

Of course, books do not get written without the support of family. My partner Stewart, especially, has been amazing and has held the fort for a very long time. He will probably know nearly as much about leprosy as I do! My late mum was always a great support too and always seemed to have faith in me. Animals too come into the equation. Thank you to our late dogs Cassie and Joss and to our current dog Misty. Finally, many thanks to my friends and colleagues who have stuck by me through the hard times, especially friends in my local community in the Yorkshire Dales.

Introduction

> Leprosy is a gravely misunderstood disease. . . . Even today clinical knowledge of leprosy runs far ahead of social acceptance and individual understanding. For the vast majority the terms leper and leprosy still produce images of people living apart from society, in . . . colonies, where blindness and deformity are rampant and where approach to the outside world is announced with the cry of unclean, unclean.
>
> Hudson and Genesse (1982, 997)

Leprosy is an infectious disease of humans that is caused by *Mycobacterium leprae,* a slow-growing bacterium that is hard to contract, and the more recently discovered *Mycobacterium lepromatosis.* The type of leprosy a person has depends on the strength of their immune system. A high resistance to the bacterium leads to tuberculoid leprosy and a low resistance to lepromatous leprosy. Leprosy is curable and therefore should be a disease of the past. However, it is still little understood by most people, even in the information age. This may explain why it remains a stigmatized infection. Thankfully, in recent years, scholars have been evaluating how the history of leprosy has led to misunderstanding, sensationalism, and misrepresentation. Leprosy remains an infection surrounded by myths that have developed over the hundreds and even thousands of years it has been present in the world. It is viewed as declining in frequency today, although it is still present in many communities around the world. Leprosy can be considered a "special" disease in clinical medicine because of the socioeconomic consequences related to its propensity to disfigure, generate stigma, and severely affect the lives of people who have it.

A Neglected Disease

People often think that leprosy is not a major concern today, unlike malaria, HIV, and other infectious diseases. However, along with sixteen other diseases, leprosy is classed as a neglected tropical disease. In May 2013, the

World Health Assembly adopted a resolution about this group of diseases that urged member states to

> ensure country ownership of prevention, control, elimination and eradication programs;
> expand and implement interventions and advocate for predictable, long-term international financing for activities related to control and capacity strengthening;
> integrate control programs into primary health-care services and existing programs;
> ensure optimal program management and implementation;
> achieve and maintain universal access to interventions and reach the targets of the roadmap.[1]

To a large extent, this resolution is being adopted around the world. However, it has proved difficult to raise the funds needed to fully manage leprosy and its consequences for the people who have it; fund-raising for other infectious diseases such as malaria and HIV is much easier. Leprosy is also less frequently portrayed in the media and is the focus of less research compared to other diseases. As I was writing this book, the monthly feeds from the PubMed website often showed that six times more papers were published on tuberculosis, another mycobacterial disease, than on leprosy. This is probably largely because leprosy is seen as a declining disease. Tuberculosis, in contrast, is seen as a reemerging disease, one of the many we see in the third epidemiological transition we find ourselves in, when infectious diseases are becoming more resistant to antibiotic treatment. Other researchers have documented this decline in research on leprosy (e.g., Cairns Smith 1996). Schoonbaert and Demedts (2008), who also relied on the PubMed database, found that although 19,201 articles were published on leprosy over the period 1950 to 2007, the number of papers published each year peaked in the 1950s. As might be expected, a large proportion of these were written by authors in places where leprosy remained a challenge. Four journals, the *International Journal of Leprosy, Leprosy Review, Indian Journal of Leprosy*, and *Leprosy in India* published over a third of the studies Schoonbaert and Demedts identified.

Leprosy has often attracted attention in academia from medical historians and bioarchaeologists. The former group usually works with descriptions and depictions of leprosy in texts and artwork while the latter group documents evidence for leprosy in archaeological human remains (see chapters 5 and 6 of this volume). Medical historians tend to focus on evidence in particular countries or during specific periods of time or on certain aspects of the history of leprosy, such as diagnosis and treatment. In bioarchaeology, most work has been on individual skeletons that have revealed evidence of leprosy from archaeological sites in a specific place and time period. Some have also written about the diagnostic criteria used to recognize leprosy in human remains and about the social aspects of people's experience of having leprosy in the past.

Goals of the Book

Published work on leprosy is quite fragmented. The main purpose of this book is to synthesize knowledge about it from a variety of disciplines. In this book, I consider past knowledge of leprosy in light of current understandings. My primary goal is to critically evaluate what we know today in order to better understand the experience of people with leprosy in the past. As I was doing the research for this book, it became clear to me that many people had a limited understanding of leprosy. Thus, a second goal of the book is to dispel myths about leprosy. Finally, I hope that the contents of this book will better inform the wider public about this infection and ultimately help those who have leprosy have better and more fulfilling lives than many do today.

A good starting point for a book about leprosy is a list of the myths about the disease. Unfortunately, many people accept these as facts and they have become ingrained in societies around the world. I found the following ideas to be the most prevalent among beliefs about the disease:

Leprosy is easy to contract.
Leprosy can be passed from one person to another rapidly.
Leprosy can be transmitted via sexual intercourse or by touching.
Leprosy can be inherited.
Leprosy cannot be cured.
Leprosy is described in the Bible.
When a person has leprosy, the fingers and toes "drop off."
Leprosy is a tropical disease.
Leprosy is not a problem for people today.
In the past, all people with leprosy were segregated from society.

The chapters that follow will explore these statements in the context of recent knowledge and understandings. As a starting point, and as there did not appear to have been much research done on knowledge of leprosy in the West, where the frequency of the disease is low, I conducted a survey of people in parts of the English population to assess their knowledge of leprosy. I collected data relating to these myths with a questionnaire that 270 people completed in 2012 and 2013 (see Appendix 2). Most people who filled out the questionnaire knew that leprosy is an infection, that it is most frequently seen in Asia, that it affects the nerves, and that it is curable. There were mixed understandings about what pathological organism causes the disease, how it is contracted, what the key predisposing factor was, whether fingers and toes "drop off," and how people with leprosy were treated in the past. The overwhelming majority thought it was described in the Bible (nearly 80 percent), but see chapter 3. As Weymouth (1938, 22) said over eighty years ago, "Leprosy, to the average man, is merely a disease referred to in the Bible, and now, fortunately, no longer with us." This belief is clearly still prevalent today, although leprosy is indeed still with us. The data I collected from this survey show that much work needs to be done to improve the general public's knowledge of leprosy. Graciano-Machuca et al. (2013) made a similar point after conducting a survey of university students in India.

The Word "Leper"

In this book, I refrain from using the word "leper" except when referring directly to a quote from another source. When I do use it in my own writing, I put it in quotation marks to emphasize that I do not regard the word as useful or helpful for describing people with leprosy, past or present. Historian Luke Demaitre (2007, xii) also avoided this word: "The word *leper* has been shunned consistently—except when it is part of a modern quotation—and consciously because it amounts to a slur (in English notably more than in other Western languages)." In his book *Images of Leprosy*, Boeckl (2011, 7) avoided the word "except in quotations from scripture and other sources" (see also Brenner and Touati forthcoming). I learned from my questionnaire that most respondents associated the word "leper" with negative words and ideas. Definitions of "leper" provided in freely accessible online websites, the websites that people seeking information and definitions might visit, fit well with the data I collected with the questionnaires. This evidence shows that among the general public, not much has changed over the past few decades; Silla (1998, 10) also noted that people in the Western world view leprosy as a metaphor for "deformity and stigma or a curiosity from the biblical/medieval past." Further, it is often the case that authors of clinical papers talk more about the disease or the affected part of the body than about the person with the disease. A prime example of this is the article by Rohatgi et al. (2016); they titled it "The Story of a Deformed Leprous Foot." Those who objectify a person with leprosy seem to suggest that the disease is a more important entity than the person's identity (which of course includes characteristics beyond their infection). My hope is that others, especially those working in bioarchaeology, will also abandon the word "leper" in the future.

However, the signs are not good, as I noticed in September 2016 at a bioarchaeology conference. The word also continues to be used in clinical medicine (Grzybowski et al. 2016), in bioarchaeology (Magilton, Lee, and Boylston 2008; Roffey and Tucker 2012), in the media, and even on signposts (see Figure I.1). Indeed, an almost contradictory footnote appeared in a bioarchaeology paper regarding the word:

> The term "leper" is employed throughout this article in the context of its traditional use in historical sources and previous scholarly works, and as a term peculiar to the Medieval and pre-modern periods. In this sense it refers to individuals affected, or perceived as having, leprosy (as applied to the anachronism "Hansen's Disease"). It is acknowledged that the term has been used pejoratively in modern contexts and it is one of the aims of this article to challenge the root of such misconception. (Roffey and Tucker 2012, 170)

Figure I.1. Ripon "leper" chapel. Photo by Charlotte Roberts.

I have heard the word "leper" used in the media regularly. For example, in 2015, a presenter on BBC Radio 4's *Today* program said that a particular politician might as well be a "leper" in the Middle Ages and a Sunday *Times* magazine reporter, with no thought or reflection, glibly described the appeal of having children to be analogous to having leprosy.[2] In popular culture, many musicians and bands have recorded songs with titles that include the word "leper," for example REM ("New Test Leper"), Weird Al Yankovic ("Party at the Leper Colony"), Dimmu Borgir ("Lepers among Us"), Metallica ("Leper Messiah"), and Frightened Rabbit ("The Modern Leper"). Examples of negative treatments of people with leprosy include *The Thankful Leper* (Hinkle 2008) and *The Grateful Leper: Tales of Two Birds* (Forson 2012). In *The Man with Leprosy* (McDonough 2007), a children's book, nobody is pleasant to the title character. Although the book purports to be a resource for engendering sympathy and caring for those who are sick, the book uses the term "leper" extensively. Leprosy has also been depicted in pejorative ways in films such as *Ben Hur* (1959), *Papillon* (1973), *The Life of Brian* (1979), *The Fog* (1980), and *Braveheart* (1995). Of particular interest is the animated film *Pirates!* (2012), which originally had a scene where the "arm of somebody with leprosy fell off." Following campaigning by various interest groups such as LEPRA, Aardman Productions withdrew the scene.[3] In other contexts, festivals such as the Leper Festival in Taddiport, Devon, England, promote the use of the word "leper," even though it raises money for LEPRA.

Why is it that the word "leper" is still used today and often in a derogatory way? At the International Leprosy Congress in Havana in 1948 and in Madrid in 1953, "it was agreed that the word 'leper' is offensive and opprobrious" (Kalisch 1973, 481). As Richards (1977, viii) noted nearly forty years ago, "the term 'leper' is now deprecated by those who have devoted their lives to the relief of suffering from this disease, and by the World Health Organization. . . . [While] yesterday's victims of leprosy were 'lepers,' with all that the name implies[,] today's and tomorrow's are 'leprosy patients.'" However, he was also of the opinion that not using the word would "diminish the significance of events" (ibid.). I do not agree. "Leper" is not seen very often in the clinical literature today and clinicians refrain from using it. Indeed, at the International Leprosy Congress in Brussels in 2013 and again in Beijing in 2016, the word "leper" was used only rarely in the context of presentations (and then usually in the context of leprosy's history). When it was used, comment was swiftly made with a reminder that it was inappropriate. The phrases used at the congress to refer to people with leprosy included leprosy patient, person with leprosy, people affected by leprosy, leprosy-affected patients, persons and people, leprosy cases, Hansen's disease sufferers, multibacillary sufferers, paucibacillary sufferers, and individuals with leprosy. However, and quite rightly, one person with leprosy at the congress suggested that calling somebody with leprosy by their actual name was the most appropriate thing to do. It is important to describe people experiencing leprosy as "a person with, or affected by, leprosy," thus emphasizing the person rather than the infection.

Experiences of leprosy have been thus described using many terms and phrases. "Hansen's disease" (named after Gerhard Armauer Hansen, the Norwegian doctor who discovered the bacteria responsible for leprosy) is used in some parts of the world in an effort to avoid the stigma associated with the words "leper" and "leprosy." This is especially so in Brazil, where the government passed a law in the late 1980s that made it illegal to use the term "*lepra*" in government documents; the suggested term was "*Hansieníase*" ("Hansen's disease") (White 2005). White found differences in how Brazilian patients understand the terms "*Hansieníase*" and "*lepra.*" Seven of forty-three patients did not associate the two terms, but some thought that the purpose of using the term "*Hansieníase*" was to conceal the truth of the disease the person actually had. Nevertheless, health care workers felt that using "*Hansieníase*" and saying that the disease was curable was the best solution because "*lepra*" was the term that had been used in Brazil before leprosy could be cured. Recent discussion about what to call this infection on the Leprosy Mailing List shows that this debate is far from finished.[4]

STIGMA AND LEPROSY

The association of stigma and leprosy is very important in any consideration of leprosy, both past and present. As Grange and Lethaby say, "a particular stigma has become associated with the disease, and those suffering from it have often had to endure the additional burdens of ostracism and isolation. Sadly, this stigma has led to many patients avoiding medical attention until after irreversible damage has occurred" (2004, 271). Three main groups of people may be stigmatized: those with physical deformities, especially of the face, those with blemishes of character (e.g., a history of a mental disorder, drug addiction, or epilepsy), and those with a "tribal" stigma of "race," nation, or social class. All three can be encompassed in the experience of leprosy today, and stigma is generally a common problem for people with skin diseases (Dimitrov and Szepietowski 2017).

Of particular importance is how the term "leper" attracts stigma. "Leper" suggests the idea of being rejected by society. As knowledge about leprosy has grown since the discovery of the leprosy bacillus in the late nineteenth century, people's understanding has changed. Leprosy is now curable and many affected people are no longer isolated from their communities. There are also official guidelines now for reducing stigma. The International Federation of Anti-Leprosy Associations (ILEP) stresses that "leper" is a derogatory, outdated, and offensive word and that it is often associated with someone who has been rejected, ostracized, or regarded as an outcast.

Each person who contracts leprosy will experience it differently, and concepts of leprosy, stigma ("leprophobia"), and "disability" are important for understanding the experience of people with this infection, both today and in the past. While stigma directed at people with leprosy is very much driven by a misunderstanding of the infection and often by fear, a person with accompanying disability can become a target for stigma and ostracism. The level of stigma can vary according to the severity of impairment and disfigurement (e.g., see Roosta et al. 2013). Stigma is described as a component of the many factors affecting quality of life for people with leprosy (Lustosa et al. 2011), and it is clear that education of people is key to ending stigma. While so much is known now about *M. leprae*, for example through the sequencing of its genome, it can be difficult in the developed world to understand that other parts of the globe can view it differently. Disease treatment in Western societies today is based on current medical thinking and knowledge, and while the public perception of the intricacies of disease causation can vary considerably in non-Western societies, there is yet another layer of "knowledge" based on traditional beliefs.

In understanding concepts of disease, studies in medical anthropology play a crucial role in showing how different contexts shape belief systems (McElroy and Townsend 2009; Wiley and Allen 2013). Studies in this field also represent people who may live similar lives to those who lived in the past, disregarding the gaps in time and space. Concepts of leprosy in different communities are relevant to the control of leprosy and to whether stigma develops and how people with leprosy were and are treated.

How Frequent Is Leprosy and Where Does It Occur?

Over thirty years ago, the World Health Organization (WHO) noted that an appreciable proportion of patients with leprosy develop physical impairments and that it was the most stigmatized disease in many countries (World Health Organization 1985). This is still true today. Even though many countries still carry a high "burden" associated with leprosy, in many countries, "leprosy . . . struggles to stay high in the political agenda" (World Health Organization 2016, 6). While over 16 million people have been "cured" of leprosy since 1991 (World Health Organization 2013, 2016), many still endure the debilitating physical and mental effects of this infection. This is especially so if they have not been diagnosed or treated or if treatment was given too late in the progression of the disease. The stigma associated with leprosy still prevents much self-reporting, diagnosis, and early treatment. Thus, changing the image of leprosy is key to success in managing the infection (International Leprosy Association and Netherlands Leprosy Relief Association 2011a–d).

Historical Reports of Frequency

Nearly forty years ago, Fine (1982) reported that 11 million people were affected by leprosy. "Cases" were recorded mainly in tropical and subtropical regions. He highlighted seven zones:

- The Old World tropical and subtropical areas (90 percent of people with leprosy worldwide lived in sub-Saharan Africa and southern Asia).
- The Mediterranean Basin, where prevalence was generally low.
- Northern Europe, where the prevalence had declined precipitously. The last recorded person with leprosy in Britain was in 1798 on the Shetland Islands off the north coast of Scotland. In Norway, although about 8,000 "cases" were registered in the latter half of the nineteenth century, the last known native "case" of leprosy was recorded in 1950 (Irgens 1980).[5]
- Northern Asia, where leprosy was present in parts of Russia and China, although its extent and distribution was unknown.
- South and Central America including the southern United States. It is generally believed that leprosy was not present in these regions until contact with Africans, the Portuguese, Spaniards, and the French began in the late fifteenth century AD but that it remained endemic with a low prevalence (Fine 1982, citing Motta 1981).

Northern United States and eastern Canada, where French, German and Norwegian immigrants introduced leprosy in the eighteenth and nineteenth centuries. It persisted in certain places and in certain family groups, then disappeared after a few decades (Aycock 1940; Feldman and Sturdivant 1976).

Pacific Islands and Australia. Leprosy was introduced in Hawaii in 1820 and in Australia in the mid-nineteenth century (Wade and Ledowski 1952; Humphrey 1952; Worth 1963).

Fine also noted the variety of types of leprosy affecting regions, likely because of genetic and environmental factors and cultural differences. Rates of (low-resistance) lepromatous leprosy were higher (above 20 percent) in Europe and the Americas than in Asia (5–20 percent). This type of leprosy was the least frequent in sub-Saharan Africa (< 5 percent), possibly because hypopigmentation in the skin was more easily seen on darker skins and people were diagnosed there during the early stages of the disease. However, the World Health Organization (1985) emphasized that the same standard diagnostic methods had to be used in all countries if frequency rates were to be compared reliably on a global scale.

Although several studies have suggested that leprosy was a rural disease, few datasets are available to test that hypothesis. In areas where leprosy had the lowest rates of prevalence there was obvious clustering, for example in Norway in the nineteenth century (Irgens 1980) and among both blacks and whites in some parishes in the state of Louisiana in the United States (Feldman and Sturdivant 1975). Fine noted that studies of clustering had traditionally focused on the household or the family but had not standardized for household size, age, or familial relationships. The main problem in the early 1980s, Fine believed, was "sorting out the extent to which such clustering is due to shared environment, shared genes, or contact with infectious cases" (Fine 1982, 172). This remains a challenge today.

In 1985, the World Health Organization reported that it was difficult to estimate the number of people affected by leprosy because reporting was incomplete and irregular and many people with leprosy were not recognized and/or were misdiagnosed. It estimated that the total number of people with leprosy was 11.5 million, although only 5.3 were officially registered, likely because the disease had not been diagnosed for various reasons. Four million of the registered "cases" were reported from India. It also reported that leprosy was most prevalent in the tropical and subtropical areas of Africa and southern Asia, which was "considered to be the original source of leprosy" (World Health Organization 1985, 11–13; see also Monot et al. 2005).

The 1991 World Health Assembly

In 1991, the World Health Assembly, the World Health Organization's governing body, adopted a resolution to eliminate leprosy as a public health problem by the year 2000 (prevalence of less than one "case" for every 10,000 people at the global level). By 1992, the number of registered "cases" had declined to 2.7 million, but that was probably only half of the real number (Noordeen 1993). By 1996, the number was 1.3 million (World Health Organization 1998a; Fine and Warndorff 1997). Multidrug therapy is probably the reason for this sharp decline. By the end of 2000, "elimination" of leprosy had been achieved. The World Health Organization then established more radical targets of reaching elimination at the national and subnational levels. These targets have helped governments initiate and maintain high levels of political commitment in countries where leprosy is endemic. By the first decade of the twenty-first century, the goal had shifted from eliminating leprosy to reducing the burden of the disease. This is probably because of the recognition that "leprosy was not going to be eliminated in the near future" (Rodrigues and Lockwood 2011, 465). This change occurred because budgets for leprosy were cut as some countries felt they had "eliminated" the disease and because other countries had manipulated their leprosy statistics. Another reason for the new emphasis on the burden of leprosy was that as the number of "cases" declined, many researchers came to believe that studying the disease was no longer urgent (Burki 2010).

Where Leprosy Occurs Today

In 2019, the World Health Organization published a report on the latest figures for leprosy from 161 member states and territories (Table I.1). The data, which were collected in 2018, were drawn from the number of new "cases" detected (per 100,000 of the population) and the prevalence of leprosy (per 10,000 population). Around 71 percent of the 208,641 new "cases" were identified in the South-East Asia Region. The highest numbers of new "cases" were in Brazil (28,660), India (120,334), and Indonesia (17,017) (see Figure I.1). The registered prevalence of leprosy was 184,238 at the end of 2018 (World Health Organization 2019).

Leprosy figures may not be entirely accurate for a number of reasons. For any country, the number of recorded "cases" of leprosy may be incomplete because of the methods of data collection used. Such data may not

Table I.1. Registered prevalence and number of new "cases" detected during 2018 by World Health Organization region

WHO Region	Registered "Cases"		New "Cases"	
	N	Rate per 10,000 population	N	Rate per 100,000 population
Africa	22,865	0.21	20,590	1.93
Americas	34,358	0.34	30,957	3.08
Eastern Mediterranean	5,096	0.07	4,356	0.62
Southeast Asia	114,004	0.58	148,495	7.49
Western Pacific	7,876	0.04	4,193	0.22
Europe	39	<0.0	50	0.01
Totals	184,238	0.24	208,641	2.74

Source: World Health Organization (2019).

represent the entire country. It may not be recorded reliably or it may not be crosschecked. Some areas of the world do not provide any data too. Frequencies also do not represent the "cured" people who are affected with leprosy-related impairments that developed before they were cured and impact everyday life. Thus, frequency data can overlook or underreport the daily experiences of people with leprosy.

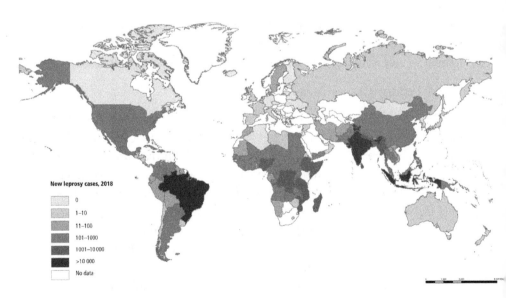

Figure I.2. Global map of total new "cases" detected. Source: World Health Organization (2018).

Are Leprosy Frequency Figures Being Manipulated?

Early in the twenty-first century, scholars thought that recording the number of new "cases" of leprosy that had been detected would be a better way of looking at the overall incidence of leprosy than measuring the prevalence rate as a percentage of the population (Saunderson 2008, citing the Leprosy Association Technical Forum 2002). The rationale was that new "cases" represented people in the community who had the disease and were transmitting it. These individuals also provided better data about the potential for transmission. The number of new "cases" has remained almost static over the last twenty years, remaining virtually the same since 2013 (World Health Organization 2013, 2015, 2016, 2017, 2019). One of the most commonly used proxies for ongoing transmission is the leprosy rate in children less than 15 years old. These data show that children are continuing to contract leprosy.

Apparent success in bringing about a decline in the number of new "cases" might paradoxically lead to a decrease in financial support for diagnosis and treatment. This potentially leads to delays in treatment, the possibility that more people will become infected before the transmitting person has been diagnosed, and impairment in more people. The need to educate physicians about leprosy is ongoing, particularly about the fact that it may be an alternative diagnosis in people who present with dermatitis or peripheral nerve disease (Golfurushan et al. 2011).

Eliminating Leprosy

Mathematical modeling of the potential decline in the incidence and prevalence of leprosy suggests that the disease will be around for at least several decades (Meima et al. 2004), although more recently it has been proposed that country-level elimination could be reached by 2020 (Blok, de Vlas, and Richardus 2015; Blok, de Vlas, Fischer, and Richardus 2015). Although there has been a considerable decline in active leprosy over the past three decades, in many countries a decline was seen even before multidrug therapy was introduced. This suggests that improved living conditions might have contributed (as Davies et al. 1999 have suggested for tuberculosis). The decline in leprosy may be partially attributable to improving socioeconomic conditions, a development that was present in many countries long before the World Health Organization began its elimination program in 1991 (Meima et al. 2004).

Today, the perception that leprosy no longer exists is widespread, but reported frequency rates may be leading to this misconception. Little research is being done on the disease and there has been a decline in sources of funding, even though it is clear that leprosy and its associated impairments are not going to disappear quickly. If we do not fully understand the epidemiology of this infection, how can health workers monitor and record frequency rates in populations that may be considered at low risk of contracting the disease? As Salipante and Hall note, "Until the modes and sources of transmission are well understood, it is unlikely that we will be able to interfere with transmission or be able to eradicate leprosy" (2011, 1512). Fortunately, advances in biomolecular science are enabling researchers and physicians to generate data that make it possible to detect leprosy in the earlier stages of the disease and to do research on modes of transmission.

Why Is Leprosy of Interest to Bioarchaeologists?

If leprosy is a declining disease today, why would a bioarchaeologist be interested in studying it? The study of disease in bioarchaeology (palaeopathology) is crucial for learning about the health and well-being of our ancestors and invaluable for taking a long view of the origin, evolution, and history of disease, in this case leprosy. Palaeopathology provides the opportunity to give skeletons a voice that can help us understand the experiences of the once living people they represent. Were their physical and social experiences the same as those who have leprosy today? Were people always stigmatized and ostracized and were they placed in institutional care? Diagnosis of leprosy in skeletons from archaeological sites is challenging: a very low percentage of people with untreated leprosy are affected in their bones. However, new methods of analysis using detection of ancient DNA of the leprosy bacteria in bones and teeth are beginning to tell us more about the bacteria's evolution and how the disease "migrated" as humans moved across the globe in the distant past.

We might also ask whether it is important for bioarchaeologists to corroborate or disprove the historical record. The short answer is yes, if appropriate. Any research requires checks and balances, and as history and archaeology complement each other, it can be the case that one may disprove a theory of the other. Historical records are especially valuable for bioarchaeologists when they are contextualizing their bioarchaeological data, alongside archaeological data that provide information about the lived environment of people and populations with diseases. Historical records also provide us with additional evidence of diseases that do not affect the skeleton (e.g., plague, malaria), although with DNA analysis those conditions are now being detected when the DNA of the infecting pathogen is preserved. At times archaeological evidence for disease in skeletons may be earlier than documentary evidence, as is the case for leprosy in some parts of the world. Of course, in prehistory there are no written documents per se, and that is why archaeological evidence per se is key to understanding that period.

While health care professionals, historians, and bioarchaeologists have written much about leprosy, no book has attempted to bring together this evidence from these different disciplinary perspectives to explore leprosy as a biological and social disease from a global perspective. This is also the first time that all the global evidence for leprosy detected in skeletons has been synthesized in an effort to understand the experience people around the world had of leprosy in the past. This book fills gaps in knowledge of this disease in the past and helps to dispel the many myths linked to leprosy, some that have been promoted by historical writings over many years. Even as leprosy declines in frequency around the world, it is still very much a part of many people's lives, from health care professionals to the people who are affected. If there was one reason to keep this disease alive in people's minds, it would be the need to continue to educate everyone about the past and present of leprosy and thereby try to promote a "new face for leprosy" (see Kumar, Lambert, and Lockwood 2019).

Structure of the Book

This book is written with the aim of taking a holistic view of leprosy and as comprehensive an approach to the available literature as possible.

Chapter 1 considers the biology of leprosy and theories surrounding transmission to humans, showing that it affects the skin and the motor and sensory nerves that connect the brain and spinal cord to the human body and control the functions of sensation and movement. The cause of leprosy, *Mycobacterium leprae,* was identified in the late nineteenth century, but it was not until the first decade of the twenty-first century that *Mycobacterium lepromatosis* was also identified as a cause of leprosy in humans. While many theories for its transmission have been described, the key route for the bacteria to spread from human to human is believed to be through exhalation of droplets containing the bacteria, which are inhaled by another person. Even though the bacterium is ultimately responsible for infection, intrinsic (e.g., age, genetic predisposition) and extrinsic (e.g., the environment) risk factors affect its occurrence. Increasingly, environmental sources of the bacteria are being identified, such as in water and soil. The sequencing of the genomes of the two leprosy bacteria has increased our understanding of the bacteria, showing susceptibility and resistance genes, identifying bacterial strains specific to geographic regions of the world, and contributing to our understanding of leprosy's origin, evolution, and history. Information about bacterial strains can also tell us about the link between the bacteria and the migration of people. This has been corroborated with historical evidence relating to colonialism and the slave trade. The only other animals that are natural hosts for the causative bacteria are wild nine-banded armadillos (found in the southern United States and South America) and red squirrels in Britain and Ireland, the latter a recent find. Nonhuman primates may contract leprosy too, but there is no convincing evidence of them being infected in the wild. The transmission of leprosy to humans from armadillos has also been suggested by several authors, a process that is facilitated by humans keeping them as pets and hunting and consuming them.

Chapter 2 considers how leprosy affects the body from the point when the bacteria is inhaled. There can be a long incubation period, probably due to the slow multiplication rate of the bacteria. Thus, a person with the infection can be symptomless but could be passing the bacteria to other humans. This has implications for affected people being infective but "invisible" and thus undetected and untreated. Impairment may occur along with the development of a social milieu that surrounds people with leprosy-related impairment. Thus, the disease may become evident many years after the initial infection. The bacteria like the cooler areas of the body. They directly affect the facial area, including the nose and mouth. Skin lesions and sensory and motor nerve damage lead to early signs that may be used for diagnosis, but *Mycobacterium leprae* has a broad immune spectrum (and classification system) that encompasses all the leprosy types. A high resistance to the bacteria leads to tuberculoid leprosy (paucibacillary—fewer bacilli) and a low resistance to lepromatous leprosy (multibacillary—greater numbers of bacilli). Sensory nerve damage can lead to trauma-related ulceration of desensitized hands and feet and subsequent infection that tracks to the bones. Motor nerve damage can cause muscle wasting and contractures of the fingers and toes. The eye musculature can also be affected, which may lead to blindness. Protecting the hands and feet from damage is an important part of care for people with leprosy today. Involvement of the autonomic nerves changes the nature of the sweating reflex in the area of skin lesions, and brittle and dry skin that develops into cracks and fissures can ensue. Potentially a person with leprosy can develop many external signs of the infection that are visible to their community.

Chapter 3 considers diagnosis, treatment, and prognosis. While a range of diagnostic tests is available, infrastructure in different areas of the world may enhance or detract from early diagnosis. It is also clear that many factors may affect whether a person is diagnosed or not, including the ability to travel to a clinic. Stigma related to leprosy in a person's community might be an important factor that prevents a person from seeking diagnosis. One other challenge is to diagnose people with subclinical or asymptomatic leprosy. The presence of specific signs and symptoms such as skin lesions, thickened nerves, anesthesia of the skin, and bacilli in skin smears is used for diagnosis. A combination of several antibiotics (multidrug therapy) is used for treatment, which has been free since 1995 and generously supported by drug companies. Treatment aims to not only use chemotherapy to stop the infection but also manage reactions, and the person is educated to cope with nerve damage. Finally, but equally important, social and psychological rehabilitation of the patient is addressed. Of course, if diagnosis and treatment is not achieved in a timely manner, impairment and disability may develop as a result of nerve damage. Alongside exploring diagnosis and treatment today, in this chapter I also delve into the past to consider what diagnostic tests were used for people who were suspected of having leprosy and what treatments were available and used. Some of the diagnostic methods seem to fit with our modern perception of medicine today (e.g., examining the patient's outward signs and their urine and blood), but they were not of course using quite as sophisticated methods as we do today. Yet some diagnostic tests seem to us to have been quite nonsensical and irrational but would have been appropriate to the time and place. The same applies to treatment in the past: some evidence seems perfectly appropriate (e.g., cleaning and treating skin lesions and bathing in healing waters), but applying hot irons to the body (cautery) or eating a diet composed of strange constituents appear very strange in our world today.

Chapter 4 considers how leprosy affects the skeleton because this is the basis for diagnosis in archaeological contexts. However, because leprosy affects the skeleton in only 3–5 percent of untreated people, recording it in skeletons provides us with information about just the tip of the iceberg of leprosy in the past. Bearing in mind that a person's immune response will determine what type of leprosy they contract and what bone changes will occur, it is interesting to note that lepromatous leprosy (the low-resistance form) is mostly what is reported in archaeological skeletons. Bone changes may be the result of the direct and indirect effects of *M. leprae,* but the diagnostic signs are recognized mainly because of the seminal research of Danish doctors Vilhelm Møller-Christensen and Johs Andersen. Diagnosis in archaeological human remains may be done using macroscopic, histological, radiological, or biomolecular methods or a combination of several of these. Leprosy affects the bones of the face, hands, and feet. The main direct evidence of leprosy is rhinomaxillary syndrome, which affects the facial bones. Indirect evidence (i.e., evidence that is not pathognomonic, or specific to, leprosy) in skeletal remains is the result of *M. leprae* affecting the sensory, motor, and autonomic nerves. Other possible leprosy-related changes are osteoporosis; inflammation of the lower leg, forearm, and hand and foot bones; damage to the eye sockets and ribs; involvement of the facial bones (maxillary sinuses) and ear canals; oral health problems; and bone trauma as a result of falls.

The safest macroscopic diagnosis of leprosy in skeletons is focused on the facial bones. However, differential diagnoses for all the bone changes should be considered because other disease processes can cause them. Biomolecular analysis of the ancient DNA (aDNA) of *M. leprae* is a relatively recent development for diagnostic purposes, and exploring bacterial strains is telling us information about the relationship between the origins and movements of people with leprosy in the past. There are many confounding factors in the diagnosis and interpretation of leprosy in archaeological human remains using any method, and inferring impairment and disability in people with leprosy is very challenging.

Chapter 5 considers the bioarchaeological evidence for leprosy. The evidence is scattered across the world but there are concentrations in certain places. These data show that leprosy has had a history of several thousand years. This picture will change as more evidence is uncovered. Three continents (Africa, Asia, and Europe) have clear evidence of leprosy. Evidence is lacking in some parts of the world because of various factors, such as a lack of work to recover evidence. In the past, leprosy was an Old World disease that focused on the northern and eastern hemispheres. The earliest evidence in skeletal remains is in Britain, Hungary, and Turkey and possibly in Iran and Sudan. Britain, Denmark, Hungary, and Sweden have the most evidence, and Northern Europe has the most evidence overall. No bioarchaeological evidence has been confirmed in the Americas, which is a surprising finding. However, there is evidence in more recent documentary data for leprosy in the Americas, perhaps suggesting that colonization had a part to play in its appearance there.

Leprosy in nonadult skeletons is rare and, of all the preserved bodies (e.g., mummies) of any age that have been examined, only one has been diagnosed with leprosy (from Africa). Another surprising finding is that the majority of skeletons with leprosy are buried normally for the time period, culture, and location. This evidence supports revaluations of the treatment of people with leprosy by scholars of historical texts in more recent years. However, the bioarchaeological evidence does not corroborate the historical evidence for the frequency of leprosy. In addition, some bioarchaeological evidence predates the historical data, which shows one of the values of studying this disease in human remains. Modern genomic data for leprosy is helping researchers evaluate the origin, evolution, and spread of leprosy, and the ancient DNA evidence is being compared with the modern data. Surprisingly, considering the many historically documented leprosaria, relatively few leprosy hospital cemeteries have been excavated. The historical evidence suggests that leprosy declined in the fourteenth century. The chapter also explores and evaluates the many possible reasons for this decline.

Chapter 6 considers the origin, evolution and history of leprosy. It includes some of the earlier discussions of the historical evidence for leprosy, the epidemiological transitions in relation to leprosy, ancient DNA evidence in relation to Monot et al. (2005, 2009), and theories about the decline of leprosy. Chapter 7 concludes the book by looking at a future for leprosy, past and present, addressing the myths about leprosy and documenting the overall findings of the study alongside considering some of the limitations of the data.

1

The Biology of Leprosy Bacteria and How They Are Transmitted to Humans

> The health burden associated with leprosy has improved significantly following the introduction of multidrug therapy in 1982 by the World Health Organization; however, this condition is still far from being eradicated, with more than 200,000 new cases detected each year for the past five years worldwide.
>
> Massone et al. (2012, 999)

It would be impossible to estimate the number of people leprosy has affected throughout time or even in recent times because of the nature of the data we have about the frequency of the disease. The accuracy of our knowledge of how many people have the infection today depends on the accuracy of reporting and our data about the frequency of leprosy in the past is incomplete because it comes from fragments of data from human remains randomly discovered in mortuary contexts. Despite these limitations, the published literature clearly demonstrates the huge impact leprosy has and has had on the lives of individuals and on communities. Even today, many people with leprosy are affected by stigma, various forms of impairment, and restrictions on their lives.

In 2016, the World Health Organization (WHO) published *Accelerating towards a Leprosy-Free World: A Global Leprosy Eradication Strategy for the Period 2016–2010*. This was not the first time the WHO had formulated a plan for eliminating leprosy (e.g., see World Health Organization 2005, 2009a), but this time the plan added a focus on the "human and social aspects affecting leprosy control" (World Health Organization 2016, viii). In this document, the WHO acknowledged that those working to eliminate this infection globally face the challenges of delays in detecting new people with leprosy, the continuing stigma associated with the disease, and the need for more work on preventing transmission.

Leprosy is often neglected in considerations of worldwide health challenges. According to the WHO's 1991 standard (a leprosy prevalence of less than one "case" for every 10,000 people at the global level), elimination of leprosy was achieved at the end of 2000. This has led to the idea that it has been *actually* eliminated in most countries. For example, the United Nations' Millennium Development Goals that were agreed in 2000 did not mention leprosy.[1] However, many issues in the agreement lists *are* related to leprosy (Rodrigues and Lockwood 2011), such as eradicating poverty, achieving universal education, and promoting equality for the sexes (women with leprosy are particularly disadvantaged). Leprosy has by no means been eliminated and in some areas of the world where it is endemic, some patients experience the effects of reduced resources for diagnosis and treatment (Scollard et al. 2006). Many people who have been "cured" of leprosy are left with impairments and are stigmatized by and ostracized from their communities. Their quality of life can be very poor. Clearly leprosy is still a challenge for those experiencing it, and for health care workers managing the challenge.

Since 1995, multidrug therapy featuring antibiotics has been the preferred treatment for leprosy. This treatment is effective and free. While it is easy to understand why this has become the treatment of choice, it often comes at the expense of a more holistic approach to care. Scollard et al. (2006, 339) go so far as to say that "leprosy is best understood as two conjoined diseases." The first is a mycobacterial infection "that elicits an extraordinary range of cellular responses in humans" and the second is the peripheral neuropathy that may have "severely debilitating physical, social and psychological consequences." Multidrug (antibiotic) treatment addresses only the first aspect of leprosy, although early treatment may prevent the second.

There have been advances in knowledge of leprosy in recent years, but challenges remain. Several aspects of the disease are still poorly understood, and to date no highly effective vaccine has been developed. The elimination of leprosy through clinical research and various policies "does not appear likely, probably due to a complex mixture of social, economic and biological factors that cannot be resolved in the laboratory alone" (Scollard et al. 2006, 339). This twenty-first-century statement reveals the failure of focused efforts to eradicate the disease. While the World Health Assembly (1991) set the goal of eliminating leprosy by the year 2000, Nshaga et al. (2011) concluded that this goal was unrealistic for several reasons. First, while multidrug treatment with antibiotics can cure a person who has leprosy, it does not prevent that person from transmitting leprosy to others. Second, no reliable diagnostic tests exist that can identify people with the infection who are not showing any signs or symptoms. Third, there is no vaccine against leprosy and, finally, reservoirs of infection exist in other animals that can be transmitted to humans.

What Is Leprosy?

In 1982, Fine observed that "there are few diseases that have a richer cultural heritage than leprosy, or are steeped in as many myths" (161). As late as the 1980s, leprosy was one of the less understood of the major infectious diseases. Although clinicians have a better understanding of the infection today, they still do not understand its full nature. Leprosy is a chronic bacterial infectious disease within the genus mycobacteria. It is caused by *Mycobacterium leprae*, a rod-shaped bacillus that mainly affects the skin and the peripheral nerves. The peripheral nervous system provides a connection between the central nervous system (brain and spinal cord) and the rest of the body, such as the muscles. It also transmits signals from the senses (such as touch and smell) to the brain. It consists of sensory, motor and autonomic nerves. In recent years, a newly discovered bacterium (*Mycobacterium lepromatosis*) that causes leprosy has been discovered in people in Brazil, Canada, India, Mexico, Myanmar, and Singapore (Han, Seo, et al. 2008; Han et al. 2009; Han, Sizer, and Tan 2012; Han, Sizer, et al. 2012; Han et al. 2014; Han and Silva 2014). This bacterium has molecular similarities to *M. leprae* (Gillis, Scollard, and Lockwood 2011). Some evidence has emerged regarding the effects of *M. lepromatosis*. For example, there are reports of a person in Canada who experienced a "leprosy-like illness" associated with it (Jessamine et al. 2012) and there is documentation that patients in Brazil and Mexico harbor both *M. leprae* and *M. lepromatosis*. In Mexico, leprosy caused by *M. lepromatosis* is more common than leprosy caused by *M. leprae* (Han, Sizer, et al. 2012), and in Brazil, patients infected with *M. lepromatosis* have exhibited tuberculoid leprosy, the high-resistance type (Han et al. 2014). Han and Silva (2014, 6) suggest that *M. leprae* and *M. lepromatosis* and their last common ancestor "most likely evolved, both in time and space, with humans." More research is needed to better understand this newly discovered cause of leprosy, but there have been developments in genomic analysis (see below).

M. leprae can induce infection in a variety of forms, ranging from low-resistance lepromatous leprosy (multibacillary, or many bacilli, and more than five skin lesions) to high-resistance tuberculoid leprosy (paucibacillary, or fewer bacilli, and from one to five skin lesions). Tuberculoid leprosy is the milder form and is characterized by skin lesions and nerve damage that occurs early in the infection. Lepromatous leprosy is more severe and is characterized by multiple skin lesions, involvement of the peripheral nerves, and further complications such as damage to the eyes and to the bones. Infections from this bacterium occur today in tropical and warm countries where there is poverty (Britton and Lockwood 2004). While its primary effects are on the peripheral nerves, *M. leprae*'s secondary effects are seen in the skin and other tissues of the body, including the eyes, the testes, the kidneys, the bones, the upper respiratory tract, and the lining of blood vessels (Jopling 1982). Our understanding of leprosy has changed considerably in recent years because of the sequencing of the *M. leprae* and *M. lepromatosis* genomes. This has enabled scholars to learn more about the biological aspects of the bacteria and how the person affected responds to the infection. Sequencing has also provided data that enables health care workers to manage leprosy better.

The Discovery of the Leprosy Bacillus

A better understanding of *M. leprae* and leprosy began with Gerhard Henrik Armauer Hansen's discovery of the leprosy bacillus in 1873 in Bergen, Norway (Hansen 1880; Harboe 1983) (see Figure 1.1). Hansen had observed people with leprosy in Norway through his work as a provider of medical care for fishermen. In 1871, he received a grant from the Norwegian Medical Society that enabled him to explore the causes of the disease. Although Hansen traveled around western Norway to study families with leprosy, he was not convinced leprosy was inherited. Instead, he theorized that leprosy was infectious because all the people with leprosy he encountered had been in contact with others with the disease. He published a report in 1872 that described it as a chronic disease. Two years later, he presented his classic description of the leprosy bacillus. He believed that what he described as brown rod-shaped "bodies" he had observed in the tissue fluid of a patient with leprosy were the cause of leprosy (Meyers et al. 1984).

Hansen tried to reduce the incidence of leprosy in Norway by hospitalizing people who were affected and focusing on the most infectious patients. He also drafted proposals for laws related to leprosy. As a result of Hansen's work, Bergen was an international center for research on leprosy for many years in the late nineteenth and early twentieth centuries. In 2001, the archives of Hansen's work on understanding leprosy in Norway entered UNESCO's Memory of the World Register.[2]

The bacillus *M. leprae* is a slightly curved or straight rod (Drabik and

Figure 1.1. Gerhard Armauer Hansen. Credit: Bergen City Museum/Leprosy Museum.

Drabik 1992; Scollard et al. 2006; Rojas-Espinosa and Løvik 2001). It is described as acid-fast, which means it cannot be decolorized by an acid after staining. This characteristic is very useful for identifying *M. leprae*. The characteristic is present because of its impermeable cell wall, which contains lipids (fats and waxes) that are specific for *M. leprae* (Rojas-Espinosa and Løvik 2001). The dominant lipid in the cell wall is PGL-I (phenolic glycolipid-I). This is what gives leprosy its immunological specificity. PGL-I reproduces intracellularly and almost exclusively in the human body. *M. leprae* can survive and multiply in the Schwann cells of peripheral nerves (the cells that produce the myelin sheath around nerves) and in macrophages (types of white blood cells) (Bennett, Parker, and Robson 2008). When *M. leprae* damages the nerves and their surrounding tissues, the host mounts an immune response *to M. leprae* antigens.

The Genome and Leprosy

An organism's genome consists of a set of DNA molecules that "contain the biological information needed to construct and maintain a living example of that organism" (Brown and Brown 2011, 14), in this case *M. leprae*. The data the molecules hold are contained in genes, or genetic material. The four individual units in a DNA molecule are called nucleotides and the biological information in a DNA molecule is characterized by the sequence of its nucleotide (adenine [A], thiamine [T], guanine [G], and cytosine [C]). A DNA sequence variation occurs when a nucleotide in the genome differs between members of a species in a population.

The complete genome sequence of the leprosy bacillus was published in 2001 (Cole, Eiglmeier, et al. 2001). An extreme case of reductive evolution was noted in the genome ("shrinking" relative to its ancestor), and it was suggested that this probably explains why the bacillus cannot be cultured in vitro (Schurr et al. 2006). Less than half the genome contained functional genes (a limited set of essential genes; Young 2001), but pseudogenes, or imperfect copies of a functional gene, were very common. The loss of functional genes means that *M. leprae* has been deprived of many metabolic activities and that it can only survive in specialized niches. Thus, it has very specific growth requirements (Grange and Lethaby 2004; Scollard et al. 2006). The bacillus was described as an "efficiently streamlined pathogen and . . . an indolent microbe on a slippery slope of evolutionary decay" (Young 2001, 1640). Young argued that the sequencing of the leprosy genome could help us understand how it is transmitted, be used to develop tools to study the immune system for assessing leprosy exposure, and play a part in improved treatments and renewed research efforts on leprosy. Sequencing has also helped us understand its evolution.

More recently, sequencing of near-complete *M. lepromatosis* genomes has been reported (Han et al. 2015; Singh et al. 2015). This research has established that *M. leprae* and *M. lepromatosis* derive from a single common ancestor and diverged about 13.9 million years ago (Singh et al. 2015) and that they cause similar pathological conditions in the human body. *M. leprae* is the more recent of the two. *M. lepromatosis* is also associated with Lucio's phenomenon, one of the leprosy reactions characterized by diffuse lepromatous leprosy and seen more commonly in Central America. *M. lepromatosis* may therefore have originated in that part of the world.

Since the *M. leprae* genome was sequenced, these biomolecular data have been used to track transmission, identify hot spots for leprosy, and highlight genes that reflect susceptibility and resistance (e.g., Fava, Dallmann-Sauer, and Schurr 2019). There has also been an increasing number of studies of *M. leprae* DNA in bioarchaeology (see Chapters 5 and 6).

The Types/Subtypes/Strains of *M. leprae*

Understanding how leprosy is transmitted includes considering relevant risk factors, the closeness of contact between people, and the bacterial load of the affected person. Today research using geographical information systems (GIS) is increasingly being used to detect hot spots for people with leprosy in association with socioeconomic indicators (Jim, Johnson, and Pavlin 2010; Queiroz et al. 2010; Paschoal et al. 2013; Sampaio et al. 2013). The occurrence of and risk for leprosy in communities are also being tracked using molecular strain typing. Molecular typing is a process that attempts to identify and differentiate between members of the same species, here being used to document the strains of *M. leprae* affecting people with leprosy (Salipante and Hall 2011). To distinguish between strains of bacteria, researchers look for single nucleotide polymorphisms (SNPs; genetic sequence variations/genotype in humans) and variable number tandem repeats (VNTRs, or a location in a genome where a short SNP sequence is organized as a tandem repeat; variations in length [number of repeats] are seen among individuals). Both VNTRs and SNPs are mutations that occur when DNA replicates (Rodrigues and Lockwood 2011), meaning that two identical replicas of DNA from one original DNA molecule are produced. Researchers distinguish between strains of bacteria by comparing the SNPs they possess. Molecular typing has led to the compilation of a freely available Leprosy Susceptible Human Gene Database (Doss et al. 2012), the first of its kind. It integrates leprosy- and human-associated genes and corresponding protein sequences and aims to be a platform for understanding SNPs in leprosy.

In 2001, through the sequencing of the *M. leprae* genome, researchers identified a number of VNTRs that they suggested could help discriminate between strains (or subtypes) of *M. leprae* (Cole, Eiglmeier et al. 2001; Cole, Supply, and Honoré 2001). This has made it possible for researchers to track the transmission of specific *M. leprae* strains in people with leprosy (Curtiss et al. 2001). Once specific strains have been identified, various methods are used to infer evolutionary relatedness and statistical analysis is conducted (phylogenetics). As an example, one study focused on five VNTR loci and twelve strains of *M. leprae* that had been isolated from patients in Brazil, Mexico, the Philippines, and the United States, plus strains from wild armadillos and a sooty mangabey monkey. Although the researchers conducting this study found diversity at four VNTRs, no particular genotype (genetic makeup) was associated with different VNTR regions or with the origin of the patients. Nevertheless, the researchers were optimistic about the future potential of genotyping. They felt that it would become an effective way to contribute to monitoring *M. leprae* strains for evolutionary change and that this knowledge could eventually lead to improvements that could limit the transmission and spread of leprosy (Truman et al. 2004).

Some researchers believe that identifying VNTRs in *M. leprae* is useful for tracking the transmission of leprosy within communities and that knowledge of small degrees of variation in SNPs and VNTRs will be beneficial when looking at wider questions related to global spread (Matsuoka 2009). However, this theory is not accepted universally. For example, Monot et al. (2009), who used VNTRs and SNPs to compare the stability of two markers of genomic biodiversity in *M. leprae*, found no variation in the SNP profiles of *M. leprae* (patients from seven different locations), although the VNTR profiles in these samples varied quite a lot. Monot et al. saw variation in different samples containing *M. leprae* from the same patient. This suggests that the VNTRs identified in this study were too unstable and had too much variability to be used as a good epidemiological marker for leprosy. In addition, while this research team found that SNPs correlated with geographical location, others have found this was not the case for some VNTRs (Monot et al. 2009). Therefore, while some researchers argue that VNTR loci are very useful for strain typing (Gillis et al. 2009), Salipante and Hall (2011) question the utility of using SNP loci to look at strain relationships. They suggest that VNTRs, which evolve rapidly, are better for exploring strain evolution and change over short periods of time and that SNPs and relatively stable VNTRs are better for looking at evolution over deeper time. One example of the latter strategy is a study that explored SNPs and VNTRs in samples from Thai patients with leprosy (Phetsuksiri et al. 2012). After comparing their data with data from elsewhere, the researchers concluded that *M. leprae* in Myanmar and Thailand has a common historical origin.

A number of studies have used molecular epidemiology to look at sources of infection and transmission chains within households and between countries (Matsuoka et al. 2004; Xing et al. 2009; Shinde et al. 2009; Lavania, Jadhav, et al. 2013; Aye et al. 2012). For example, it was found that infected people do not necessarily contract leprosy from contact with other members of their household and are likely to contract the disease from sources outside the home (Matsuoka et al. 2004). For example, in one study the same strains of the *M. leprae* were found in patients from the same house and in patients living near the house (Lavania et al. 2011). In such cases, person-to-person or environment-to-person transmission mechanisms are likely responsible (Cardona-Castro et al. 2009; Salipante and Hall

2011). In a Chinese study, similar strain types were found within family groups and in neighboring townships, indicating that where housing is located and who people with leprosy socialize and work with are important factors in leprosy transmission (Weng et al. 2011).

Although local studies are the norm, it has been possible to use the results of such work to think about the broader picture of transmission of different leprosy strains over larger regions. For example, 93 percent of *M. leprae* strains in skin samples of people with leprosy in rural and urban Maharashtra, India, belonged to one particular SNP and the Maharashtra strains were similar to southern Indian strains (Kuruwa, Vissa, and Mistry 2012). Many variables may be responsible for variations in strains, such as socioeconomic and cultural factors (Salipante and Hall 2011), but while researchers have cross-sectional data on *M. leprae* strains for a single point in time in some specific regions or countries, the amount of longitudinal data is small. Studies of the movement of the *M. leprae* with people from around the world have also been relatively rare until the last few years, although a series of papers using Gillis et al.'s VNTRs was published in 2009 (*Leprosy Review*, vol. 80). More research focusing on *M. leprae* ancient DNA could provide the deep-time longitudinal data needed to explore the evolution of the bacteria.

A landmark paper on the origin and evolution of leprosy has been particularly helpful in developing a much broader perspective on the *M. leprae* organism (Monot et al. 2005, Figure 1.2). Monot et al. established that all present-day "cases" of leprosy are attributable to a single clone (a group of identical cells that share a common ancestor). In other words, they are derived from the same mother cell. They also found that the dispersal of *M. leprae* could be tracked by analyzing rare SNPs. In this study, the research team wanted to learn whether all strains of *M. leprae* had undergone similar events and sought to determine their level of relatedness. They analyzed seven strains of *M. leprae* they had obtained from people with leprosy in Brazil, Ethiopia, India (two), Mexico, Thailand, and the United States. While the genomes of these seven strains were practically identical, the geographical origin of the people was correlated with different SNPs. To increase the possibility of detecting SNPs, Monot and colleagues analyzed selected genes, noncoding DNA sequences (components of an organism's DNA that do not encode protein sequences), and pseudogenes (an imperfect copy of a functional gene) in the Brazilian strain. They chose this strain because of its geographical remoteness and because at the time Brazil had the second highest leprosy burden in the world. They found five SNPs in

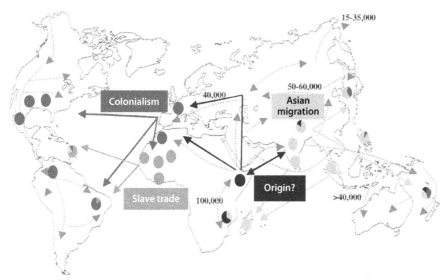

Figure 1.2. The origin and spread of leprosy using DNA data. The original figure was in color; the reader is advised to access it for further detail. Source: Monot et al., "On the Origin of Leprosy," *Science* 308 (2005): 1040–1042. Reprinted with permission from the American Association for the Advancement of Science/RightsLink.

this strain. When they analyzed the rest of the strains, they found only three SNPs in two or more of the strains. They found the remaining two SNPs only in a strain from Tamil Nadu, India.

Monot et al. (2005) then wanted to document the distribution of the SNPs in *M. leprae* throughout the world. They analyzed 175 clinical and laboratory samples from people from twenty-one countries on five continents (Africa, Asia, Europe, and North and South America) and found four SNPs (genotypes/types). Type 1 is associated predominantly with Asia, the Pacific region, and East Africa; Type 3 is associated with Europe, North Africa, and the Americas; and Type 4 is associated with West Africa and the Caribbean region. SNP Type 2, the most rare type, was detected only in Ethiopia, Malawi, Nepal/Northern India, and New Caledonia. Based on their findings, they suggested that leprosy originated in East Africa or the Near East and spread as populations from Europe and North Africa migrated to West Africa and the Americas over the last 500 years. This is contrary to the idea that leprosy was taken across the Bering Strait to the Americas from Asia. They also outlined two possible scenarios for the spread of the infection and the development of SNPs. In scenario 1, SNP 2 originated in East Africa or Central Asia and gave rise to SNP1, which was then carried eastward to Asia. SNP 3 was disseminated westward to the Middle East and Europe. SNP 3 gave rise to SNP 4 in West Africa and then spread to countries linked to West Africa by the slave trade. In scenario 2, SNP 1 (Asia/Pacific, East Africa) was the originator of SNP 2 (Ethiopia, Malawi, Nepal/Northern India, and New Caledonia), then SNP 3 (Europe, North Africa, and the Americas), and finally SNP 4 (West Africa and the Caribbean).

In 2009, Monot et al. published a second study that analyzed the same 175 samples of *M. leprae* to better understand the worldwide distribution of *M. leprae* SNPs. They found four SNP permutations (SNPs 1–4) but could not associate a specific VNTR pattern with particular SNPs, although there was a correlation between geographical origin and SNP profile. In this research, Monot and colleagues identified additional SNPs that enabled them to identify sixteen subtypes (or strains) of the four main genotypes of *M. leprae*. They argued that these sixteen subtypes have "limited geographic distribution that correlate with the patterns of human migration and trade routes" (Monot et al. 2009, 2). This 2009 research provided a more nuanced appreciation of the global distribution and spread of leprosy and provided new data that generated two new suggestions about the global origin and spread of leprosy. They proposed a northern and southern route for the introduction of leprosy to Asia. The northern route spread SNP 3 strains from the Eastern Mediterranean and Turkey through Iran, China, Korea, and Japan via the Silk Road. SNP 1 strains were conveyed along the southern route from the Indian subcontinent to Indonesia and the Philippines.

Data such as these can provide testable hypotheses about the movement of people with specific bacterial strains of leprosy and the routes of transmission of those strains. Identification of different strains in people that are not usually seen in a region suggests longer-distance movement of populations, especially today as people move more frequently, more rapidly, and farther from their place of birth than they did in the past. Now that the *M. leprae* genome has been fully sequenced, researchers are beginning to understand the bacterium better. It is hoped that this new knowledge will eventually contribute to the elimination of the disease.

Theories of How Leprosy Is Contracted

Leprosy is viewed as a contagious disease that is transmissible to other humans. It has low pathogenicity and is not highly infectious. The precise mechanism of leprosy transmission is not clearly understood (Scollard et al. 2006; Turankar et al. 2012; Bratschi et al. 2015). People who are close to others with leprosy (especially those with multibacillary leprosy) are more likely to contract it (Yawalkar 2009). However, over 90 percent of people who are exposed to leprosy develop only a subclinical infection and not the disease. In a recent review of publications on the subject, Bratschi and colleagues concluded that although "no study unequivocally demonstrated the mechanisms by which bacteria travel from one case to another, . . . the nose, the oral cavity and the skin have been tentatively identified as entry and exit points of *M. leprae*" (Bratschi et al. 2015, 151). Although research on transmission has focused on the nasal mucosa (the lining of the nasal cavity), more recent studies have shown that the oral mucosa also harbor the bacilli (Martinez et al. 2011; Taheri et al. 2012).

It is generally accepted that *M. leprae* is transmitted via droplets exhaled from the nose and mouth of infected people (Noordeen 1993; Curtiss et al. 2001). Because the bacteria remain inside the cells of the body's tissues, most people are not infectious and do not develop clinical leprosy (Rodrigues and Lockwood 2011). Leprosy is not highly transmissible, but people who have lepromatous leprosy and are excreting large numbers of bacilli from their nose are highly contagious to others if they are not treated (four to eleven times more contagious than patients with tuberculoid leprosy). The bacteria can also cause subclinical infection in people they are in contact with. Less than 5 percent of people with subclinical leprosy infection develop signs of the disease (Dayal et al. 1990).

Fine (1981) suggested six scenarios for people who have been exposed to *M. leprae*, bearing in mind the difference between infection (that is, the bacteria has entered the body) and disease (that is, body cells are damaged by the bacteria and signs and symptoms have developed). People may become infected by the bacteria and then be cured with no obvious manifestations of leprosy; they may become infected and their infection may resolve but they may still have obvious disease; they may harbor the infection but have no recognizable disease (subclinical infection); they may develop overt leprosy; they may have the infection but move along the disease's immune spectrum according to how resistant they are to the bacteria; or the disease may completely resolve.

All of these scenarios probably occur in people with leprosy today and likely did so in the past. Fine (1981) posited that all of these scenarios might relate to particular genotypes. The questions of how and to what degree genetic factors determine how a host responds to *M. leprae* (or *M. lepromatosis*) exposure or infection is beginning to be answered now that the leprosy genome has been sequenced.

Droplet Infection

As noted above, leprosy bacilli in droplets exhaled from the upper respiratory tract are the key source of infection for humans (e.g., see Araujo et al. 2016). As early as the 1940s, researchers noticed that the nasal area was affected in people with lepromatous leprosy (Rogers and Muir 1946). By the late 1960s, researchers also understood that viable leprosy bacilli could occur in nasal discharges in large numbers even when obvious signs and symptoms were not present and that leprosy bacilli could persist in the nose even after prolonged drug therapy (Rees et al. 1967). The nose is affected at an early stage in people with lepromatous leprosy and bacilli can be detected in the spongy bones of the nasal passages (turbinate bones). When an immunologically vulnerable person inhales *M. leprae* through the nose, the bacilli multiply in the nasal cavity and then bind to macrophages and to the Schwann cells that keep peripheral nerves alive (Lockwood 2004). Untreated patients with leprosy can discharge as many as 100 million bacilli per day; a single nose blow of somebody with lepromatous leprosy can contain up to 20 million bacilli per milliliter (Waters 1981; Sreevatasan 1993). Sneezing or coughing can project bacilli from the upper respiratory tract at distances of up to 30–50 centimeters (Figure 1.3). Researchers also believe that it is possible that people can discharge bacilli during normal breathing (Bedi et al. 1976). The evidence clearly indicates that the upper respiratory tract is significant in the transmission of leprosy. Experimental research continues on *M. leprae* in the oral and nasal cavities of people with leprosy (e.g., Lavania et al. 2013; Morgado de Abreu et al. 2014). Lavania and colleagues (2013) found a higher frequency of *M. leprae* during the monsoon season than during the summer and the lowest frequency during the winter. They also discovered that while few people harbored *M. leprae* in every season, in areas where leprosy is highly endemic people are likely exposed to the bacteria regardless of the season.

Huang (1980) emphasizes the relevance of immune system strength as a factor in the transmission of leprosy. In order to survive, viable leprosy organisms must be released into the environment from the person, enter a new host, distribute themselves through the body of the new host, and produce an illness. Any of these variables can affect their rate of transmission, including the strength of the immune system of a potential host. However, only some of the people who are exposed to leprosy acquire the disease (Noordeen 1993). It is not yet clear which factors differentiate the people in a population who contract leprosy from those who are exposed to *M. leprae*

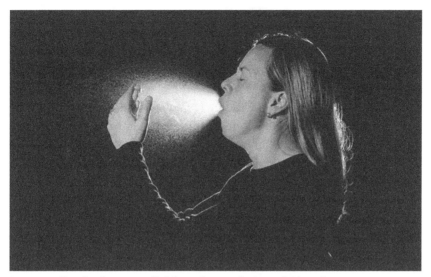

Figure 1.3. Droplets being expelled from the mouth. Scottish News Services.

but do not develop the disease (subclinical infection), but susceptibility and resistance genes may play a part. Unfortunately, researchers did not begin to focus on developing methods of diagnosing subclinical infection in leprosy until fairly recently (e.g., Martinez et al. 2011).

Since the late 1950s, researchers have believed that long-term continuous contact is not necessary for transmission of leprosy and that repeated intimate contact (or even a one-off contact under ideal conditions) may be enough (e.g., Badger 1959). While intimate and prolonged close contact is related to leprosy transmission particularly in families, a short period of contact can be enough to transmit the infection for people who are susceptible (Dayal et al. 1990; Bratschi et al. 2015). However, some people with leprosy have had no discernible contact with others who have the disease (Fine 1982). While the intensity of contact will differ, proximity of contact with a person with leprosy likely correlates with how many bacilli are inhaled. Close contact may be related to overcrowding in the home. Crowding is also linked to other factors such as poverty, but the relationship between leprosy and poverty could just as easily work the other way around: the presence of a member of a household with leprosy could lead to lowered socioeconomic status for a family and poverty. All this research suggests that the patterns of contact of uninfected members of a household or a community with people with leprosy are not the sole determinants for who will contract the disease.

Skin-to-Skin Contact

Because skin lesions are characteristic signs of leprosy, researchers have attempted to learn what role the skin plays in the transmission of the disease. Newell (1966) believed that the skin was the site where the bacilli entered the body. Ridley, Jopling, and Ridley (1976) found that the fingers were the sites with the highest bacterial load and suggested that skin-to-skin contact may be a key route of transmission for leprosy. Huang (1980) was of the opinion that the broken skin of a person with leprosy might enable leprosy bacilli to escape onto its surface and into the environment but that it was unlikely that leprosy bacilli penetrated intact skin. However, a decade later, Dayal et al. (1990) found some evidence that bacilli can infiltrate open skin. Fine (1982) argued that unclothed body parts were the likeliest points of skin-to-skin contact between people, being the most likely areas to be subject to trauma, and the areas most exposed to insects, and that they were relatively cool areas that are favorable for the proliferation of *M. leprae*. Despite this research, the prevailing opinion is that the skin is not significant in transmitting leprosy; it does not excrete bacilli and bacilli are rarely found on it (Lockwood 2004, 29.3).

Insects

Researchers have also explored the role of insects in the transmission of leprosy through experimental work, particularly using those that feed on blood (e.g., Kirchheimer 1976). In one experiment conducted where leprosy was highly endemic, researchers allowed laboratory-bred mosquitoes (*Culex fatigans*) and bedbugs (*Cimex hemipterus*) to feed freely on people with lepromatous leprosy. While it was found that the insects often contained leprosy bacilli and some bacilli were viable for up to forty-eight hours after the blood meal, they also found that the frequency of bacilli in mosquitoes, bedbugs, head lice, scabies, and mites was the same in homes where people with leprosy had not been treated and in randomly selected homes where no people had leprosy (Narayanan, Manja, and Bedi 1972; Narayanan, Manja, and Kirchheimer 1972). In fact, the frequency of *M. leprae* in mosquitoes and bedbugs was actually higher in the randomly selected houses. On the other hand, recent research has suggested that ticks may act as a reservoir for *M. leprae* and as vectors for spreading the infection (Ferreira, Oliveira, et al. 2018).

Sreevatasan (1993) found that bacilli can persist in the proboscis of mosquitoes that have fed on the blood of people with leprosy for at least 144 hours but that they last for a shorter time in the gut (96 hours) and in feces (72 hours). However, the proboscis needs to reach beyond the epidermis to transmit leprosy to humans. In people with leprosy, the average depth from the skin surface to the bacilli was 200–400 micrometers and the average thickness of human epidermis is 100 micrometers. The average length of a mosquito proboscis is from 15 to 2,300 micrometers. When the proboscis is long enough, a mosquito can therefore access blood-containing bacilli. As insects ingest only small quantities of blood, the number of bacilli they transmit will also be small, and for *M. leprae* to be viable and multiply in another human host the insect would need to inject a large number of the bacilli into a human.

Geater (1975) studied houseflies (*Musca*), bluebottles (*Calliphora*) and biting flies (*Stomoxys*) to see whether they transmitted leprosy; all of these flies have been observed feeding on skin lesions of people with leprosy. Geater found that the legs, mouthpieces, abdominal walls, and stomachs of these flies were heavily infected one hour after feeding and that a small proportion had bacilli present up to three days later. The flies were also capable of infecting the surfaces on which they subsequently fed. However, Skinsnes (1977) did not consider flies to be a significant means of leprosy transmission and Fine (1982, 174) concurs: "Arthropod transmission . . . appears technically feasible, but what *can* happen does not necessarily coincide with what *does* happen." On the basis of these data, the current thinking is that insects are not major players in leprosy transmission (Lockwood 2004), even though they "cannot be completely ruled out" (Yawalkar 2009, 25).

Gastrointestinal Route

The gastrointestinal system is not known to be a route of transmission of leprosy for humans. However, bacilli can be ingested (they have been found in the breast milk of mothers with lepromatous leprosy) and flies may transmit bacilli to food (Pedley 1967, 1968; Huang 1980; Saha, Sharma, and Siddiqui 1982; Brubaker, Meyers, and Bourland 1985; Jopling and McDougall 1988; Bryceson and Pfaltzgraff 1990). However, Rodriguez (1926) and Duncan et al. (1983) found no evidence of bacilli in samples of breast milk from mothers with leprosy they studied. Duncan and colleagues hypothesized that bacilli could be present in breast milk only in people with advanced leprosy involving the nipples and the milk ducts. This transmission route therefore remains a possibility.

Infection In Utero

Perceptions that leprosy might be transmitted from an affected mother to her unborn child (congenital leprosy) may originally derive from the historical association of leprosy with the Bible, sin, and sexual intercourse. Extensive research has been carried out to explore whether mothers can transmit leprosy to their developing fetuses. Duncan (1985b) observed that women of low socioeconomic status who had inadequate diets could transmit *M. leprae* to a fetus. She also found that men with advanced lepromatous leprosy could harbor large numbers of *M. leprae* on their genitalia and have leprous urethritis, suggesting that leprosy may spread through sexual intercourse. Drutz et al. (1972) looked at transmission through the placenta. People with lepromatous leprosy can have up to 10^5 leprosy bacilli in a milliliter of their blood. Because the placenta is highly vascular, damage to its integrity might enable *M. leprae* to enter the fetus from the mother via that route. Melsom, Duncan, and Bjune (1980) found an increase in *M. leprae* antibodies in umbilical-cord blood that was significantly higher in babies of mothers with lepromatous leprosy and tuberculoid leprosy than it was in a control group. They believed that this was an indication that the fetus had produced antibodies, a possible sign that *M. leprae* or *M. leprae* antigens can transfer across the placenta. Melsom, Harboe, and Duncan (1982) also found that the number of antibodies was higher in some infants of mothers with lepromatous leprosy. Several researchers have noted *M. leprae* in the placentas of some mothers with leprosy (Duncan 1982; Duncan and Oakey 1982; Duncan et al. 1983; Bryceson and Pfaltzgraff 1990), and *M. leprae* bacilli can be transmitted via amniotic fluid (Melsom et al. 1981; Melsom, Harboe, and Duncan 1982). However, after a baby is born to a mother with lepromatous leprosy, it is exposed to bacilli from excretions from the mother's nose and mouth, thus increasing their risk of contracting leprosy in this way. The evidence considered here suggests that in utero infection of the fetus is relatively unlikely but not out of the question.

Environmental Sources

Living conditions are naturally part of the environmental risk factors that can predispose people to leprosy (see below). The environment includes the altitude and latitude where people live and seasonal, temporal, interannual, climatic, and precipitation effects, including temperature and humidity. All of these can contribute to different rates of disease occurrence (Fine 1982; Lafferty 2009; Turankar et al. 2012). However, while these data are relevant to understanding leprosy occurrence, they can present contradictions. For example, *M. leprae* can survive in both humid and arid environments (Desikan 1977).

In the mid-1980s Jopling (1985) described forty-one mycobacterial species, including *M. leprae,* but noted that most were not harmful to humans. Mycobacteria are a genus of acid-fast bacteria. Most live in water (rivers, ponds, swimming pools, water storage tanks, taps) and some live in soil, in animal droppings, and in vegetation. A few mycobacterial species cause leprosy in rodents, birds, and fish and some cause the infection in humans, as seen above. The few that infect humans are opportunistic when they are pathogenic. This occurs in special situations, such as when they infiltrate a skin abrasion or when a person's immune response is impaired. The number of mycobacteria identified has increased in recent years; Brosch et al. (2000) described seventy species and Grange (2008) identified over 100 species. Daffe and Draper (1998) note that mycobacteria have a lipid-rich wall that resists desiccation, which is helpful when analyzing them.

In some parts of the world where leprosy is still a challenge, there is increasing evidence that the environment remains a source of *M. leprae* (Fine and Warndorff 1997; Tadesse Argaw et al. 2006). *M. leprae* may survive in soil or vegetation and on dead or decaying organic matter (Kazda 1981). This might explain why most people who contract leprosy do not report contact (knowingly) with anyone with the infection. At least one biomolecular study has indicated that other sources of the infection exist in places where large numbers of people with leprosy concentrate around specific geographic features such as water tanks (Salipante and Hall 2011). This finding supports the idea that *M. leprae* can survive outside the human host under favorable conditions, such as in a hot and humid climate.

The bacilli can survive outside the human body for days or months in droplets from nasal secretions and in most soils at room temperature (Dayal et al. 1990; Desikan and Sreevasta 1995). *M. leprae* have been identified in soil around houses of people with leprosy in India (Stearns 2002; Lavania et al. 2006; Mohanty et al. 2016) and in water (Lavania et al. 2014; Arraes et al. 2017). For example, Lavania and colleagues (2008) studied *M. leprae* DNA detected in soil samples that were collected from a range of places, including from fifteen villages where leprosy was endemic. One-third of the samples was positive for DNA-specific *M. leprae*. A comparison of *M. leprae* presence in soil samples from residential areas of people with leprosy and from non-leprosy residential areas showed a lower presence of *M. leprae* in the non-leprosy areas. Turankar et al. (2016) also noted viable *M.*

leprae in soil in India, where samples were collected from around a leprosy hospital and in a resettlement village inhabited by both people who had been cured and people with active leprosy. Analysis of soil samples and skin samples from people with leprosy, again in India, has also shown the same genotype of *M. leprae* (Turankar et al. 2012). Thus, these more recent studies are pointing toward the environment as a key factor in the transmission and contraction of leprosy. Furthermore, Tió-Coma et al. (2019) found *M. leprae* DNA in soil samples collected near houses of people with leprosy in Bangladesh, from armadillo holes in Suriname, and from habitats of red squirrels in Britain.

Prior to the most recent environmental data, the material that was most intensively studied as a transmitter of *M. leprae* was sphagnum moss in Norway (Kazda et al. 1986). Kazda, Irgens, and Muller (1980); Kazda, Irgens, and Kolk (1990); and Irgens (1981) had suggested that sphagnum bog vegetation was a source of the bacteria that caused leprosy. In Norway, sphagnum moss, which needs high humidity, is found only in the country's northern and western areas. Researchers found noncultivable acid-fast bacilli in samples from areas of western Norway where there had been a high frequency of leprosy in the nineteenth-century population. During the period when leprosy was endemic on the west coast of Norway, many people walked barefoot over wet ground, and Hansen and Looft (1895) suggested that sores acquired from damage to the soles of their feet provided a route for leprosy to enter the body. Sphagnum and other moss vegetation, both of which thrive in the high humidity of the Atlantic climate, surrounded farms in this part of Norway. Thus, in the nineteenth century, farms located under southern slopes in northern and northwestern Norway were perhaps the ideal environment for mycobacteria to multiply in sphagnum moss due to the heat of the sun and high humidity in the summer (Hansen and Looft 1895). The water supply from sphagnum moss bogs probably increased the risk of leprosy during Norway's epidemic of 1851–1885, but this does not necessarily mean that all sphagnum moss areas harbor the leprosy bacilli (Kazda 2000). Kazda found mycobacteria in sphagnum in Africa, the Americas, Europe, and New Zealand, but the proportion of samples containing mycobacteria ranged from 31.9 percent in Sweden to 79 percent in Ireland (the mean was 41.2 percent). However, while Kazda found twenty-six species of mycobacteria, he found no *M. leprae*. Chakrabarty and Dastidar (1989; 2001–2002) concluded that the microclimate of sphagnum, rather than the particular taxonomic unit of the vegetation, was the key factor in mycobacterial density. They suggested a correlation between leprosy distribution around the world and sphagnum and argued that because only one-third of people who contract leprosy acquire it from other people with leprosy, other nonhuman sources must be responsible (2001–2002). The data from their soil analysis support this hypothesis; in their soil samples, they found the phenolic glycolipid-I antigen, which is unique to the cell wall of *M. leprae.*

Polluted surface water is another possible source of the leprosy bacteria (Kazda 2006). Kazda hypothesized that people could be exposed to waterborne *M. leprae* through the water they drink and through clothes washed in that water. He noted that in humid tropical climates, where it is often impossible to naturally dry clothes, people wear these items while they are still damp. Kazda believed that *M. leprae* could gain access to human hosts through broken skin, including the skin of bare feet, and when washed dishes are placed on contaminated ground. The research of Matsuoka et al. (1999) supports this hypothesis. They found that in Indonesia, leprosy was more frequent in people who used water sources for bathing and cleaning that contained *M. leprae* than in those who did not. They also found that *M. leprae* bacilli, some of which were viable, were present in well water in villages where people with leprosy lived (Matsuoka et al. 2013).

Other environmental organisms have been studied by some authors. Lahiri and Krahenbuhl (2008) and Wheat and colleagues (2014) have focused on the relationship between amoeba (one-celled organisms/protozoa) and leprosy bacilli. Wheat and colleagues investigated whether it is possible that pathogenic amoebae could be host cells for *M. leprae,* thus protecting the bacteria from environmental decay. Lahiri and Krahenbuhl used cultures of the soil amoeba *Acanthamoeba castellanii* and found that in more than 90 percent of the cultures, the amoeba ingested *M. leprae* and the ingested bacilli remained viable for at least seventy-two hours. However, Wheat et al. found that *M. leprae* could survive much longer (up to eight months). It was also recently found that *Acanthamoeba* in soil and water samples from around houses in West Bengal appeared to provide protection for *M. leprae* (Turankar et al. 2019). Additional studies are now being done to see if amoebae also play a role in transporting leprosy bacilli through broken skin or through the nasal mucosa.

Other researchers have used geographical information systems to explore the relationship between environmental factors and leprosy prevalence. For example, in Ethiopia, Tadesse Argaw et al. (2006) studied the presence of leprosy in over 100 health institutions in association with the photosynthetic activity of plants, levels of rainfall, temperature readings, and the amount of moisture in the climate. For each health clinic, the research team took into account elevation, longitude and latitude, land use, soil types, amount and types of infrastructure, political boundaries, and population data. Their results showed that leprosy might be associated with specific environmental features, for example thermal-hydrological regimes. While it is possible that *M. leprae* in the environment can be transmitted to humans, even when *M. leprae* are found in environmental sources, it does not necessarily follow that they are viable bacteria or that they are infectious reservoirs (Salipante and Hall 2011).

Genetic Predisposition

There has been much debate about whether people who contract leprosy have a genetic predisposition to the infection and if this relates to susceptibility and resistance and/or the type of leprosy that develops (Bryceson and Pfaltzgraff 1990). In the nineteenth century, Norwegians Danielssen and Boeck (1848) were chief protagonists of the hereditary theory of leprosy transmission, but they also acknowledged that it was possible for leprosy to occur "spontaneously." The idea of heredity in the occurrence of leprosy was advocated until it was challenged at the end of the nineteenth century, when Hansen finally identified the bacillus (Duncan 1985a; Scollard et al. 2006). Following identification of *M. leprae*, researchers believed that familial patterns of transmission were the result of transfer of the bacteria through close contact with family members (Spickett 1962a). Support for the idea had also been shown on the Shetland Islands in Scotland because people who had leprosy there came from only a few families. In the 1960s, Spickett maintained that there was still confusion in the scientific/medical community in relation to inherited leprosy and leprosy that was contracted because of a predisposition. However, some took the fact that the prevalence rate of lepromatous leprosy almost never rose above 1 percent anywhere to mean that there was a constant prevalence of a genetically determined defect in all populations (Newell 1966). Spickett also noted that "Caucasian" groups had a higher frequency of leprosy than "Negroid" groups, but the evidence for this claim was weak (Spickett 1962a, 1962b). He argued that because there were gene clusters in families, genetically determined traits must cluster too. While clusters had been found and family clustering had been studied in the mid-twentieth century, there were methodological difficulties in studies. Although much research has looked at the differences between the frequency of leprosy and the expectation of how often leprosy would occur in families, a large proportion of such studies failed to account for family size. In addition, many family cluster studies dealt with households rather than families. This meant that people in those studies were not necessarily related (Spickett 1962a, 1962b). Because age at infection correlates with family size, comparisons between family studies must also consider age distributions. It is now generally accepted that there may be a genetic component to contracting leprosy, but there is still debate about this topic (Rodrigues and Lockwood 2011). Researchers do not know how and when leprosy is inherited and which leprosy-associated genes are important in this process (Yang et al. 2012), and scientists have also not fully explored how *M. leprae* proteins contribute to the biology and pathogenesis of leprosy (Wiker, Tomazella, and de Souza 2011; Parkash and Singh 2012).

Scholars have studied both the present and past of leprosy and have benefited in many ways from the genetic sequencing of *M. leprae,* as we have already seen (Cole, Eiglmeier et al. 2001). Some have explored the mechanisms of leprosy transmission while others have focused on the origin and evolution of the bacteria (e.g., see Zhang, Liu, et al. 2009). Overall, this work has enabled us to understand that immunity to *M. leprae* is controlled at two levels: 1) genetic determinants of overall susceptibility and the type of clinical leprosy that develops; and 2) resistance to the organism (Scollard et al. 2006; Fernando and Britton 2006; Mazini et al. 2016). The data from strain typing have also provided some understanding of gene variants in leprosy that are involved in the immune response to the bacterium and the subsequent different expressions of the infection along its spectrum (Alter et al. 2011; Rodrigues and Lockwood 2011; Cardoso et al. 2011). Fine (1981) has suggested that the degree to which a person is susceptible must reflect an underlying genotype consistent with the response of the individual. However, questions of how and to what degree genetic factors determine how a host responds to exposure to leprosy or to infection were not answered in any serious way until fairly recently.

For example, the human leucocyte antigen (HLA) and the major histocompatibility complex (MHC), both proteins, regulate human immune systems. Studies have identified several genes in the MHC region that may be associated with susceptibility to leprosy, and leprosy has clear phenotypes that are associated with the genetic characteristics of the person affected (Sauer et al. 2016), such as eye color and disease history. HLA-linked genes also influence the development of lepromatous leprosy and tuberculoid leprosy. These genes appear to determine which type of the disease develops. HLA-DR2 and DR3 (antigens of the human leukocyte blood cell that induce an immune response to fight infections) seem to be linked to susceptibility to leprosy in general (see Siddiqui et al. 2001; Zhang, Liu, et al. 2009; da Silva et al. 2009; Hsieh et al. 2010; Lavado-Valenzuela et al. 2011). In recent research focusing on medieval skeletons from five archaeological sites in Denmark, Krause-Kyora et al. (2018) found a significant association of an HLA Class II region (DRB1*15, 01) and susceptibility to lepromatous leprosy. In combination with DQB1*06, 02, it is common in people in Europe today and is associated with inflammatory disorders (such as ulcerative colitis) and multiple sclerosis. It also protects against Type 1 diabetes.

It is clear from published research that people can have genes that make them more likely to contract leprosy, that influence progression from infection to overt disease, and that determine the likelihood of developing leprosy reactions (sudden-onset acute aggressive inflammation in skin lesions). Skamene and colleagues (1982) were some of the first authors to generate data about genetic determinants of overall susceptibility to leprosy. The macrophage protein 1 gene (NRAMP1), which is associated with natural resistance to infections, is found on chromosome 1 in mice. NRAMP proteins influence pathogen viability and/or replication within macrophages. In 1998, Abel et al. were the first to associate this gene with susceptibility to leprosy. The genes that influence the immune response to *M. leprae* occur at both innate (through the genes that control overall susceptibility or resistance to the infection) and acquired levels if innate resistance is insufficient and an infection has been established (Scollard et al. 2006). Acquired immunity is primarily mediated by the functioning of the T-lymphocytes within the immune system when there are antigen-presenting cells. T-cells protect the body from disease by killing infected cells. What mechanism lies behind how NRAMP1 influences susceptibility to *M. leprae* is unknown, but not all studies find this association (see Mira et al. 2003).

Genetic indicators have been used to explore leprosy susceptibility in families. Schurr and colleagues (2006, Vietnam) found an association between leprosy and chromosome 6q25-q27, but only in people with tuberculoid leprosy. They suggested that there is a susceptibility gene in people with all forms of leprosy and a specific susceptibility gene for each type of leprosy (see also Wang et al.'s 2012 study of Han Chinese in southwest China). People who are genetically susceptible to leprosy but have no genetic risk factors for the advancement of *M. leprae* infection to clinical disease constitute the unknown reservoir that makes continued transmission possible in endemic communities. Bakija-Konsuo and Mulić (2011) identified an allele (a variant form of a gene) protective for leprosy in the PARK2 gene in people from Mljet Island in Croatia, where those with leprosy had been segregated in the past. Of particular interest are genetic variants of NOD2, the gene that is linked to inflammatory bowel disease (Cho 2001; Behr and Schurr 2006; Grant et al. 2012). This gene recognizes bacterial molecules, stimulates immune reactions, and regulates the innate immune response. It also influences the risk for leprosy and the type of the disease a person contracts and can affect susceptibility to leprosy and the development of leprosy reactions (Zhang, Huang, et al. 2009; Berrington et al. 2010).

Two amino acid chains, TAP 1 and TAP 2, that are present in the protein of all cells are also associated with the tuberculoid type of leprosy (Rajalingam, Singal, and Mehra 1997). The tumor necrosis factor further plays a major role in nonspecific inflammation and innate resistance. It is one of the most powerful stimulants of cell-mediated immunity and is generally associated with resistance to *M. leprae*. Its gene could be responsible for the different clinical types of leprosy. Toll-like receptors (TLRs) are another type of molecule on cell surfaces that have been found to be important in the recognition of pathogens (Kang, Lee, and Chae 2002). TLR2 controls the production of cytokines (substances that immune-system cells use to signal to other cells), cell signaling, and other aspects of resistance to *M. leprae* (ibid.). While TLRs are very important for pathogen recognition (in this case the immune response to mycobacteria), they are also necessary for the production of interleukin 12 (IL-12), a pro-inflammatory cytokine that is responsible for activating cells and destroying tissue in leprosy reactions (Underhill et al. 1999; Brightbill et al. 1999; Kang and Chae 2001). Polymorphisms (genetic variations resulting in several different forms of individuals among the members of a single species) in TLR2 are associated with a higher risk of a particular type of reaction (reversal) in leprosy (Bochud et al. 2009; see also chapter 2, this volume). Mutations in TLR2 confer susceptibility to severe infection by mycobacteria, and TLR4 polymorphisms have been associated with susceptibility to leprosy. Finally, when Schuring et al. (2009) studied the association between a polymorphism of TLR1 N248S and susceptibility to leprosy, they concluded that TLR1 function may affect both progression from infection to disease and the course of the disease. Studies of cytokine gene expression in leprosy lesions also provide a detailed picture of the immunological characteristics of lepromatous leprosy and tuberculoid leprosy (Scollard et al. 2006). These confirm that tuberculoid leprosy is a response to delayed hypersensitivity and cellular immunity and that lepromatous leprosy develops when immune recognition occurs but the infected person is unable to develop cellular immunity to *M. leprae*. In recent experimental work, exposure of human immune cells to *M. leprae* affected the expression of inflammatory cytokines, indicating that the bacilli can cause a detectable immunological shift (Crespo et al. 2016).

Intrinsic and Extrinsic Factors Related to Leprosy Occurrence

Intrinsic traits that a person has little control over (e.g., biological sex) and extrinsic factors that define the characteristics of that person's life can all contribute to the occurrence of leprosy. Discussions about genetic predisposition to leprosy are also relevant/related to ancestry and ethnicity. Indeed, in a recent study, it was found that the ancestry of some groups of people in Brazil might be a factor in leprosy risk. For example, European ancestry can increase risk, while African ancestry decreases risk (Pinto et al. 2015, 8). The former is explained by the high number of people who came to Brazil from Portugal, increasing the "frequencies of alleles of susceptibility on Brazilian populations." A related study found that European ancestral composition was common in Andean Colombian populations with leprosy, while in the Atlantic area African ancestry predominated (Cardona-Castro et al. 2015). Both studies thus show the part migration can play in the occurrence of infectious diseases (see Cunha et al. 2015).

Intrinsic Factors

Underlying genetic susceptibility and resistance and the immune response to leprosy are additional to the intrinsic factors discussed here that determine the appearance and outcome of leprosy (Britton 1993, 508), discussed already. Essentially a person is born with a set of genes that are intrinsic for that person, although gene therapy is increasingly used to treat disease (Dunbar et al. 2018). However, two other important intrinsic factors, age and biological sex, may also be related to and affect immune system strength.

Age

The age at onset of leprosy is variable and is related to a variety of factors, including where a person works and other environmental variables (Guha et al. 1981). However, it can be difficult to access data about such factors because patients may not be aware of the actual onset of their infection, they may hide skin lesions if they are aware of their significance, and in the early stages of the disease they may have no sickness or disability. Contemporary data on frequency rates by age are therefore presented as "age at detection" rather than as "age at onset" (Yawalkar 2009, 22). People may also not access diagnostic tests or may evade being diagnosed.

Some studies have shown that people can first be affected by leprosy when they are very young. However, the majority are 20–39 years old at diagnosis (Sehgal et al. 1977). The first appearance of leprosy in people is much more common in adolescents and adults than in infants and young children, although debates remain about whether leprosy can be contracted at very early ages (Rao et al. 1972; Noussitou, Sansarricq, and Walter 1976; Fine 1982; World Health Organization 1985; Bryceson and Pfaltzgraff 1990; World Health Organization 2009b). However, in areas where leprosy is endemic, it may increase in children at times of stress such as during growth spurts (Rao et al. 1972). It is also more common in children living with adults who have leprosy (Fine 1982; Melsom, Duncan, and Bjune 1982; Dave and Agrawal 1984). The presence of leprosy in children indicates that there is a continuation of transmission within a community and that transmission is occurring within households (Dave and Agrawal 1984; Butlin and Saunderson 2014). Young children rarely present with lepromatous leprosy; leprosy in children tends to be tuberculoid or a borderline or indeterminate type (Kishmore, Kamath, and D'Silva 1985; Kant and Mukherji 1987). Some of the youngest patients with leprosy have been reported from Japan (Nakajo 1914, 2½ months old), and Martinique in the West Indies (Montestruce and Berdoinneau 1954, 3 weeks old), and occasional studies have found higher than expected rates for lepromatous leprosy in children (e.g., Palit and Inamadar 2014). Frequencies rise for individuals 10–20 years old; the peak age of onset is 20–30 years. Frequency is high again for individuals 30–50 years old, and then it decreases in the older age groups (Kumar, Mathur, and Rao 1973; World Health Organization 2009a, 2009b). It has been suggested that the decline in later life could be attributable to recovery (mainly for people with tuberculoid leprosy), selective mortality (mainly for people with lepromatous leprosy), or poor identification of "cases" (Doull et al. 1942).

While leprosy in infants and young children is more rare than in adolescents and adults, control in this vulnerable group today is a key challenge, as noted in South Asia (Sachdeva et al. 2011). Early detection and prophylactic measures are of paramount importance in preventing impairment and disability in later life (Sachdeva et al. 2011; Butlin and Saunderson 2014). However, because the incubation period of leprosy can be long, leprosy in an infant one year old or less is considered likely to be uncommon. This was confirmed by a large study of the Leprosy Registry at the former Armed Forces Institute of Pathology in Bethesda, Maryland, in the United States; published reports and questionnaire data from people working with patients with leprosy were collected (Brubaker, Meyers, and Bourland 1985). However, because newborn infants can possess *M. leprae*-specific IgM (immunoglobulin) antibodies, intrauterine tolerance is possible in some infants. However, as cell-mediated tolerance can be induced in neonatal mice, early intrauterine exposure to antigens of *M. leprae* could also suppress the cell-mediated response to infection by *M. leprae* in humans; this could promote leprosy in infancy (Carnaud, Ishizaka, and Stutman 1984).

Miscarriages can be common in women with leprosy today and the fetus may grow more slowly in utero. In addition, newborn babies may have a low birth weight and develop at a slower rate than normal (Duncan 1980, 1985b; Lewis 2002, 2008). While leprosy is rare in children less than two years of age, frequency rates can increase in the age categories of 5–9 and 10–14 years (Noussitou, Sansarricq, and Walter 1976). This is supported by data from a large study of children 0–14 years old with leprosy. Ninety percent of the cohort of 1,000 was over 6 years of age and the majority of that cohort had tuberculoid leprosy with a high rate of nerve involvement (Nadkarni et al. 1988). Today it can be common for children under 15 years of age to have leprosy in leprosy-endemic countries; a majority of these children are located in Southeast Asia (37% of the global total for 2018: WHO 2019).

Many studies have focused on leprosy in pregnancy, on leprosy in young infants and children, and on the mechanisms of transmission (e.g., Duncan, Melsom, et al. 1981; Duncan, Pearson, and Rees 1981; Duncan et al. 1982; Duncan, Miko, Howe, et al. 2007; Duncan and Oakey 1982; Duncan 1980, 1985a, 1985b, 1993). For example, pregnancy has been long associated with the first appearance of leprosy or with a worsening ("downgrading") of the disease. After childbirth, leprosy may develop in women or become more severe; this is probably associated with suppressed cell-mediated immunity (Tajiri 1936; Hardas, Survey, and Chakrawarti 1972; Duncan et al. 1981; Duncan 1993). Leprosy reactions can also occur in pregnancy and during lactation, both because of hormonal imbalance and as a result of immunosuppression (Arora et al. 2008)

It has been noted that Type 2 reactions peak in late pregnancy and both Type 1 and Type 2 can be experienced into the postpartum period. Nerve inflammation (neuritis) affects women with leprosy who are pregnant or lactating, which can lead to loss of sensory and motor functions. It may be that leprosy onset or leprosy reactions seen in pregnant women occur because many women in leprosy-endemic countries spend a large part of their adult lives either pregnant or lactating or because pregnant women have more contact with health services and are therefore more likely to be diagnosed (Fine 1982).

Can babies be born with congenital leprosy? If this possibility does exist, it is very rare (Butlin and Saunderson 2014). This has been the subject of debate for many years and of course is relevant to ideas about how leprosy is contracted. It is also relevant to the developmental origins hypothesis (the Barker hypothesis), which states that health problems in early life can predispose people to poorer health in later life (Barker 1994). There have been reports of leprosy in children less than one year of age, and Brubaker, Meyers, and Bourland (1985) have reported that these children were exposed to *M. leprae* in utero. However, in order to prove that a child was infected in utero, it would be necessary to differentiate leprosy onset from infection contracted immediately after birth from the mother. Bacilli could be transmitted through the placenta or via amniotic fluid (Duncan et al. 1983), and some studies have proved this to be the case (e.g., Melsom et al. 1981; Duncan et al. 1982), as noted above. Once an infant is born, breast milk may also be a route of transmission (Pedley 1967, 1968). While children may miss being diagnosed because early skin lesions are overlooked, the routes of transmission of leprosy and the context in which children live are of course highly variable but relevant to whether they contract leprosy.

Biological Sex and Gender

Before discussing the impact of biological sex on leprosy occurrence, a comment about the difference between sex and gender should be given. This is because the terms are confused so often and gender is constantly used as a catch-all word to refer to biological sex, certainly in the media. Sex is biological and gender is a social construct (see Walker and Cook 1998; Springer, Hankivsky, and Bates 2012). A person's biological sex may influence whether they contract leprosy, but how they live and work—social constructs that relate to gender—will also influence whether they contract the infection and how their disease progresses. Thus, gender could be considered an "environmental" factor in leprosy occurrence, but it is discussed here together with biological sex. For example, a woman working in agriculture who has leprosy will be more likely to develop injuries to desensitized hands and feet and consequently develop ulcers and infection of hand and foot bones than a woman who is working in a more sedentary job. Health-seeking behavior may also vary between men and women, thus affecting the course of the infection. In Western societies, men might be less likely to seek medical care, but in developing countries the opposite can occur. In those countries, women's mobility may be limited to such a degree that they cannot easily seek care (Courtenay 2000). However, sex differences in health and in reported symptoms seem to be to be more related to psychological differences between the sexes than to differences in biology (Gijsbers van Wijk, Huisman, and Holk 1999), thus showing again how complex studying the relationship between sex, gender, and health can be.

It is well known that biological sex differences exist in the occurrence and experience of some diseases (e.g., see Pollard and Hyatt 1999), including leprosy, and that physiological, socioeconomic, and cultural factors are also very important. Women have a stronger immune response to *M. leprae* than men do (as they do to infections in general), have a lower incidence rate, and experience less severe clinical forms (Ulrich et al. 1993). They are also better buffered against environmental risk factors (Stini 1985; Stinson 1985; Ortner 1998). Men are thus believed to be more susceptible to leprosy. In some studies, female rates are generally lower than male rates for all leprosy types. This suggests that women have a greater resistance to the infection except in stressful periods of their life (e.g., pregnancy), which may of course be related to age (Rao et al. 1972: Tamil Nadu, India). Higher rates in adult men have also been noted, especially for lepromatous leprosy, as has a male predominance following puberty (Fine 1982 and Britton and Lockwood 2004, respectively). Essentially, men and women and boys and girls, experience different risks over their lives that predispose them to health problems. Some diseases affect women more than men and some affect both equally (Le Grand 1997).

However, achieving accurate frequency rates for males and females with leprosy can be very challenging and subjective. These data ultimately contribute to global figures. Finding and documenting people with leprosy can be active or passive (Le Grand 1997). Passive detection includes voluntary reporting, referral, and notification. However, passive detection is not efficient because of the stigma linked to leprosy in many parts of the world that prevent voluntary reporting. Active detection includes general reporting and contact reporting. The former involves house-to-house inquiry or gathering populations into a central place. This may be more accurate, as it should ensure that all women are seen, but when the community is invited to a central place, females may not attend. Contact reporting involves a survey of all contacts of people with leprosy, but it only identifies a small proportion of new "cases" in a specific area. Actual detection of people with leprosy is generally lower for women, due to factors related to both biological sex and gender and to the fact that women do not often voluntarily report leprosy (Kumar et al. 2004). For example, women in India have been identified in some studies as isolated, living in poor conditions, and generally more poorly educated about leprosy. These factors put them at a high risk for leprosy and make it less likely that they will be diagnosed (Morrison 2000; Thilakavathi, Manickam, and Mehendale 2012). However, in some African countries detection rates are similar for men and women (for example Le Grand 1997: Kenya), although female rates can be higher (as they are, for example, in Burkina Faso; see Tiendrebeogo et al. 1996).

Exposure to disease is clearly affected by a person's occupation, socioeconomic status, and lifestyle. Men in areas of high leprosy frequency might be infected because they have more mobility and more opportunities to have contact with infected people than women do (see also World Health Organization 1985). Women often work within the home, which may reduce the risk of infection, but that lower risk could change as more women in developing countries are beginning to work outside the home. Women with leprosy can also experience more stigma, isolation and rejection from their families and more restrictions than men (Rao, Garole, et al. 1996; Rao, Khot, et al. 1996), and in some regions of the world they may have reduced access to health care services (Santow 1995). Factors related to this include lack of decision-making power, poor awareness/literacy, lack of money to pay for travel costs to health care facilities (or less awareness of them), low mobility, and being restricted to the home environment (e.g., see Crook et al. 1991, India). In a study Peters and Eshiet (2002) conducted, women were less aware of the signs and symptoms of leprosy and the available health services, and men or their mothers often controlled their access to health care. The time between observing symptoms and suspecting that they were leprosy related and the time between suspecting leprosy and seeking diagnosis was also longer for women. As a result of these delays, more leprosy-related deformity is documented in women. Some studies in Africa show that women have little power and are dependent on men, even for making decisions about their health requirements (Morrison 2000, study groups in northwest Botswana, Ethiopia, Tanzania; see also Varkevisser et al. 2009; Mull et al. 1989; Rao, Garole, et al. 1996; Rao, Khot, et al. 1996; Kumaresan and Maganu 1994). In some cultures male health care staff are not allowed to examine women, even though leprosy clinics may only have male staff.

As unmarried girls and women cannot expose their bodies to male health care staff to enable diagnosis, they may not be properly examined or diagnosed. Indeed, "it is improper to examine females completely in some cultures" (Fine 1982, 169; see also Butalia 1992), and social stigma may encourage concealment of leprosy in young girls and young women. Women may also be conditioned to ignore health problems. Although women are generally more compliant with treatment than men, they may not take medication because of the potential side effects (Mull et al. 1989, Pakistan). In India, husbands of women diagnosed with leprosy have insisted on separation because of various late nineteenth- and twentieth-century marriage acts (Kaur and Ramesh 1994), but in 2019 India removed leprosy as a grounds for divorce. Some women in India voluntarily withdraw from the household to spare their families "contamination" (Vlassoff, Khot, and Rao 1996). In contrast, women with leprosy in Africa are well integrated into communities (Kumaresan and Maganu 1994, Botswana).

In summary, the effects of biological sex and gender on leprosy frequency rates are extremely complex, but the data are not consistent across cultures or even within cultures (Schäfer 1998). However, an awareness of the potential factors that contribute to reported frequency rates across the world is important when interpreting the differences between men and women, and other genders, as is a consideration of the various lifestyle factors that might affect whether men or women contract leprosy. However, educating schoolchildren about leprosy may lead to better awareness in their parents, especially their mothers.

Extrinsic Factors

Many extrinsic factors may put a person at risk of contracting leprosy. These include their indoor and outdoor living environment, the food they eat, the extent of their mobility, which in turn relates to who they come into contact with, the type of work they do, and whether the health care they have is effective at diagnosing and treating leprosy. The intrinsic factors described above are intertwined with the extrinsic factors described below, as is social status. For example, the quality of a person's diet may be linked to their status; this will directly affect the strength of their immune system and resistance or otherwise to leprosy. Status may also be linked to the quality of care accessed; higher-status people may receive better care (Heijinders 2004). Social status also influences each person's conceptualization of leprosy. Considering the differences in all of these factors within communities over time and space, the task of disentangling the effects of each component that contributes to how a person contracts leprosy is very complex. In addition, people and communities are differentially exposed to risks for leprosy but may or may not be able to mitigate those risks.

The Environment

Environmental factors in relation to the presence of leprosy bacteria have already been discussed separately above, but indoor and outdoor living conditions naturally include many variables that place a person at risk for developing leprosy. For example, poor living conditions could compromise their immune system and predispose them to leprosy or to its progression to a more severe form. Other factors include the number of people living in a house, the structure of the house (for example, whether ventilation is adequate), and whether a person is living in a rural or an urban area. This also extends to how much social contact people experience and if that contact puts them at risk from contracting leprosy. As already discussed, biomolecular analysis has shown that people living in the same household or village are more likely to have the same strains of leprosy, thus emphasizing that household and community interactions are important risk factors. The same genotype of the leprosy bacteria has also been found in people with leprosy and in soil samples near where they live. It is clear that the combination of living in an urban environment with high population density and crowding in residences can promote leprosy (Cabral-Miranda et al. 2014).

Another relevant and increasingly important finding is the association between leprosy and vitamin D deficiency. The vitamin D receptor in the body's tissues and cells is important in vitamin D function (Kato 2000), and the alipoprotein E (ApoE) is linked to fat metabolism and to *M. leprae* infection (Wang et al. 2017). (An alipoprotein is a protein that combines with a lipid to form a lipoprotein.) One of ApoE's alleles (E4) is associated with high levels of vitamin D in the body. Low levels of this receptor have been found to influence how leprosy develops in people who are experiencing leprosy reactions (Mandal et al. 2015). In a review of the subject, Lường and Nguyên (2012) illustrated that vitamin D also plays a role in whether a person actually contracts leprosy. Exposure to ultraviolet light, which leads to vitamin D synthesis in the skin, may inhibit bacterial growth. Living in crowded conditions or being exposed to particulate pollution can lead to lack of exposure to ultraviolet light, as might be the case if a person is working for long hours indoors or wearing clothing that prevents exposure to ultraviolet rays. Therefore, preventing synthesis of vitamin D in the skin ultimately affects normal bone metabolism and reduces resistance to leprosy.

Of relevance here is the finding of an association between higher rates of vitamin D deficiency in people with tuberculosis who were not exposed to ultraviolet light because they lived on the lower levels of a high-rise building in Hong Kong. The study suggested that lack of access to ultraviolet light may have been a factor in their vitamin D deficiency (Lai et al. 2013). This shows how spatial data (i.e., where and how people live) can provide much more nuanced interpretations of why people contract specific diseases, as can latitude.

The presence of crowded living spaces varies by country, region, and culture. Exposure to crowding will also vary according to biological sex, gender, age, and status. Some authors who have considered crowding and its effect on leprosy transmission find that overcrowding in the home and closeness of contact may be related to transmission of leprosy (Fine 1982) but find a weak relationship between sleeping space and leprosy rates. People with leprosy sleeping close to those without leprosy did not result in a higher risk of infection transmission from infected to uninfected. However, crowding can be confounded by many other factors and could be secondary to having leprosy. For example, leprosy could lead to lowered socioeconomic status and crowded living conditions (Ojha, Chaudhury, and Choudhary 1984).

Diet

A well-balanced and sufficient diet will strengthen and maintain the resilience of the immune system of a person (Scrimshaw 2000), which in turn will create a situation where resistance to *M. leprae* may prevent them from contracting leprosy (see Dwivedi et al. 2019). The amount of food a person eats and how nutritious it is will affect how frequently and severely they experience an infectious disease and their rate of recovery. Infections can also have an effect on particular nutrients; for example, infection can lead to lower protein levels (Scrimshaw 2000), and helminth infections may coexist with vitamin deficiencies in people with leprosy (Fairley et al. 2019). While food poverty is a very important risk factor for people with leprosy, more work needs to be done to establish the underlying processes that contribute to it.

Deficiencies of specific nutrients are known to be risk factors for leprosy (Oktaria et al. 2018). For example, protein calorie malnutrition (PCM) affects cell-mediated immunity. This is the body's major defense mechanism against the capability of a pathogen to grow and reproduce inside the cells of a host, especially pathogens that cause chronic infections such as *M.*

leprae. PCM coexists with leprosy in most developing countries. Severe PCM before contact with *M. leprae* also enhances the chances that leprosy will develop beyond the subclinical stage. PCM after infection may then moderate the illness. Thus, low protein intake can increase susceptibility to leprosy (see Fine 1982 on protein deprivation and depressed cell-mediated immunity responses). Paradoxically, in the early twentieth century, eating fish (especially rotten fish) was linked to a predisposition to leprosy. This perhaps derives from the experience of leprosy in Norway, where it survived until the 1950s (see Hutchinson 1906). As Norwegians consumed a high proportion of fish in their diet, eating fish became associated with leprosy. This was also noted in a review of nutrition and leprosy (Foster et al. 1988). While fish liver oils have high levels of both vitamins A and D (O'Keefe 2000, 386), paradoxically a high level of the latter in the body can protect against leprosy (see above). However, the presence of sufficient quantities of vitamins A, C, D, and E and the B groups have been recorded as particularly influential on the progression of leprosy in a person: leprosy was more likely to progress rather than decline in severity. As Dwidevi et al. (2019) have indicated, more emphasis on the diet in the treatment of leprosy is needed.

Lower levels of vitamin A and zinc have also been associated with depressed immune systems (Good, West, and Fernandes 1980; Smith et al. 1976). The liver is affected in all forms of leprosy and zinc is needed to mobilize vitamin A from the liver. Lower levels of vitamin A and zinc and have been found in people with leprosy, particularly those with lepromatous leprosy (Agarwal et al. 1973). This may be due to deficient nutrition or because leprosy leads to a defective liver and subsequent poor protein synthesis. While Agarwal et al. (1973) found no difference in iron levels, they found that hemoglobin levels in patients with lepromatous leprosy were significantly lower but that serum ferritin levels (iron storage) were not. Anemia due to iron deficiency can be common in people with lepromatous leprosy (Lapinsky 1992; Sen et al. 1991; but see Oktaria et al. 2018), and co-infections may be one contributory factor to the development of anemia (see Fairley et al. 2019).

It is recognized that the interaction between leprosy and malnutrition is complex and poorly understood. Malnutrition and undernutrition obviously modulate infectious and immune processes (e.g., see Edelman 1979; Ryrie 1948). Undernutrition in particular can lead to a low body mass index and predispose people to leprosy (Rao and John 2012). Nevertheless, some people with infections may benefit from malnutrition. For example, iron deficiency anemia in people with infections prevents pathogens from accessing the iron they need to survive and replicate (see Stuart-Macadam 1992). As Nesse and Williams suggest (1994, 30), "Iron is a crucial and scarce resource for bacteria, and their hosts have evolved a wide variety of mechanisms to keep them from getting it." Cholesterol has also been described as a risk factor for the development of lepromatous leprosy. Ell (1985) suggests that this was the case for the medieval upper classes because they ingested more of it. In a clinical study (Gokhale and Godbole 1957), higher serum lipid levels were found in patients with leprosy and total body and serum cholesterol were reported to be altered, especially in people with lepromatous leprosy. If cholesterol is necessary for the survival of *M. leprae*, then people in endemic areas who eat diets high in cholesterol could be at risk.

Clearly, the quality and quantity of a person's diet, their eating habits, and even food security should be key considerations of any leprosy control program (Teixera et al. 2019).

Socioeconomic Status

Poverty is a predictor of many problems for human society, including diet quality and occurrence of disease (Wilkinson and Pickett 2009). Poverty leading to undernutrition and a less diverse diet is related to socioeconomic status and may be more severe for people with leprosy than it is for poor people who do not have the disease (Raju and Rao 2011, India; Wagenaar et al. 2015, Bangladesh). However, there is no hard evidence linking poverty, poor nutrition, and leprosy (Lockwood 2004; Sommerfelt, Hadfield, and Meyers 1985, India). Socioeconomic status relates to many factors, not just the quality and quantity of food, but a higher risk for leprosy has been noted in studies of people in poorer areas (Cury et al. 2012). A combination of socioeconomic, demographic, environmental, and behavioral factors is clearly important when assessing this outcome. Kerr-Pontes et al. (2006, Brazil) found that a low level of education, having experienced food shortage, bathing weekly in open water, and changing bed linen infrequently were all significantly associated with leprosy.

Other studies have confirmed the association of people with leprosy and lower socioeconomic status (e.g., see Singh et al. 2009). It is logical to assume that in a general sense, poverty would predispose to leprosy, even on the grounds of affecting immune response, but the association is complex. Poverty and poor diet may also lead to stress, which can affect the immune response to disease (Selye 1950), although some researchers claim that the relationship between stress and infectious disease is weak (Sapolsky 2004; Masland et al. 2002). However, the developmental origins theory of health and disease posits that stress and disease in early life can predispose to health problems in later life, which may include infections (Barker 1994). It is important to remember that each person is an individual and has specific risks during their life course that may or may not lead to disease. Indeed, as Wilkinson and Pickett (2009) have indicated, if people were more equal globally, then disease (including leprosy) would not be as much of a problem. As they also show, infectious disease tends to be the one health problem that declines as countries become more affluent.

Mobility

People have been on the move for thousands of years and for many reasons (King 2010). Travel presents risks to the health of those who travel in many ways, varying according to the individual traveler and the places encountered. Experiencing environmental changes such as increased altitude, higher or lower temperature and humidity, and being exposed to different foods, different water quality, different pathogens, and poorer sanitation can all lead to health problems (Zimmerman, Kiss, and Hossain 2011).

When people move, they can transmit the diseases they have to people that have not experienced them before, but they may also become targets for discrimination and become poor (Acevedo-Garcia and Almeida 2012; Murto et al. 2013, Brazil). Migrant groups may also become exposed to new diseases as they travel (e.g., Singh and Chauhan 2018, *M. lepromatosis*) and when they reach their destinations. Much has been written about the effect of travel on the spread of other infectious diseases to new places, such as tuberculosis (e.g., Albert and Davies 2008). An increasing social science literature on place and health highlights the strong link between health and where people reside (e.g., the journal *Health and Place*, dedicated to the study of all aspects of health and health care related to place or location).

However, very little published research specifically links an increase in leprosy in a place to infected people moving there or with people moving to a new region that is endemic for leprosy contracting leprosy because their immune system had not previously been exposed to the infection (but see Massone et al. 2012, Italy; Kwan et al. 2014, Malaysia). Nevertheless, people have migrated and do migrate to new areas with leprosy and have taken and do take leprosy with them (Yu et al. 2015; Aftab, Nielsen, and Bygbjerg 2016; Ramos, Romero, and Belinchón 2016; see also Roa and Morris 2006 on the possibility that leprosy was taken to the Americas 12,000–13,000 years ago). Because the incubation period is very long, it may not be easy to detect leprosy in an immigrant, but transmission to an uninfected person needs long-term close contact. The impact of migration today on leprosy transmission and occurrence does not seem to be as significant as it is for other infectious diseases such as tuberculosis, in spite of recent data that has identified increased mobility of people with leprosy from endemic to non-endemic areas as a risk for others (World Health Organization 2016). Indeed, a study of infectious disease morbidity associated with travel and migration of Europeans within and external to Europe in 2008 found that only one person of nearly 7,000 travelers studied had contracted leprosy (Field et al. 2010). Nevertheless, native populations today have identified immigrants with leprosy as a health risk. This has stoked anti-immigration agendas and has led to the misrepresentation of people with leprosy, particularly by the media. This in turn has led to an increased fear of the infection (White 2010). This has particularly been the case when risks to military personnel have been considered (Chambers, Baffi, and Nash 2009). While the US military has reported a decline in the number of their personnel with leprosy over the years, the incidence of leprosy in military personnel has been documented at specific points in time as being much higher than for the general US population (e.g., Schlagel, Hadfield, and Meyers 1985). One key concern is that clinicians are not necessarily familiar with leprosy in non-endemic countries and that diagnosis and treatment may be delayed.

As people move rapidly, more distantly and more frequently in the future, there is no doubt that travel-related risks for transmission and contraction of leprosy will be more prominent.

Animals and the Bacillus

Unlike tuberculosis, which may be transmitted to humans by many animal hosts (Pfeiffer 2008), leprosy is essentially a human disease, with rare exceptions. Because *M. leprae* depends on host cells to survive, researchers have looked at armadillos and other animals as potential reservoirs that may pose a risk to humans (Salipante and Hall 2011). However, before discussing the different animal model studies in leprosy, the next section considers leprosy in species other than humans more broadly.

As early as the nineteenth century, Armauer Hansen tried but failed to transmit leprosy to other animals by inoculating rabbits, cats, and monkeys. Later, guinea pigs, dogs, rats, and hamsters joined the list (Meyers et al. 1984). An isolated report of a leprosy-like disease was noted in a rat, but the bacillus could not be cultivated (Dean 1905). Later, in the 1960s, the footpads of mice and nine-banded armadillos were successfully inoculated with *M. leprae*. This was the beginning of research that still continues. In 1984, Meyers and colleagues issued an urgent call for animal models for leprosy; ideally, it needed to be an immunologically unaltered animal that had experienced the full spectrum of the disease (Meyers et al. 1984). Meyers and colleagues developed their ideas one year later (1985), suggesting that the best animal model for leprosy was one that duplicated the maximum number of leprosy features and was genetically closest to a human. Even though lepromatous leprosy had been found in armadillos, mangabey monkeys, and chimpanzees, no indigenous reservoir for the latter two could be identified anywhere in the world.

In a more extensive paper on leprosy in wild and domestic animals, Rojas-Espinosa and Løvik (2001) recorded experimental transmission to more than fifty animal species, but the definition of success varied. For example, the rabbit was the first species used in experiments as an animal model for leprosy, but genuine transmission has been documented in a range of other animals: five nonhuman primates (white-handed gibbon [*Hylobates lar*], rhesus monkeys [*Macaca mulatta*], African green monkeys [*Cercopithecus aethiops*], mangabey monkeys [*Cercocebus atys*], and chimpanzees [*Pan troglodytes*]), in three species of armadillos (*D. novemcinctus, D. hybridus,* and *D. sabanicola*), in normal mouse footpads, in rats, in Korean chipmunks (*Eutamias sibiricus asiaticus*), and in experimentally immunosuppressed mice and rats. Murine leprosy (*M. lepraemurium,* which affects rats and mice) was abandoned as a model for leprosy in humans in the 1940s but is still used to understand host susceptibility and resistance to infection by intracellular mycobacteria similar to *M. leprae* (Rojas-Espinosa 1994). Stefansky (1902) first described "mouse leprosy" when he found it in rats in Ukraine, and Marchoux (1912) reported it in England. However, through improved knowledge of leprosy and *M. leprae,* the idea that murine leprosy was a reservoir for human leprosy was eventually rejected.

The first evidence of leprosy in cats was reported in New Zealand (Brown, May, and Williams 1962), and it has been reported elsewhere since then. The characteristics of the cat leprosy bacillus were identical to *M. lepraemurium* (Hughes et al. 1997). More recently, feline leprosy caused by *M. lepraemurium* has also been diagnosed using molecular techniques. The location of the skin lesions was compatible with rodent bites and the owners had seen the cat hunting and eating rats (Courtin et al. 2007, Kythira,

Greece). Similarly, the histological features of dermatitis in red squirrels in Scotland have been likened to the features of lepromatous leprosy and genetic analysis has displayed a similar DNA sequence to *M. lepromatosis* (Meredith et al. 2014). In a more recent study of red squirrels from England, Ireland, and Scotland with and without lesions related to leprosy, both *M. leprae* (Brownsea Island, off the south coast of England) and *M. lepromatosis* (England, Ireland, and Scotland) have been detected (Avanzi et al. 2016). Furthermore, when *M. leprae* and *M. lepromatosis* strains were considered, the two closest relatives to the *M. leprae* strain (3I) found in red squirrels on Brownsea Island had both been identified in medieval skeletons from Denmark (St. Jørgen, Odense 625) and England. The latter derived from skeleton 8 excavated from the St Mary Magdalen leprosy hospital cemetery in Winchester, Hampshire, a location very close to Brownsea Island. Strain 3I is currently endemic in wild armadillos in the United States. Even more recently, more evidence of red squirrel leprosy has been found on the Isle of Wight, also off the south coast of England.[3] However, in a wider study that included samples from squirrels from Europe, including red squirrels, no DNA evidence of leprosy was found (Schilling et al. 2019; Tió-Coma et al. 2019).

The development of animal models for leprosy has been challenging. One obstacle has been the poor quality of the *M. leprae* inoculum (Scollard et al. 2006). However, a large colony of *M. leprae*–infected mice was established at the National Hansen's Disease Program laboratories in the state of Louisiana in the United States that was used to produce, characterize, and provide viable *M. leprae* for research.[4] The research confirmed that the *M. leprae* bacillus likes cooler temperatures: 4° C (39.2° F) for storage and 26–33° C (78.8–91.4° F) for metabolic activity (Truman and Krahenbuhl 2001). Another obstacle was the failure of researchers to recognize the prolonged growth cycle of *M. leprae* and to acknowledge its preference for cooler temperatures.

Nine-Banded Armadillo

Armadillos are uniquely and highly susceptible to *M. leprae* (Kirchheimer, Storrs, and Binford 1972; Storrs et al. 1974; see also review by Oliveira, Deps, and Antunes 2019). Wild nine-banded armadillos (*D. novemcinctus*) are found only in the New World and are natural regular hosts of *M. leprae* (Truman 2005; Hamilton et al. 2008). They also have a low body temperature (30–35°C, or 86–95° F), which favors the survival of *M. leprae* (Figure 1.4). Armadillos were originally chosen as a model for leprosy because of

Figure 1.4. A nine-banded armadillo. Credit: Greg Lasley Nature Photography.

their low body temperature and long life span (Rojas-Espinosa and Løvik 2001). They are also the only immunologically intact animal that regularly develops fully disseminated *M. leprae*. They have been the hosts of choice for several decades for in vivo (within a living organism) propagation of *M. leprae*. Approximately 65 percent of all armadillos experimentally inoculated with *M. leprae* will develop disseminated infection. Armadillos can upgrade or downgrade their response to the bacteria (Scollard et al. 2006). In addition, the microscopic appearance of the affected tissues of armadillos is identical to humans.

The first armadillos were experimentally inoculated with *M. leprae* in 1968 (Walsh et al. 1988). Three years later Kirchheimer and Storrs (1971) reported that they had developed lepromatous leprosy. Seven-banded and eight-banded armadillos have also been artificially inoculated with *M. leprae* and some developed leprosy. A more detailed report of the effects of inoculation in *Dasypus novemcinctus mexicanus* shows that after fifteen months, lesions appeared at the inoculation sites that resembled invasion of nerves in humans (Kirchheimer et al. 1972). In autopsy, widespread evidence of leprosy was seen and was said to be much more severe than in humans. After 1975, naturally acquired leprosy was observed in 100 newly captured armadillos, primarily from Louisiana. The infective changes were identical to those seen in the experimentally inoculated armadillos (Walsh et al. 1981). However, while many of the features of experimentally induced leprosy in the armadillo resemble leprosy in humans, there are differences. In armadillos, it is fatal in 1.5 to 3 years, nearly all of them have lepromatous leprosy, and deformities related to nerve damage are not reported (Meyers et al. 1985). In a DNA study, Frota et al. (2012) reported that the six-banded armadillo (*Euphractus sexcinctus*) also harbored viable *M. leprae*. In support of the theory of transmission from mother to baby, milk and mammary gland tissue of armadillos have been found to contain acid-fast bacilli; bacilli were discovered in the placentae of three pregnant armadillos with lepromatous leprosy and in the spleens of fetuses from them (Job, Sanchez, and Hastings 1987). Researchers suggest that humans are exposed to leprosy-infected armadillos through handling, hunting, and preparing them for eating (da Silva et al. 2018; Kerr et al. 2015; Frota et al. 2012; Lumpkin, Cox, and Wolf 1983).

Today, armadillos are found in Arkansas, Louisiana, Mississippi, and Texas and their geographic range is expanding (Sharma et al. 2015). There are also a few reports of leprosy in armadillos in Central and South America (Deps, Antunes, and Tomimori-Tamashita 2007; Deps et al. 2008). Researchers have identified the same *M. leprae* 3I strain in armadillos and in people with leprosy in the United States who have been in contact with armadillos but not with other humans with leprosy (Truman et al. 2011). Strain 3I, as described above in relation to red squirrels, is a European strain that could have been transmitted to armadillos by humans who migrated to the New World after the fifteenth-century contact period (Monot et al. 2005). However, armadillos appear to have existed for some time in the New World and hunter-gatherers were likely catching and consuming them back to 7,000 years BC (Frotini and Vecchi 2018). Whether they harbored leprosy at that time is unknown.

While humans from outside the New World may initially have infected armadillos after they migrated, armadillos first reached Louisiana from Mexico around 1926 (Truman 2005; Truman and Fine 2010). Leprosy-like disease was first reported in a wild armadillo in 1975, but some scholars suggest that humans had been affected by leprosy in Louisiana for around 150 years at that time, suggesting that the armadillo could not be held responsible (Walsh et al. 1975; Walsh, Storrs, and Meyers 1977). The World Health Organization (1985) was also of the opinion that armadillos were not an important source of the infection in the United States because there was no conclusive evidence at that time.

However, because armadillos can harbor the infection, humans exposed to these animals may be at risk. The source of leprosy in armadillos in Texas and Louisiana remains unknown, but it may have come from untreated patients with advanced untreated lepromatous leprosy before drug therapy was available. Indirect contact with clothing and bandages from people with leprosy, perhaps during the pre-treatment era, is also a possibility (Meyers et al. 1984). Transmission may also occur via insect bites from the nose and ears of wild armadillos; insects can carry *M. leprae* and the majority of an armadillo's diet in the wild consists of insects. It is also important to note that in the United States in the late 1980s approximately two-thirds of native-born patients with leprosy were inhabitants of Louisiana or Texas, the same states where armadillos were being found with naturally acquired leprosy (Walsh et al. 1988).

Bacilli are likely excreted from armadillos via the gastrointestinal tract, skin nodules, the lungs, and the urinary tract, but conclusive evidence of animal-to-human transmission was lacking (Truman 2005). Nevertheless, in 2011 whole-genome sequencing, SNP typing, and VNTR analysis that compared *M. leprae* from thirty-three wild armadillos to *M. leprae* from thirty-nine US patients with leprosy found a unique *M. leprae* genotype in 85 percent (28) of armadillos and 64 percent (25) of the humans. The patients lived in areas where exposure to armadillos was possible (Truman et al. 2011). If armadillo genotypic strains of *M. leprae* are the same as those in humans, then direct transmission from animal to human could be possible, as also seen in a recent study by Bonnar et al. (2018). In this research, the European 3I strain, also affecting armadillos, was found in a Canadian man who had only ever visited Florida. A report of three native-born people with leprosy from the Mississippi Delta with a history of exposure to armadillos, no history of travel, no contact with each other, and no contact with patients with leprosy supports this possibility (Abide et al. 2008; see also Lane, Walsh, and Meyers 2006). This is a major advance in understanding leprosy in armadillos and their relationship to humans. While keeping armadillos as pets may expose humans to leprosy, only low rates of transmission of *M. leprae* from pet armadillos to their owners has been reported (Curtiss et al. 2001), and some studies are mixed in their conclusions about armadillo-to-human transmission (e.g., Clark et al. 2008).

Thus, on the basis of the data so far, armadillos seem a plausible explanation for leprosy in some people in the United States today, especially so with increasing support from biomolecular research.

Nonhuman Primates

The evidence for *M. leprae* infecting and multiplying in nonhuman primates is less clear. Early attempts to infect monkeys and chimpanzees were repeatedly unsuccessful (Nicolle 1905). Later attempts saw better success and a range of nonhuman primates were infected through inoculation with *M. leprae*. While there have been no scientifically accepted reports of infection in the wild, leprosy as a zoonosis in nonhuman primates has been cited as a source of infection in humans in geographic areas where human leprosy was endemic and nonhuman primates were prevalent (Walsh et al. 1988).

The white-handed gibbon was successfully infected in the 1970s (Waters et al. 1978), and the first report of naturally acquired leprosy in a nonhuman primate (a chimpanzee) was published around the same time (Donham and Leininger 1977). The chimpanzee had been imported to the United States; researchers suggested that it had naturally acquired leprosy in Sierra Leone (Leininger, Donham, and Meyers 1980). The disease was "leprosy-like," consisting of thickening of the skin, including the ears, eyebrows, nostrils and lips, and a skin rash. Many acid-fast bacilli were found in nasal swabs and skin lesions. Infection may have occurred via contact with patients with leprosy in Africa; human leprosy was endemic in Sierra Leone at that time (Walsh et al. 1988). Meyers et al. (1985) reported leprosy in a mangabey monkey (*Cercocebus atys*) imported to the United States from Africa in 1975. It was housed securely in a research institute in Louisiana and the first signs of leprosy appeared in September 1979. *M. leprae* was identified in its body tissues, it developed paralytic deformities of the hands and feet, and its face was affected. Its response to antibiotic treatment used for humans was favorable (dapsone at the time), but it was partially resistant to it, which suggested that it had contracted leprosy from a human. Another mangabey monkey that had been housed in the same cage as the previous monkey also acquired leprosy (Gormus et al. 1988). Signs of leprosy appeared seven years after detection in the first monkey and may represent the first reported natural transmission of leprosy among nonhuman primates. Rhesus monkeys (*Macaca mulatto*) and African green monkeys (*Chlorocebus sabaeus*) have also been studied by inoculating them with *M. leprae*; leprosy developed in some of them (Meyers et al. 1984). Recently, Honap et al. (2018) studied the complete genomes of *M. leprae* from three naturally infected nonhuman primates. This research confirmed that *M. leprae* can be transmitted from humans to nonhuman primates and between nonhuman primate species, but no one has yet observed nonhuman primates that contracted leprosy in the wild.

CONCLUSIONS

This chapter has considered the nature of the bacterial causes of leprosy, *M. leprae* and *M. lepromatosis,* and research on their genomes. Paucibacillary leprosy is the high-resistance form of leprosy and multibacillary leprosy is the low resistance form. It is clear that genomic research, including the documentation of different strains of the bacterium and identifying susceptibility and resistance genes, is providing knowledge that is helping researchers track transmission and identify areas in regions of countries that remain challenges for management. While researchers have concluded that leprosy is transmitted through the exhalation and then inhalation of bacteria-laden droplets, other reported mechanisms exist, including contact with environmental sources of the bacteria. Leprosy in children is rare and males are affected more than females. The wide range of intrinsic and extrinsic factors that make people more or less susceptible to leprosy provides a complex picture to navigate when thinking about why any particular person contracts the infection. These factors are also individual to that person. The wild nine-banded armadillo and red squirrel are natural endemic hosts for *M. leprae,* but while nonhuman primates may be affected there is no evidence of them being infected in the wild.

2

How Leprosy Affects the Human Body

> Clinical leprosy is remarkable for the variety of its manifestations.
> Fine (1982, 164)

Leprosy mainly affects the skin, the nerves, and the upper respiratory tract. Some of the earliest signs are skin lesions, which can have a variety of appearances, and damage to the sensory (numbness, tingling, or pain) and motor nerves ("bent" fingers and/or toes, i.e., shortening or stiffening of muscles or skin that results in decreased movement and range of motion). Leprosy also affects other parts of the body, such as the eyes and the bones. Its effects on the body vary according to the strength of the immune system against the bacteria (resistance) but, as with any disease, each person infected will experience the disease differently.

Infection with *M. leprae* leads to a cell-mediated immune response (CMI) to the bacteria, meaning that the infected body cells are detected and destroyed. Lymphocytes (T-cells) are responsible for CMI. Both T-cells and macrophages (large white blood cells) are key to the body's recognition and response to *M. leprae* (Lockwood 2004). Cell-mediated immunity also determines what type of leprosy develops (Godal 1978; Adams 2012). When the bacteria enter the body, they produce antigens that enter the lymph nodes or the spleen through the lymphatic system or the bloodstream, and these initiate the immune response.

In lepromatous leprosy, or the type of leprosy where people have the least resistance to the bacteria, the bacilli target the cooler areas of the body: the eyes, the face, the mucus membranes of the upper respiratory system (i.e., the thin layer of cells that produce mucus), the hands, the feet, the skin, and the peripheral nerves (Bennett, Parker, and Robson 2008). Lesions that develop in the mouth also occur more frequently in the lower-temperature areas (Scheepers 1998).

INCUBATION PERIOD

Once infected, the bacillus multiplies in the skin of the host. Bacilli also replicate in the Schwann cells, the principal supporting nerve cells that wrap around the nerve fibers in the peripheral nerves. This leads to nerve damage. Recent research on Schwann cells has found that *M. leprae* have the ability to differentiate and reprogram Schwann cells into a stem cell–like state (Masaki et al. 2013). This promotes the bacteria's spread around the body, including to the skeleton.

The bacteria have a long generation time of eleven to thirteen days. This is the time it takes for the number of bacteria to double themselves, from receipt of the infection by a host to the maximum infectivity of that host (Bennett, Parker, and Robson 2008). However, most pathogens can reproduce themselves in minutes (Bryceson and Pfaltzgraff 1990) and the related mycobacterium *M. tuberculosis* divides every twenty hours (Fine 1982). The bacteria thrive best at 27–30°C (80.6–86.0° F). The incubation period (time between infection and when a person has clinically detectable alterations in the body) can also be long, but opinions differ. The long length of the period is likely because of the bacteria's slow multiplication rate (Shepard and McRae 1965). The length of the incubation period and length of time it takes for symptoms to develop are relevant to the timing of diagnosis and treatment. For example, an incubation period of six months to twenty years has been recorded for American servicemen. The medians were two to five years for tuberculoid leprosy and eight to twelve years for lepromatous leprosy (Brubaker, Binford, and Trautman 1969). While Suzuki, Udono, et al. (2010) report an incubation period of 39 years in a chimpanzee, the average incubation period for humans is around five years.[1]

One other problem is defining the time between infection and disease. Differentiating between reactivation of a subclinical infection (with no signs/symptoms) and reinfection can be difficult (Fine 1984). It is possible that a subclinical or dormant infection is fairly common but that development into the disease is rare (Lockwood 2004). There is no test to identify people with subclinical infection or to determine when their immune system has responded. It is probable that only 10 percent of people with subclinical leprosy outwardly display the disease.

Classification of *M. leprae*

The nature and strength of a person's immune system is important for determining what infections a person will contract, which clinical signs they will display, and which symptoms they will experience. The quality of a person's diet plays an important part in that process (Scrimshaw 2000; Passos Vázquez et al. 2014). The lepromin test is used to monitor cell-mediated immunity (CMI) and T-cell function in leprosy. The CMI in patients infected with *M. leprae* with multibacillary infection (lepromatous and borderline lepromatous) is defective, but in paucibacillary infection (tuberculoid and borderline tuberculoid leprosy), it is strongly reactive (Binford, Storrs, and Walsh 1976). However, immunity due to subclinical infection is the most common outcome following *M. leprae* infection.

Initially the Madrid system was used to classify types of leprosy, but it relied heavily on clinical evidence and proved inadequate (Davison, Kooij, and Wainwright 1960). In the mid-1960s, Ridley and Jopling (1966) developed a classification system based on histopathology (microscopic analysis of diseased body tissues). This system is used today (see Figure 2.1). It groups people with leprosy into high-resistance, low-resistance, and borderline groups. At the high-resistance end of the immune spectrum (tuberculoid leprosy), the person's cellular response to *M. leprae* is vigorous and there is limited disease development (Lockwood 2004). Few bacilli are found in the dermis (the skin layer below the epidermis, or outer layer) or the nerves, and skin tests are strongly positive. At the opposite end of the spectrum (lepromatous leprosy), the bacilli multiply uncontrollably and the person develops extensive skin lesions and a strong antibody response to the mycobacterial antigens (proteins) (Britton 1993). There is no response to the lepromin skin test. People classified with leprosy in one of the three borderline groups—borderline tuberculoid, mid-borderline, and borderline lepromatous—tend to have a progressive reduction in the cellular response, frequent nerve and skin lesions, a large number of bacilli, increasing antibody levels, and a tendency to downgrade to a lower-resistance type of leprosy.

Some of the terminology has changed in order to create a more simplified classification system so that everyone shares an understanding of the type of leprosy people are describing. In 1982, the World Health Organization recommended dividing people with leprosy into two groups—multibacillary and paucibacillary—based on a bacterial index (World Health Organization 1982; Moschella 2004; Table 2.1). Patients with mid-borderline,

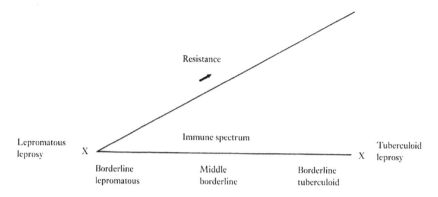

Figure 2.1. Ridley and Jopling's leprosy classification system. Credit: Redrawn by Yvonne Beadnell, Durham University, after Ridley and Jopling (1966).

borderline lepromatous, and lepromatous leprosy were classified as multibacillary and those with primary neuritic leprosy, tuberculoid leprosy, and borderline tuberculoid leprosy were classified as paucibacillary. This simplified system assigns people with one to five skin lesions to the paucibacillary group and people with more than five skin lesions to the multibacillary group. These categories roughly correlate with the effectiveness of cellular immunity and bacterial load and were introduced to simplify and standardize diagnosis and treatment. In indeterminate leprosy, the host response is not differentiated enough to allow classification (Ridley and Jopling 1966). It is characterized by a single skin lesion two to four centimeters in diameter. The lesion has a poorly defined border and loss of color, the person has minimal or no loss of sensation, and bacilli are rarely seen in biopsies of the skin lesion. This is the earliest detectable form of leprosy. Lesions in 50–75 percent of patients with indeterminate leprosy heal spontaneously; the rest progress to a classic form of leprosy. In primary neuritic leprosy, there is asymmetric nerve involvement and no skin lesions; this form of leprosy can develop into tuberculoid leprosy (Yawalkar 2009, 42).

Many health care providers use this system as they make decisions about the best treatment options (Bennett, Parker, and Robson 2008; Rodrigues and Lockwood 2011).) At the same time that it introduced the two classification categories of multibacillary and paucibacillary, the World Health Organization introduced two antibiotic treatment regimes, one for people with multibacillary leprosy and one for people with paucibacillary leprosy (World Health Organization 1982, 1998b). Rifampicin is combined with dapsone to treat paucibacillary leprosy for six months, and rifampicin and

Table 2.1. Types of leprosy

Type of leprosy	Degree of resistance to *M. leprae*
Multibacillary leprosy: Lepromatous (>5 skin lesions)	Very poor/none
Multibacillary leprosy: Borderline lepromatous, mid-borderline lepromatous	Poor
Paucibacillary leprosy: Tuberculoid, primary neuritic, and borderline tuberculoid (1–5 skin lesions)	Fair

Source: Yawalkar (2009, 28).

clofazimine are combined with dapsone to treat multibacillary leprosy for twelve months. A onetime treatment for single lesion paucibacillary leprosy consists of rifampicin, ofloxacin, and minocycline.

While patients can move along the immune spectrum either way, the majority of patients today lie in the borderline categories (Scollard et al. 2006; Bennett, Parker, and Robson 2008). Untreated patients also have unstable immunological responses and tend to move to the low-resistance end (also known as downgrading, or decreased cell-mediated immunity). This instability can increase or decrease as leprosy develops in their bodies. When patients are treated with chemotherapy, they move toward the high-resistant end (improved cell-mediated immunity), also known as "reversal" or "upgrading" (Godal 1978; Bennett, Parker, and Robson 2008). The closer people are to the center of the spectrum (borderline), the more liable they are to move, unlike the polar groups at each end of the spectrum. Some movements are associated with acute inflammation, also called a reaction.

The Ridley-Jopling system of classifying leprosy was key to understanding the relationship between the bacillus that causes leprosy in humans and recognized that the immunological response of the host was related to *M. leprae*'s classification (Britton 1993).

Skin Lesions

The most common early sign of leprosy, apart from loss of feeling (anesthesia), is one or several skin lesions. Lesions are often the presenting sign for a patient (Lockwood 2004). Thickened nerves can also lie in the region of skin lesions and may also be an early sign. The skin is of course a very important part of the body because its appearance has ramifications for social and political life (Millard and Cotterill 2004). It is well known that disorders of the skin may indirectly lead to psychological problems. If the face is affected, as it can be in leprosy, major distress and lowered self-esteem can occur because the face is a very visible part of overall body image.

However, many health problems lead to skin lesions (Yawalkar 2009), not just leprosy, and the appearance of skin lesions for the same disease can vary among different groups, especially given differences in skin colors. This makes diagnosis challenging (Gawkrodger 2004). Leprosy is classed as a "great imitator" of other diseases that affect the skin (Kundacki and Erdem 2019). Clinically, leprosy can resemble other skin diseases, such as contact dermatitis and vitiligo (Anderson et al. 2007), and correct diagnosis of leprosy using skin lesions can be compromised by local cosmetic practices that conceal lesions, occupations that cause disfigurement of the hands, and birth marks (Bryceson and Pfaltzgraff 1990). Thus, diagnosis today using skin lesions can present many challenges. Nevertheless, some characteristics differentiate leprosy from all the diseases or other conditions that cause skin lesions (Lockwood 2004). The different types of leprosy also have recognizable skin lesions (Table 2.2).

Lesions in tuberculoid leprosy are large reddish, copper-colored, or purple plaques (French "plate") with sharp raised outer edges that slope toward a flattened center (Ridley and Jopling 1966; Lockwood 2004). They are dry, hairless, and sometimes scaly, and they lack sensation except when they are on the face. The number of lesions is usually few and they are asymmetrically distributed. They are common on the face, buttocks, and limbs, all of which are cooler areas of the body. However, they can occur anywhere, except the scalp, the armpits, the groin, and the perineum, where the skin temperature is higher. In borderline tuberculoid leprosy, skin lesions are macules (flat discolored area on the skin) or plaques that resemble those of tuberculoid leprosy, but they are smaller, less dry, and more numerous. Their outer edges are less clearly defined, hair growth is less affected, and the associated thickened nerves are not as prominent as they are in tuberculoid leprosy. In mid-borderline leprosy, lesions are intermediate in number and their size is somewhere between that of lepromatous leprosy lesions and tuberculoid leprosy lesions. They have moderate anesthesia. They are either irregular plaques with vague outer edges and an oval hypopigmented center (one that is lighter than overall skin tone) or raised oval or circular bands with well-defined outer and inner edges. In borderline leprosy, lesions are numerous but are not truly symmetrically bilateral over all affected regions. In this type of leprosy, some plaque lesions are

Table 2.2. Skin lesions in different types of leprosy

Leprosy type	Characteristics of lesions	Arrangement	Location
Tuberculoid leprosy	Large plaques that are copper-colored or purple	Asymmetric	Anywhere except scalp, armpits, groin, and perineum
	Sharp raised outer edges		
	Slope toward a flattened center		
	Dry and hairless		
	Sometimes scaly		
	No sensation		
	Few in number; sometimes only one		
	In light-skinned people can be reddish macules; in dark-skinned people can be hypopigmented		
Borderline tuberculoid leprosy	Macules or plaques	Asymmetric	
	Smaller than in tuberculoid leprosy		
	More numerous than in tuberculoid leprosy		
	Less dry than in tuberculoid leprosy		
	Outer edges less clearly defined		
	Hair growth less affected than in tuberculoid leprosy		
	No sensation		
	Smaller than in tuberculoid leprosy		
Mid-borderline leprosy	Intermediate in number and size between borderline tuberculoid and borderline lepromatous leprosy	Asymmetric	
	Moderate loss of sensation		
	Irregularly shaped reddish plaques with vague outer edges and an oval hypopigmented center or raised reddish oval or circular bands		
	Well-defined edges		
Borderline lepromatous leprosy	Numerous	Asymmetric	
	Some plaques large		
	Some plaques punched out		
	Loss of sensation in parts		
	Some nodules dimpled in center		
	Hair growth affected		

Leprosy type	Characteristics of lesions	Arrangement	Location
Lepromatous leprosy	Multiple macules (in early stage) and papules (raised area of skin tissue)	Bilateral and symmetrical	Face, earlobes, buttocks, limbs
	Small in size		Not located in hairy areas
	Copper colored		Oral mucosa (papules on lips, nodules on palate)
	Bacilli in lesions		Uvula, tongue, and gums
	Poorly define outer edges		
	Smooth and shiny surface		
	No loss of sensation		
	Not dry		
	Hair growth normal		
	Small in size		
Indeterminate leprosy	Macules only		Face
	Usually one lesion		Extensor surface of limbs
	Hypopigmented or slight redness		Buttocks
	Diameter of a few centimeters		Trunk
	Slight impairment of sensation		
	Hair growth normal		

large and anesthetic in parts, some are nodules ("lumps" under the skin) that are dimpled in the center, and some plaques are "punched out."

As cell-mediated immunity declines, skin lesions become more numerous (Jopling 1982). In lepromatous leprosy, many bilaterally and symmetrically placed copper-colored macules, papules with distinct borders, and nodules develop (Figure 2.2). There are no associated thickened nerves and there is no anesthesia, dryness, or problems with hair growth. Skin lesions on the face, the earlobes, the buttocks, the limbs, and the oral mucosa (lips, palate, uvula, tongue, and gums) occur in this type of leprosy. In the intermediate form of leprosy, skin lesions are purely macular with hypopigmented lesions that are few in number. Sensation impairment is slight, the margins are poorly defined, and hair growth is normal.

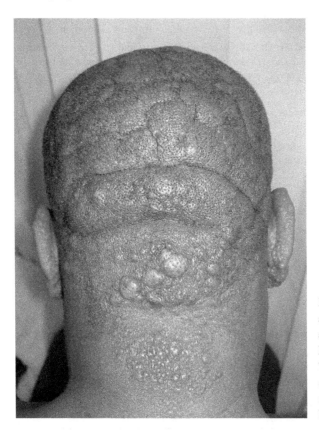

Figure 2.2. Skin lesions in a person with advanced untreated lepromatous leprosy. The skin is thickened and infiltrated with many nodules. Photo by Diana Lockwood.

Leprosy and the Nervous System

M. leprae is the only bacterium that selectively invades the peripheral nerves, leading to irreversible anesthesia in the skin and partial loss of movement or even total paralysis of the muscles of the face and extremities (Binford, Storrs, and Walsh 1976). If leprosy is not treated, nerve damage can ensue, as can impairment and disability (see Table 2.3). *M. leprae* directly affects the facial, median, ulnar, radial, common peroneal (or lateral popliteal), and posterior tibial nerve trunks, along with peripheral nerves that lie superficially in the skin (see Figure 2.3; also Yawalkar 2009, Table 2.3). The nerves become inflamed, thickened, hard, and tender due to mycobacterial antigens. The size of the nerves increases, which can lead to increased intra-nerve pressure and nerve compression (Figures 2.4, 2.5). Peripheral nerve thickening is rarely seen in any other disease (Bryceson and Pfaltzgraff 1990; Lockwood 2004).

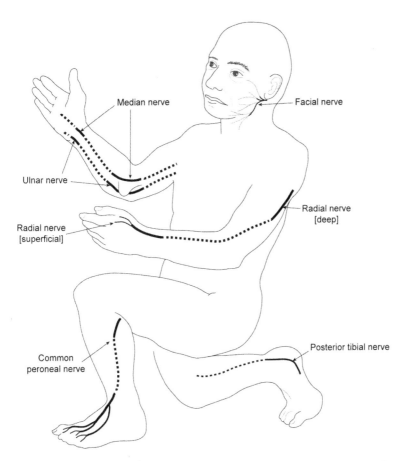

Figure 2.3. Main nerve trunks affected in leprosy. Credit: Redrawn by Yvonne Beadnell after Jopling and McDougall (1988).

Table 2.3. WHO classification of leprosy-related impairments

	1	2	3
Hands	Insensitivity	Ulcers and injuries and/or mobile claw hand and/or slight absorption	Wrist drop or clawed fingers and stiff joints and/or severe absorption of fingers
Feet	Insensitivity	Tropic ulcer and/or clawed toes or foot drop and/or slight absorption	Contracture and/or severe absorption
Eyes	Red conjunctiva	Lagophthalmos (incomplete closure of the eyelid) and/or blurring of vision and/or damage to cornea/iris	Severe loss of vision or blindness

Key: 1 = mild difficulty or possible future difficulty; 2 = moderate difficulty or need for therapeutic action to prevent severe impairment; 3 = severe impairment or impairment too severe to benefit from treatment.

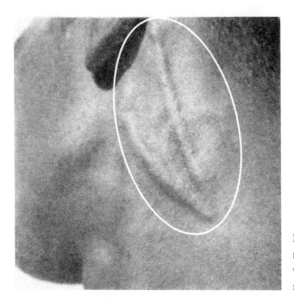

Figure 2.4. Enlarged nerve next to the ear of a person with leprosy. Source: Dharmendra (1947, figure 46).

M. leprae target the Schwann cells in the peripheral nerves because these cells provide an environment that enables the bacteria to survive. *M. leprae* probably enters the nerve via the endothelium (inner lining) of blood vessels and binds to Schwann cells in the innermost layer of connective tissue of the nerve. Cells ingest the bacilli, which can survive only for a limited time in vitro at 33°C (91.4°F). Nerves may be impaired mild to moderately in patients with lepromatous leprosy with a minimal immunological

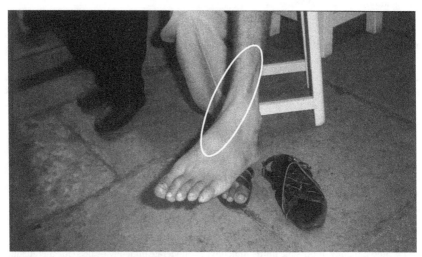

Figure 2.5. Enlarged nerve in the ankle/foot of a person with leprosy. Photo by Charlotte Roberts.

response to *M. leprae*. In all forms of leprosy, selected nerves will undergo demyelination, a process by which the myelin (insulating) sheath of white fatty tissue around certain nerve fibers of nerve cells is damaged. Invasion of the sensory nerves by *M. leprae* leads to loss of sensation in the skin, flexion contractures in fingers and toes (motor nerves), and impaired sweating and vasomotor function—dilatation/constriction of blood vessels (autonomic nerves). The sensory nerves, which are the first to be affected, are the most severely damaged of all types of nerves. Anesthesia, skin dryness, ulcers, scar tissue, and secondary infection develop, as does muscular paralysis (motor nerves) that predisposes people with leprosy to deformities and impairments in their hands and feet. This is because the extremities continue to be used after damage begins because the person cannot feel pain (Bryceson and Pfaltzgraff 1990).

In tuberculoid leprosy, asymmetrical nerve damage occurs early in the disease, earlier than it does in lepromatous leprosy. In lepromatous leprosy, damage to nerves is symmetrical and more widespread and progresses very slowly. People with borderline leprosy usually have the largest number of nerves affected and have leprosy reactions more commonly. If a person has borderline tuberculoid leprosy the nerves are affected asymmetrically, while in mid-borderline leprosy and borderline lepromatous leprosy, involvement is symmetrical. A study in India of people with a range of leprosy types found that 36 percent had nerve damage and that people with multibacillary leprosy and borderline leprosy were most affected. Over 45 percent of those with paucibacillary leprosy had one limb affected and more people were affected in the arms than the legs (mainly due to ulnar nerve involvement). None had motor nerve paralysis without anesthesia, suggesting that paralysis must be preceded by anesthesia (Karat, Rao, and Karat 1972). Calcification of nerves in leprosy is also described but is rare (Trapnell 1965; Chauhan, Wakhlu, and Agarwal 1996).

Today it is imperative to treat leprosy to prevent or stop the progression of nerve damage. Early treatment mitigates impairment and the potential associated stigma. Prevention of nerve damage is thus extremely important in managing people with leprosy. Recent advances have been made in the study of nerve injury. These include improvements in sensory nerve testing, direct examination of immunological parameters in small samples of affected nerves, and using in vitro studies to identify the molecular mechanisms by which *M. leprae* binds to Schwann cells (Scollard et al. 2006). However, because many people with leprosy lack access to early diagnosis, they continue to experience nerve damage and consequent impairment.

76 · Leprosy

Sensory Nerves

As *M. leprae* infiltrates the sensory nerves, a loss of sensation develops. In the arms, the lateral, intermediate, and medial supraclavicular nerves; the posterior, lateral, and medial antebrachial nerves; and the radial nerves are affected. In the legs, the posterior, lateral, medial and intermediate femoral and the saphenous, sural, and peroneal nerves are affected (Figures 2.6, 2.7). Loss of sensation in the face is associated with incomplete closure of

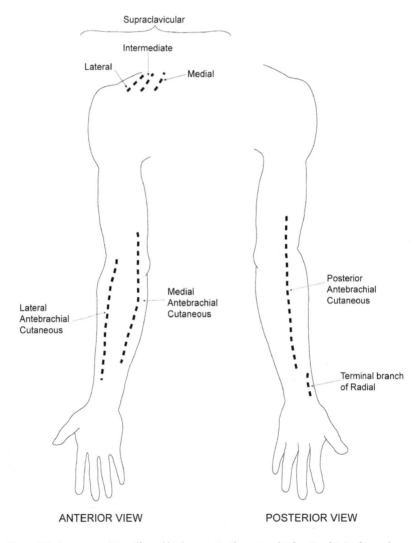

Figure 2.6. Sensory nerves affected by leprosy in the upper limbs. Credit: Redrawn by Yvonne Beadnell after Jopling and McDougall (1988, figure 2.5a).

the eyelids, especially in people with multibacillary leprosy (Daniel et al. 2013).

Because people with leprosy cannot feel pain or extremes of temperature, trauma to the affected parts of the hands and feet can occur with continued use (Table 2.4). This can lead to ulceration and even loss of all the finger and toe bones eventually (Figures 2.8, 2.9, 2.10). In effect, minor injuries can "become major disasters" (Bryceson and Pfaltzgraff 1990, 142). Patients with loss of sensation also experience a change in the way they perceive their body. Some do not regard areas that have lost sensation as part of their body, which can affect how they look after the affected parts.

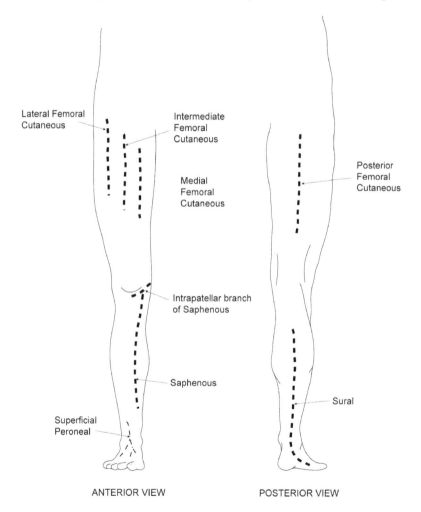

Figure 2.7. Sensory nerves affected by leprosy in the lower limbs. Credit: Redrawn by Yvonne Beadnell after Jopling and McDougall (1988, figure 2.5b).

Table 2.4. Causes and effects of injuries

Cause of injury	Effect
Repeated trauma; thrust and shear forces to foot (e.g., walking on foot with scar tissue)	Bruises, damage to tissues, blisters, ulcers
Prolonged or abnormal pressures (e.g., from poor-fitting shoes)	Bruises, cuts, necrosis, ulcers
Thorns, nails, sharp objects	Puncture wounds and cuts, necrosis
Burns	Blisters, scar formation
Friction (e.g., from poor-fitting shoes)	Blisters, ulcers
Loss of sensation of stretching and pain	Joint dislocation, soft tissue damage, unstable joint

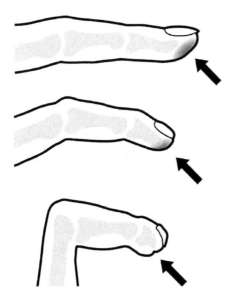

Left: Figure 2.8. Damage caused by repeated injury to an anesthetic fingertip as the finger develops deformity. Credit: Redrawn by Chris Unwin after Bryceson and Pfaltzgraff (1990, figure 10.6).

Below: Figure 2.9. How an ulcer develops in an anesthetic foot: bruising, death of tissue, and breakdown of the skin leads to an ulcer. Credit: Redrawn by Chris Unwin after Bryceson and Pfaltzgraff (1990, figure 10.7).

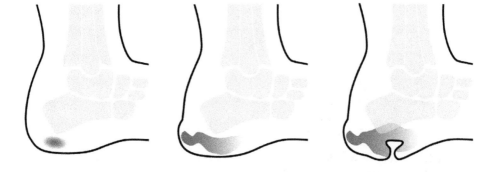

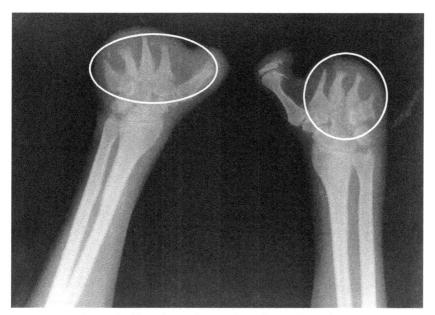

Figure 2.10. Radiograph of hand bones showing loss of bones due to leprosy. Credit: Johs Andersen.

If injuries are not attended to, infection can affect the bones underlying the soft tissues. Price (1964b, 259) described how ulcers develop on the plantar surfaces (or soles) of the feet in such a circumstance: "The crushing effect of body weight on the walking foot is added to a slipping that occurs between the bony framework of the foot and the ground beneath as each step is taken.... The slipping does not occur in a foot with competent musculature; but when sensory and motor loss is present, the loss of control by the intrinsic musculature allows the soft tissues to be ground between bone and walking surface beneath, so that tissue necrosis takes place." Necrotic (dead) soft tissue tracks to the surface of the skin, breaks through it, and causes an ulcer. Ulcers occur at pressure points on the foot. The extent of infection of the tissues and the degree of collapse of the foot architecture determines the level of impairment that ensues (Figure 2.11). The progress of a foot ulcer is primary tissue damage, plantar ulcer, collapse of the foot, and then secondary ulceration.

The soft tissue overlying the heads of the metatarsals, the base of the fifth metatarsal, the heel (calcaneum), the interphalangeal joint of the big toe, and the fleshy parts of the ends of the toes are the areas affected. Ulcers on the metatarsal heads can merge to create a long ulcer across all of the metatarsals. Ulcers at the base of the fifth metatarsal, on the heel, and on

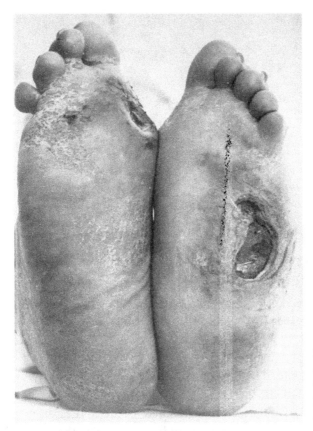

Figure 2.11. Ulceration of the foot in a person with leprosy. Source: Diniz et al. (1960, figure 40).

the head of the first metatarsal can severely affect the future of the foot's integrity. Heel ulcers may lead to damage to the ligaments of the foot because if the ulcer is not healing, the ligaments can become infected. The arch of the foot can collapse, creating friction against and pressure on the resulting new bony prominences. Long-standing ulcers that result in disuse of a person's foot over many years can lead to osteoporosis, then collapse of the foot and secondary ulcers on the lateral and medial areas (edges) of the sole of the foot. Total collapse leads to a boat-shaped foot and a centrally located ulcer. In such a case, the primary ulcers then heal because they are not subject to the pressure and strain of walking. Ultimately, ulcers cause secondary infection of the soft tissues (cellulitis) and bone, and the bone is subsequently absorbed and remodeled (Figure 2.12a and Figure 2.12b). In the soft tissues, scar formation develops and, because scar tissue is weaker and has a poorer blood supply, it can break down and ulcers can recur (Bryceson and Pfaltzgraff 1990). Sometimes amputation of a limb can be the end stage of initial ulceration.

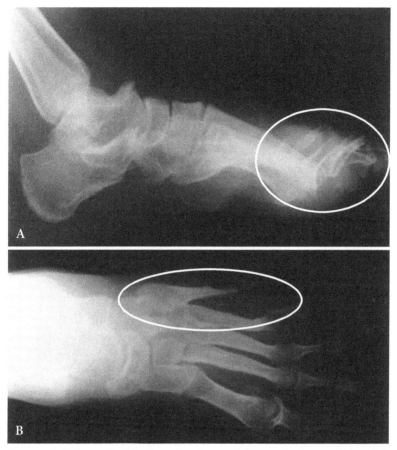

Figure 2.12. Radiograph of the foot of a person with leprosy showing A, flexion deformity, or "bending" of the toes due to motor nerve damage; and B, "thinning" of the fifth metatarsal due to autonomic nerve damage. Credit: Johs Andersen.

Motor Nerves

In motor nerve involvement, muscle atrophy (wasting) and even complete or partial paralysis can occur, including flexion (contracture) of toes and fingers (Table 2.5). Hyperextension of the metacarpophalangeal and metatarsophalangeal joints and flexion of the interphalangeal joints develop (Figures 2.13, 2.14, 2.15). In the legs, the common peroneal, lateral, and posterior tibial nerves are most affected (the latter most frequently), while for the arms it is the median, ulnar (deep), and radial (superficial) nerves. Damage to the facial nerve can also occur, leading to injury to the cornea. If the common peroneal nerve is affected, people find difficulty in dorsiflexion and eversion of the foot and the lateral border of the foot becomes

Table 2.5. Effects of *M. leprae* on the motor nerves

Nerve	Effect
Ulnar	Weakness in fourth and fifth fingers
Median	Weakness in thumb and index and middle fingers
Radial	Wrist or hand drop
Lateral popliteal	Drop foot
Common peroneal	Drop foot
Posterior tibial	Flexed/bent toes; arch collapse
Facial	Lagophthalmus; loss of sight

anesthetic and susceptible to trauma and ulcers (Lockwood 2004). Posterior tibial nerve involvement leads to paralysis and the small foot muscles contract. Paralysis then leads to muscular imbalance and the bones of the joints become positioned abnormally. When loss of sensation accompanies paralysis, this causes tissue destruction and ulcers due to exposure of the hands and feet to injury (Bryceson and Pfaltzgraff 1990). When anesthesia accompanies paralysis, the latter makes the effects of the former worse.

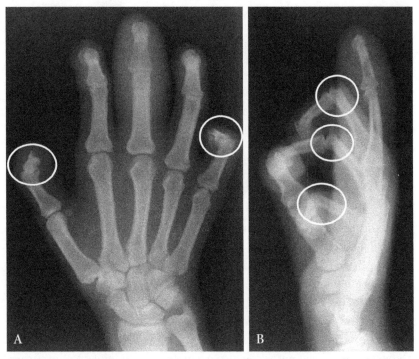

Figure 2.13. Radiograph of a person's hand showing flexion deformity, or "bending," of the fingers in a person with leprosy. *A*, postero-anterior view; *B*, lateral view. Credit: Johs Andersen.

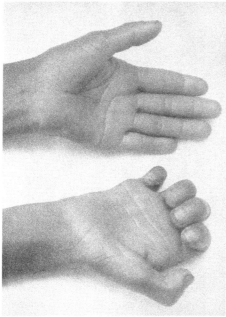

Left: Figure 2.14. Flexion deformity, or "bending," of the fingers of the hand in a person with leprosy. Source: Diniz et al. (1960, figure 36).

Below: Figure 2.15. Radiograph of the foot of a person with leprosy showing flexion deformity, or "bending," of the toes. Credit: Johs Andersen.

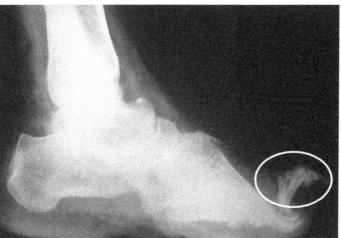

For example, contracted fingers and toes can lead to damaged desensitized hands and feet.

Involvement of the motor nerves can affect the eyes and lead to facial paralysis and defective lower eyelid closure (lagophthalmos) (Drabik and Drabik 1992) (see Figures 2.16 and 2.17). Even though lagophthalmos is a complication of leprosy, it is not known how common it is because it is often detected when a person presents with other signs and symptoms (Jaiswal and Subbarao 2010). In such cases, the orbicularis oculi muscle is

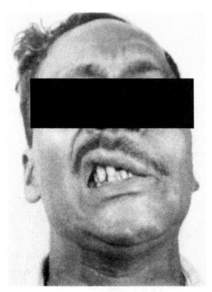 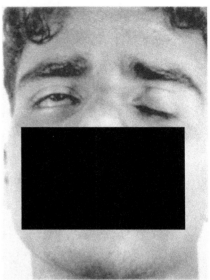

Figure 2.16. Facial palsy on the left side as a result of leprosy. Source: Dharmendra (1947, figure 90).

Figure 2.17. The effects of lagophthalmos caused by leprosy in the right eye. Source: Dharmendra (1947, figure 86).

affected; this muscle is responsible for closing the eye. Partial or complete paralysis ensues if it is damaged (Bryceson and Pfaltzgraff 1990). Facial skin lesions also increase the possibility of facial nerve damage (Hogeweg, Kiran, and Suneetha 1991). Further damage to the eye can develop in patients with multibacillary leprosy when damage to the ophthalmic branch of the fifth cranial nerve occurs (trigeminal). Loss of sensation develops in the cornea (front of the eye) and the conjunctiva (the mucus membrane covering the eye), which can affect the normal blinking function. In such cases, a dry eye, conjunctivitis, and corneal ulcers may develop. *M. leprae* can also invade the eye directly via the bloodstream (Lockwood 2004) and blindness may eventually result. Inflammation of the structure in the eye that controls the amount of light that reaches the light-sensitive retina (iris) can lead to pressure in the eye and short sightedness or blindness. Cataracts (cloudiness) may also occur in the normally transparent lens of the eye. One long-term study of patients with leprosy in the UK found that 22 percent had complications that threatened eyesight in one or both eyes (Malik, Morris, and Ffytche 2011). The matter becomes worse when a person experiences a Type 2 reaction in which iris inflammation from light exposure causes eye discomfort or pain, blurred vision, pupil constriction, and cloudiness. Preventing the eye complications of leprosy may help reduce stigma and enable those with leprosy to participate in social life and integrate better in their communities (Boku et al. 2010). Eye drops can be used to prevent dry eyes, goggles can be worn in work situations to prevent damage, and glasses can be worn for everyday activities. However, surgery may be the ultimate treatment for the resulting damage to the eyes.

Autonomic Nerves

The autonomic nerves are affected in patients with multibacillary leprosy, leading to changes in normal perspiration (parahidrosis). Damage to the autonomic nerves in the blood vessels in the skin and the subcutaneous tissues also occurs and capillary blood becomes sluggish because heat is not readily dispersed. In addition, the sweating function is lost in the area of skin lesions. Because the circulation of the blood is not as efficient, oxygen is reduced to the body tissues and healing is slower. This affects the rate at which ulcers heal (Bryceson and Pfaltzgraff 1990). As a consequence, people with leprosy can develop a very dry skin that is brittle and inflexible and cracks or develops fissures easily. This makes it possible for secondary infections to develop, especially in the feet (Drabik and Drabik 1992; Bryceson and Pfaltzgraff 1990). The longer the duration of leprosy, the more likely it is that autonomic nerve involvement occurs (Ramachandran and Neelan 1987).

Acute Inflammation in Leprosy (Reactions)

The term "leprosy reaction" refers to sudden-onset acute aggressive inflammation in skin lesions that produces specific signs and symptoms (Bryceson and Pfaltzgraff 1990; White and Franco-Paredes 2015). There are three types of reactions. Type 1, or reversal, affects one-third of people who have borderline forms of leprosy; when this happens the person's disease is upgraded to tuberculoid leprosy or downgraded to lepromatous leprosy. Type 2 (erythema nodosum leprosum) affects patients with borderline lepromatous leprosy and lepromatous leprosy (Britton and Lockwood 2004). Type 3 is called Lucio's phenomenon. Reactions usually occur after treatment has started. They are the body's response to *M. leprae* or its products in the body tissues, but they do not mean the disease is getting worse (Dayal et al. 1990). During the normal course of leprosy, the clinical state of patients fluctuates spontaneously (Scollard et al. 2006). This is because the immune response in leprosy is dynamic: up to 50 percent of all patients with leprosy have reactions while they experience the disease (Kumar,

Dogra, and Kaur 2004). Reactions are key drivers for permanent disabilities (Fava et al. 2012). The peripheral nerves and skin lesions become swollen and painful and reddish nodules appear. People also experience loss of sensation and muscle weakness, fever and malaise, swelling of the hands and feet, and deformities.

The types of reactions seem to have different underlying immunological mechanisms that are poorly understood. The factors that cause them are unknown and it is not possible to predict who will be affected or when (Scollard et al. 2006). A topic of ongoing research is how genetic factors in the host person contribute to the development of reactions (e.g., Fava et al. 2012). While reactions occur during chemotherapy that is administered to treat *M. leprae,* exposure to other mycobacteria also cause them (Godal et al. 1973; Barnetson et al. 1976; Motta et al. 2012). They can further occur after pregnancy in some women, possibly due to cell-mediated immunity recovery (Rose and MacDougall 1975). Corticosteroids and thalidomide are the usual treatments for reactions, but therapy has to be implemented rapidly to prevent irreversible deformities (World Health Organization 2000a–c).

Reactions can also occur long after leprosy has been "cured" using medication, especially in people with multibacillary leprosy (Stearns 2002), but people who have been cured may still have psychological problems due to nerve pain and consequently may have a poor quality of life (Reis et al. 2014). However, people who have been cured following completion of their treatment are removed from the World Health Organization leprosy frequency data. Because of this, it is not known how many people are experiencing and being treated for reactions (World Health Organization 2000b). Even though health care workers emphasize that leprosy is curable, White notes that "the idea that a patient is cured of Hansen's disease when [multidrug therapy] is complete is completely flawed" (White 2005, 324). This dynamic can lead people who have ostensibly been cured to mistrust medical workers; they may also experience depression and stop believing that they have been cured when they develop a reaction. Encounters with medical professionals can shape patients' perceptions of the disease at the time of diagnosis, especially if they develop negative side effects from drugs or if they experience a reaction. While multidrug therapy is of fixed duration and makes patients noncontagious, further treatment to prevent disability may be necessary once that therapy is finished.

Type 1 (Reversal)

Type 1 reactions occur in the people who develop borderline leprosy. They are called reversal reactions because after the reaction the diagnosis is upgraded to tuberculoid leprosy or downgraded to lepromatous leprosy, although the latter is rarely seen now (Scollard et al. 2006). The reaction is the result of rapid changes in immunity and delays in hypersensitivity to *M. leprae*. The reactions vary from mild to moderately intense exacerbations of existing skin lesions and to new skin lesions and edema (fluid) in lesions on the hands and feet. Progressive nerve inflammation causes sensory and motor nerve damage and lesions may ulcerate. This type of reaction may last many weeks (Godal 1978).

Type 2 (Erythema Nodosum Leprosum)

Type 2 reactions, also called "lepromatous reactions" or "erythema nodosum leprosum," occur in people with multibacillary leprosy (lepromatous leprosy and borderline lepromatous leprosy), either during treatment or spontaneously. They are not associated with alterations in cell-mediated immunity; they are a humoral antibody-mediated immune response by B-lymphocytes (Jopling and McDougall 1988). Up to 50 percent of people with multibacillary leprosy can develop a Type 2 reaction, whether therapy has started or not (Modlin and Rea 1987; Rodrigues and Lockwood 2011). In this reaction, skin and nerve lesions contain many bacilli and there is a strong antibody response. There is also an abrupt onset of very tender reddish nodules on the face, the extremities, and the trunk, bilaterally and symmetrically. These nodules target cooler regions of the skin. Fever; malaise; an aching body; inflammation and enlargement of and damage to sensory and motor nerves; paralysis of the facial nerve; inflammation of the eyes, muscles, testicles in men, and veins near nodules; and arthritis accompany this reaction in patients (Scollard et al. 2006). Severe Type 2 reactions may include swelling in some lesions; the tissues may ulcerate, die, and separate from living tissue. The reaction lasts one to two weeks, although some patients develop multiple recurrences over several months.

Lucio's Phenomenon

Lucio's phenomenon is a reaction seen only in people with untreated multibacillary leprosy in Central and South America and Mexico. There appears to be an association with this reaction and the recently discovered organism that causes leprosy, *M. lepromatosis* (Han and Jessurun 2013). Here, painful and tender patches develop in the skin that become red or purple, necrotic, and ulcerated. The lesions finally develop a black or brown crust that falls off, leaving a scar (Jopling 1982a). Acute, severe, necrotic inflammation of blood vessels can also occur. Lucio's phenomenon is rare but can be associated with high mortality and morbidity and may be accompanied by severe anemia (Scollard et al. 2006).

Conclusions

This chapter has considered how leprosy affects the human body. The bacillus has a long incubation period, which means that many people may harbor the infection but will not show any signs, thus preventing early diagnosis and treatment. The variety of forms that leprosy can take shows how important the immune system is to the development of different leprosy types. This has consequences for the experience of each person with leprosy. The range of different types of skin lesions also reflect what a dynamic disease leprosy is. The effect of the bacteria on the nervous system can lead to profound damage to the extremities in particular that can have considerable consequences for the person concerned physically, socially, and mentally. Acute inflammatory episodes (reactions) provide a challenge to people with leprosy and to health care workers who are helping to manage their patients' infection.

3

Past and Present Diagnosis, Treatment, and Prognosis

> Today leprosy is no more a dreaded disease as it is curable . . . and the disease is capable of being eliminated as a public health problem through organized case detection and treatment"
>
> Yawalkar (2009, 11)

These words concisely illustrate that it is potentially possible to eliminate leprosy in the world. This chapter considers how leprosy is diagnosed and treated today, followed by some perspectives on what methods were available for people living in the distant past.

CLINICAL DIAGNOSIS OF LEPROSY TODAY

The methods used to diagnose leprosy today vary by region of the world and by available resources and personnel. Education about leprosy for people in countries with the highest rates of the disease highlights the important features people at risk should look for and that people should go to a diagnostic facility as soon as they experience signs and symptoms of the disease. However, because of the stigma associated with leprosy, it can be difficult to persuade people to seek a firm diagnosis. It is likely that many people remain undiagnosed because they fear social stigma and eventual marginalization. This is also the case for tuberculosis (Venkatraju and Prasad 2013). In some countries, legislation actually promotes stigma against people with leprosy, for example laws related to travel restrictions (Senior 2009).

One aspect of diagnosis that can be very challenging is detecting people with subclinical (latent) and asymptomatic leprosy. As leprosy becomes less common around the world, diagnosis of people with no overt evidence may be delayed or misdiagnosis may occur (Bennett, Parker, and Robson 2008; Rongioletti et al. 2009). One study in Nepal found that people with leprosy delayed seeking advice at a health care clinic, which of course delayed diagnosis and treatment (Robertson, Nicholls, and Butlin 2000). The factors associated with delays included the age of the person, living in a rural environment, relative distance/walking time to the health facility, lack of awareness of leprosy's signs and symptoms, and the absence of another person with leprosy near where a person lived. The total travel time to a clinic in Nepal was a mean of 13.4 hours and the cost of travel was the equivalent of several days' wages. In China, delay in in diagnosis was also related to a high rate of disability and a lack of knowledge about leprosy (Zhang, Chen, et al. 2009).

If diagnosis and treatment occur early in the progress of the infection, the person with leprosy may experience a better quality of life (Bottene and Reis 2012). However, those involved with diagnosis are advised that they need to be aware of signs in people that might be misread as caused by leprosy (Bryceson and Pfaltzgraff 1990). For example, health care personnel should take into account the range of skin color for the population they are working with and be aware that some skin colors may conceal lesions. Another factor that may affect accurate diagnosis is the activities people do that might lead to trauma-related scars on the hands that could be mistaken for anesthetic-related scarring on the hands of a person with leprosy. Local cultural practices may involve mutilation of certain parts of the body that may be misinterpreted as a sign of leprosy, for example ritual removal of the anterior upper teeth or eyebrow plucking. Health personnel should also be familiar with the range of skin diseases that may be expected for a particular population in order to avoid misdiagnosis. In addition, thickened nerves, a potential early sign of leprosy, can be caused by other factors and other causes of loss of sensation should be considered, such as trauma to nerves and nerve compression. Ulceration of the feet may develop in any disease that causes lack of skin sensation (e.g., diabetes), and flexion contractures of the fingers and toes can occur in the tertiary phase of yaws (a treponemal disease) and rheumatoid arthritis.

Diagnosis is usually made when one or more of the following conditions is observed in patients (Britton and Lockwood 2004, 1213):

Hypopigmented (loss of color) or reddish patches on the skin with a definite loss of sensation.
Thickened peripheral nerves.
Acid-fast bacilli in skin smears (a sample taken from a tiny cut in the

skin and stained for *M. leprae*); these may be taken from the forehead, earlobes, chin, forearm, trunk, and buttocks.

Biopsies of other body tissue showing bacilli.

However, up to 70 percent of people can be negative for bacilli if they have paucibacillary leprosy (Ustianowski and Lockwood 2003), and lack of sensation alone is usually not used for diagnosis. In addition to skin and nerve observations, a general clinical examination is done, and a medical history is taken (Grange and Lethaby 2004).

Signs and Symptoms (Clinical Features)

People with leprosy might show and experience a range of clinical features. All the five traditionally accepted senses (sight, hearing, touch, taste, and smell) are affected, which perhaps emphasizes the impact of the infection on the person but also on the senses of the community in which they reside. A community may not want to touch a person with leprosy, they may be able to smell them if they have infected ulcers, and they may find the sight or sound of people with the infection disturbing (as might have been likely in the past).

The range of clinical features is dependent on the patient's cell-mediated immunity (Jopling 1982), the type of leprosy a person has, and whether it is treated or not. The onset of signs and symptoms can be gradual, but also may be sudden in highly susceptible people (Dayal et al. 1990). Signs and symptoms in lepromatous leprosy include thinning of the eyebrows and broadening and collapse of the nose due to destruction of underlying nasal structures (Figure 3.1). Additionally, nasal stuffiness, discharge and bleeding, loss of taste, deepening of the forehead lines as the skin thickens, and loss of the eyelashes and the incisor teeth can occur. The larynx can also be affected; the most prevalent voice complaints are hoarseness and "hawking" (clearing the throat of phlegm) (Palheta et al. 2010). Furthermore, edema of the lower legs and feet may develop due to increased permeability of the skin capillaries. This is because the capillaries contain many bacilli and this allows fluid to build up in the tissues. Damage to autonomic nerve fibers and the effects of gravity can also contribute. As a result, the legs become hard, shiny, and waxy. There may be ulceration of the hands, legs, and feet due to loss of pain and touch sensation in the skin. Pins and needles and a sensation of "ants crawling under the skin" can also occur, and people with leprosy have a relatively higher rate of restless leg

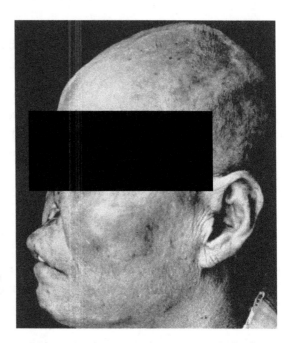

Figure 3.1. Destruction of the nasal bones as a result of leprosy. Source: Mitsuda (1952, figure 70).

syndrome than people who do not have the disease. This is a neurological disorder with many underlying causes that involve the uncontrollable need to move the legs, leading to poorer sleep quality (Dayal et al. 1990; see also Choi et al. 2012; Padhi and Pradhan 2014). Nerve pain, muscle weakness, and shortening of fingers and toes develop as the hand and foot bones are affected. Nerve pain is now considered a late development in leprosy but is reported to be quite common; this condition also impacts sleep quality and a person's ability to function normally in their community (Ramos et al. 2014). Fever, malaise, joint pain, arthritis, inflammation of the tendons and ligaments and lymph nodes, the absence of the sweating reflex, and itching before the development of skin lesions are also signs and symptoms (Bryceson and Pfaltzgraaf 1990). The majority of patients with lepromatous leprosy have enlarged lymph nodes. Damage to the eyes can also be a significant complication of leprosy.

Psychological signs and symptoms can become part of the infection for many, including for people with leprosy without a previous history of mental health problems (Ryrie 1951; Jindal et al. 2013; Rocha-Leite et al. 2014). The most common underlying factor that health care workers who manage people with leprosy cite is irrational fear. Even though there is much more educational material about leprosy today, particularly in developing countries, it is still the case that community members often have views about leprosy that are unsupported by biomedical knowledge. This impacts how people with leprosy react to their infection. Evidence shows that poor mental health and especially clinical depression has increased in people with leprosy (Singh 2012; Yamaguchi, Poudel, and Jimba 2013). It is common for people with leprosy to develop complexes, neurotic symptoms, and psychotic reactions. The likely reason for these is social stigma (Ranjit and Verghese 1980). A mental assessment of several thousand people with leprosy in southern India found that some had a high score for mental health problems. People with physical deformity had a higher rate of mental health issues than the population without physical deformity. The researchers concluded that "the psychology of crippling is a very important part of the psychology of leprosy" (ibid., 433). Another study found that single unemployed women with long-standing leprosy who were poorly educated experienced psychiatric disorders, suggesting that psychiatric care could be required for particular cohorts of affected people (Bakare et al. 2015).

Diagnostic Tests

Although many agree that the lepromin test is not diagnostic of leprosy or of exposure to *M. leprae,* it does provide a measure of an individual's ability to mount a response to the antigens present in his or her body (Scollard et al. 2006; Lockwood 2004; Dayal et al. 1990). However, it is described as a "crude, semi-standardized . . . non-specific test of occasional value in classifying a case of leprosy" (Lockwood 2004, 29–16). An ELISA test has also been used to test for the presence of *M. leprae*-specific immunoglobulin antibodies (a biochemical technique used mainly in immunology to detect the presence of an antibody or an antigen). Lepromin is prepared from lepromatous armadillo liver tissue or a lepromatous nodule (Lockwood 2004). The reaction to the injection of lepromin is assessed after 48 hours (Fernandez reaction: sensitivity of the body tissues to leprosy bacilli protein) and after four to five weeks (Mitsuda reaction; resistance of the person to the bacteria). If the reaction to the latter is strong, the person has tuberculoid leprosy. If there is no reaction, a diagnosis of lepromatous leprosy is made (Finch et al. 2002). A Mitsuda reaction correlates strongly with the Ridley-Jopling classification system. The number of positive Mitsuda reactions is also said to rise with increasing age (Fine 1982). This may be because the immune system has matured and strengthened its response to *M. leprae,* showing that there has been a cumulative effect of exposure to mycobacterial antigens. However, the Fernandez and Mitsuda reactions have been found in people with no clinical leprosy and both may be induced by the tuberculosis vaccine (Bacillus Calmette-Guerin; Shepard and Saitz 1967).

Unfortunately, there are no serological tests for a routine laboratory diagnosis of leprosy. However, some believe that the presence of antibodies to phenolic glycolipid-I which is specific to *M. leprae* and can be detected with the *M. leprae* Lateral Flow Test (MLFT), is indicative of *M. leprae* and correlates with the bacterial load of people with leprosy (Anderson et al. 2007; Grossi et al. 2008). People with paucibacillary leprosy are positive in 1–40 percent of cases; for people with multibacillary leprosy, the positivity rate is 80–100 percent. The MLFT is a very simple technique that requires no special equipment and can be used with whole blood and serum from the blood. The result is known in less than ten minutes. The MLFT is thus very suitable for use in the field. Seropositive people who are in contact with patients with leprosy are at increased risk for developing leprosy and this test can be used to identify these contacts.

However, "the major advance in the laboratory diagnosis of Hansen's disease in the last 15 years . . . has been the development of methods for the extraction, amplification and identification of *M. leprae* DNA in clinical specimens using [polymerase chain reaction] and other molecular techniques" (Scollard et al. 2006, 342). Rapid molecular type tests are now being used (Katoch 1999). This can provide rapid information directly from the patient about drug susceptibility. More recently, *M. leprae* specific proteins (cytokines) that are secreted by lymphocytes in the immune system have been used to monitor responses of the immune system. Their use is encouraged for the early detection of leprosy (Pena et al. 2011; Bobosha et al. 2011). While more work is needed to understand which proteins are associated with *M. leprae,* they may be used more in the future for diagnosis and for the development of vaccines (Wiker, Tomazella, and de Souza 2011; Parkash and Singh 2012). Genetic analysis is also helping researchers identify markers for specific strains of bacteria so they can assess exposure to *M. leprae* and trace transmission patterns. Predictions that genetic markers would be helpful in understanding leprosy have clearly become reality (Jopling 1992; Scollard et al. 2006). Up to 50 percent of people with leprosy that is not diagnosed through histological assessment may now be detected using molecular methods (Anderson et al. 2007).

However, diagnosis may be delayed in areas of the world where leprosy is not normally experienced. People could be diagnosed late in the disease in regions where medical personnel are not trained to diagnose leprosy or have any experience of the infection (Nolan et al. 2014). For example, US military personnel who were deployed abroad and returned with the infection experienced a delay in diagnosis of up to twelve months (Chambers, Baffi, and Nash 2009). The perceived relative importance of leprosy within a country is another area for consideration. There is a perception that leprosy does not affect people today in the Western world. This was the case in the UK, where there was considerable undernotification of the infection in England and Wales from 2001 to 2012 because of the belief that national surveillance was not important (Fulton et al. 2016).

Treatment of Leprosy

Today the diagnosis and treatment of leprosy is straightforward and most countries where the disease is endemic are striving to fully integrate leprosy services into existing general health services. This is especially important for the underserved and marginalized communities that are most at risk from leprosy. Members of these communities are often the poorest of the poor.[1] However, a number of factors may influence whether a person is diagnosed and treated. Public awareness campaigns about the integration of leprosy services into general health services were implemented many years ago, but the World Health Organization continues to emphasize that this needs to happen more often. Whether this has had any impact on success in treating leprosy is of interest. A survey of 3,000 people in three villages in India found that only around 46 percent were aware of the availability of treatment for leprosy in primary health care centers, although most knew they could access the treatment for free (Verma, Rao, and Raju 2011). In addition, many studies of the mechanisms for reducing stigma have emphasized that leprosy should be included in general health care programs instead of setting the disease apart (Sermrittirong, Van Brakel, and Bunbers-Aelen 2014).

The key goals of treatment are to stop the infection (chemotherapy), treat reactions, educate the patient to manage nerve damage, treat nerve damage complications, and rehabilitate the patient socially and psychologically (Lockwood 2004). Stopping the infection using chemotherapy obviously will reduce the incidence of leprosy for people who live in households with others who have leprosy (Rodrigues and Lockwood 2011). A meta-analysis and review of the use of drugs to prevent leprosy found that when a person with leprosy had this prophylactic treatment, the rate of leprosy among those they lived with was reduced (Reveiz, Buendía, and Téllez 2009). However, implementing drug therapy is not always easy in some parts of the world.

The elimination strategy of the 1991 World Health Organization Assembly centered on treatment with multidrug therapy. It included expanding multidrug therapy services to all health care facilities to ensure that all existing people with leprosy and newly diagnosed people were given this therapy, and it encouraged all patients to take treatment regularly and completely (World Health Assembly 1991; World Health Organization 2000c). Additionally, it emphasized promoting awareness of leprosy in communities to encourage voluntary diagnosis and treatment, setting targets and a timetable for activities, and keeping good records to monitor progress toward elimination. However, as Rafferty (2005, 123) notes, "It is simply not enough for the medical profession and society to treat the disease and ignore the patient as a whole person." A holistic approach is needed, as for all treatment of disease. While it is clear that health care professionals working with people with leprosy today have a better understanding of the impact of leprosy on patients and of the need to recognize how their lives can be disrupted by having the infection (Dean et al. 2019), the complete control of leprosy will not be achievable until the social stigma of leprosy is eradicated. Kellersberger (1951) said this over half a century ago and it is still true today.

Chemotherapy

Although some scholars report that herbal remedies may be effective in treating leprosy (e.g., Kumar, Sheikh, and Bussman 2011; Jana and Shekhawat 2011; Sharma et al. 2014), the treatment recommended today is six months of rifampicin and dapsone for paucibacillary leprosy and twelve months of rifampicin, dapsone, and clofazimine for multibacillary leprosy. This antibiotic treatment has been free since 1995 for all people with leprosy worldwide. The Nippon Foundation's drug fund paid for the treatment until 2000, and Novartis and the Novartis Foundation for Sustainable Development have taken over this service. Novartis has said that it will continue this arrangement until leprosy is eradicated from the world (Yawalkar 2009, 73). While relapses in infection are very infrequent, they can lead to continued transmission of the infection by people with multibacillary leprosy. Relapse may occur when only one drug is used, when a therapy is poorly implemented and/or accepted by the person, when a person with leprosy has a poor response to treatment, and when multiple skin lesions or thickened nerves are present (Kaimal and Thappa 2009).

History of Treatment

The modern period of treatment began in the 1940s, when the first breakthrough for leprosy treatment using chemotherapy occurred (Drabik and Drabik 1992). The American leprologist Guy Faget revolutionized treatment when he introduced Promin, a derivative of diaminodiphenyl sulfone (or dapsone, an antibiotic), as a new protocol for leprosy. It was first used as a therapy for people with leprosy in 1946 and was readily available from the 1950s (Yawalkar 2009, 65; Faget et al. 1943; Jopling 1992). This soon became the protocol that leprologists around the world followed, and it was used most intensively in the 1950s (Yawalkar 2009). The fifth and sixth International Leprosy Congresses (in Havana in 1948 and Madrid in 1953) endorsed it as particularly effective in treating advanced lepromatous leprosy. This was the first effective remedy for leprosy.

However, this treatment only arrested the disease and the length of treatment could extend to the lifetime of the patient. Until 1981, dapsone was the only drug used. This was problematic because this long-term monotherapy led to dapsone resistance. The first dapsone-resistant strains of the leprosy bacteria were reported at the end of the 1960s; by 1982 one-third of recently infected patients in thirty countries had resistance (World Health Organization 1982). This is when multidrug therapy was recommended.

Rifampicin and clofazimine were later added to the treatment, but people also became resistant to rifampicin and clofazimine was only partly effective (Scollard et al. 2006). Drug resistance develops as chromosomes mutate in genes that encode drug targets. These mutations develop because of errors in DNA replication. In addition, inappropriate or inadequate drug use enriches mutants. In 1981, the World Health Organization's Study Group of Leprosy for Control Programmes recommended multidrug therapy and one year later the World Health Organization officially recommended it. This was an important turning point in the history of treatment of leprosy. After that time, therapy consisted of a combination of dapsone, clofazimine, and rifampicin. In 1984, at the 12th International Leprosy Congress in New Delhi, it was made mandatory for treatment. By the time of the 13th Congress in 1988, the new protocol was a confirmed success (Drabik and Drabik 1992).

Initially, the objective of multidrug therapy was to make treatment as economical as possible and prescribe a minimum dosage with maximum efficiency; treatment time ranged from six to twenty-four months. Following various discussions, the World Health Organization suggested that it might be possible to reduce the treatment duration for all types of leprosy to six months (World Health Organization 1998a; Bhattacharya and Seghal 2002). This led to guidance that patients with multibacillary leprosy be treated for twelve months and those with paucibacillary leprosy for only six months (World Health Organization 2000b). This caused much criticism from people trying to manage people with paucibacillary leprosy around the world, and some suggested that the World Health Organization's recommended treatments might not be compatible with its aim of eliminating leprosy (Naafs 2006). However, the treatment regime remains. This also includes a one-off dose of rifampicin, ofloxacin, and minocycline for people with single-lesion paucibacillary leprosy. If leprosy bacteria are still present after two years, treatment is continued until none are detected. Following discontinuation, people with paucibacillary leprosy are observed for two years and those with multibacillary leprosy are observed for five years. This is because *M. leprae* multiplies so slowly. Rifampicin is the most important anti-leprosy drug and is included in the treatment of both types of leprosy; within a week of treatment with rifampicin, nonviable *M. leprae* bacilli will normally be present (Levy, Shepard, and Fasal 1976). Indeed, a single dose of rifampicin is recommended for preventing leprosy in people who have been exposed to those with the infection, along with the Bacillus Calmette-Guérin (tuberculosis) vaccine. The antibiotics clarithromycin, minocycline, and ofloxacin are also used.

The impact multidrug therapy has had on leprosy frequency since it was introduced is clear. By 1990, almost 50 percent of registered people with leprosy were receiving multidrug therapy, but therapy coverage was uneven across the world. Twenty-nine of 105 countries where leprosy was endemic (31 percent) accounted for 76 percent of people being treated, and 49 countries (47 percent) were treating at least half of the people with leprosy. Both patient compliance and detection of new people with leprosy through self-reporting had increased. Health care workers were better motivated and there was more community support, probably because treatment was seen as more successful. However, while multidrug therapy after one week left a patient non-infectious, one leprologist noted that "the real disappointment is that the overall rate of implementation of multidrug therapy worldwide is unsatisfactory and is particularly poor in Africa and Latin America; by 1990 56 percent of registered "cases" worldwide were receiving multidrug therapy, but only 18 percent in Africa and 24 percent in Latin America" (Jopling 1992, 10). Times have changed as more people with leprosy have been given access to free treatment. However, the validity of the numbers provided for "cured" people continues to be debated. The World Health Organization defines "cured" people as those who have completed a course of multidrug therapy (Stearns 2002). Some studies suggest cure rates for paucibacillary leprosy of between 67 and 100 percent and for multibacillary leprosy of between 38 and 100 percent (World Health Organization 2000c). Lower figures are likely attributable to people with leprosy who have poor or no access to treatment, but the challenge is greater than this.

Scholars in many disciplines, from social scientists to health care workers, have emphasized that combating leprosy with the ultimate aim of eliminating the disease requires a combination of drug therapy and education of both people who are affected and those who are not. If films, the press, television programs, and posters were used to draw attention to leprosy as a curable disease, people might be more likely and willing to access treatment on their own initiative and a healthy (and balanced) appreciation of what the infection really encompasses would be achieved (Drabik and Drabik 1992; see also World Health Organization 2016). Treatment of patients goes hand in hand with reduction of prejudice.

Noncompliance with Drug Therapy

Noncompliance with treatment is a challenge for those trying to manage, control, and eradicate any disease, and leprosy is no exception (e.g., see Lira et al. 2012). For example, a study of people with untreated leprosy who reported to the Leprosy Mission Referral Hospital in Champa, India, showed that 151 (9 percent) were from indigenous groups. These groups were accessing help, but delays in reporting leprosy made controlling transmission and management of complications very challenging. Researchers thought that this was because it was difficult to travel in a harsh environment to visit health facilities, leading to delay in diagnosis and treatment (Kumar, Kumar, et al. 2011; see also Pandey and Rathod 2011).

Women can be very disadvantaged in some countries. Some may delay visiting a hospital until their husband or guardian deems it necessary. A study of women who attended a leprosy referral hospital in India found that while they were at the hospital their chores were not done and that they were expected to continue as normal when they returned home. If their hospital stay was lengthy, their social worth diminished. In addition, they could not always follow the advice they were given to help them cope with the effects of the infection (such as using protective gloves when handling hot cooking utensils if they had skin sensation loss). These experiences made it less likely they would visit a hospital, and many of these women had leprosy-related disabilities as a result (John, Rao, and Das 2010).

Decentralization of leprosy management is one way of giving more people better access to diagnosis and treatment, particularly in cases where specific societal groups cannot easily reach a medical facility (Fuzikawa et al. 2010a, 2010b). Even though decentralization is challenging, it has led to earlier detection of leprosy and treatment of patients closer to their homes. Rao (2008) recommended encouraging patients to stay motivated, providing counseling, making frequent contact with patients, and making services more patient friendly.

Another reason for noncompliance relates to the side effects associated with the drugs used to treat leprosy. Dapsone may lead to anemia, and rifampicin, which is the most effective drug for treating leprosy, may result in liver damage (Drabik and Drabik 1992). Clofazimine can cause red urine for a few hours after taking it and can lead to pigmentation of the skin and hair (World Health Organization 2000b). However, education can help patients overcome their fear of these side effects.

Nondrug Therapies for the Management of Leprosy

Management of leprosy involves more than drug therapy, including access to counseling if people with the infection are stigmatized (e.g., Floyd-Richard and Gurung 2000). Other modes of treatment can help combat the effects of nerve damage in the body and mitigate subsequent damage to the hands and feet of those affected. Thus, one of the main foci of leprosy education is related to protecting the hands and feet from trauma and ulceration. People with leprosy are strongly encouraged to wear gloves when handling hot items and to wear shoes to avoid abnormal pressure, subsequent injury to the soles of the feet, ulcers, and ultimate deformity. Gloves and shoes are often provided to patients free of charge, as many people with leprosy may not have the resources to purchase them. Croc brand shoes have even been recommended in some regions (Figures 3.2, 3.3).

Education about self-care is also important (Sathiaraj, Norman, and Richard 2010). As autonomic nerve damage can lead to dryness and cracking of the hands and feet, patients often receive advice about how to care for their hands and feet (World Health Organization 2000b). Soaking the feet and hands every day helps soften cracks and fissures and applying cooking oil or petroleum jelly regularly can help prevent ulcers from developing. When ulcers do occur, patients are advised to clean and dress them

Above: Figure 3.2. Shoes being fitted for a person with leprosy. Credit: Leprosy Mission International.

Left: Figure 3.3. An example of more recent shoes (Crocs) used to protect the feet. Photo by Charlotte Roberts.

to prevent them from worsening, and to rest their feet. However, while patients may be aware of how to care for their faces, feet, and hands, they may not have the resources or time to do that care (Lima et al. 2018). Nevertheless, self-care groups have proven helpful; the prevalence of leprosy-related ulcers may decrease markedly among the members of such groups (Ebenso, Muyiwa, and Ebenso 2009). To prevent eye problems and even blindness, patients who are experiencing blurred vision or a discharge or pain are usually prescribed analgesic eye drops that relax the eye muscles and dilate the pupils.

Seboka and Saunderson (1996), among others, have written about the challenges of ensuring that patients comply with guidelines for the care and treatment of their hands and feet. Although the ideal situation is for patients to rest their hands and feet and to have wound care to allow ulcers to heal (Birke et al. 1992; Kazen 1993), it is difficult for many people to maintain ulcer-free feet because they need to continue to work (Krishnamoorthy 1994). Nevertheless, the shoes that health care facilities provide that reduce pressure and spread the load over a wider area of the foot offer a compromise between normal activity and complete rest (van Schie et al. 2013; Fuk-Tan-Tang et al. 2015). However, wearing the shoes these facilities provide can cause problems. First, stigma associated with specific designs of shoes may prevent people from wearing them (Govindharaj et al. 2017). When payment is required for footwear that is stigmatized, "patients are understandably unwilling to pay for such products" (Cross et al. 1996, 407; see also Iyere 1990). Thus, footwear that looks similar to normal shoes available on the market is recommended (Kulkarni, Antia, and Mehta 1990). Second, some people never wear shoes inside the house because of the custom or culture of their community. Third, in some occupations (e.g., working in paddy fields), wearing shoes is not practical. Fourth, these shoes need to be repaired about every six months, but this can be difficult for people to organize. Fifth, in some regions of the world, it is difficult to provide adequate numbers of shoes.

The All Africa Leprosy, Tuberculosis and Rehabilitation Training Centre (ALERT) in Addis Ababa, Ethiopia, provides canvas shoes to people with leprosy. This shoe is acceptable to patients and can be made in high numbers. In a study of people at ALERT whose feet lacked sensation and had developed subsequent deformity and were regularly attending the foot care clinic, selected participants were allocated either canvas shoes or standard Plastazote molded sandals (Seboka and Saunderson 1996). Researchers examined the two groups at the start of the study and then at two-, four-, six-, and twelve-month intervals. They measured the sizes of each person's foot ulcers at each visit, questioned participants, and assessed whether they were wearing their footwear. All of the foot ulcers in the group that was wearing canvas shoes decreased in size over time except in two people. People who wore canvas shoes fared better than those who wore the Plastazote sandals. Eighty percent of the people in the canvas shoe group were accepting of their shoes. In contrast, only 20 percent of the people in the Plastazote sandal group were happy with their shoes. Canvas shoes were thus more acceptable to patients and to their communities, they were more beneficial for protecting their feet from damage, and they were preferred for farm work and walking over dusty or stony ground. This study showed that with a small amount of effort, mitigating further effects of leprosy can be achieved.

The Bacillus Calmette-Guérin Vaccine

Prevention of leprosy is of course a better option than trying to cure it. As leprosy and tuberculosis are both caused by *Mycobacteria,* the idea developed that using the Bacillus Calmette-Guérin (BCG) vaccine for preventing tuberculosis in humans might prevent people from contracting leprosy in endemic countries. BCG was first used to attempt immunization against leprosy in 1939, when it was injected into lepromin-negative healthy children. A lepromin-positivity rate of 90 percent was the result (Mukherjee 1989). In the 1940s, four major field trials tested the efficacy of BCG in preventing leprosy; in Burma, 20 percent of the study participants were protected, in New Guinea 46 percent, in Uganda 80.9 percent, and in India 23 percent (Fine and Rodriguez 1990). While these results were promising, the findings were not without problems. Following additional field trials in 1970, the World Health Organization took the position that BCG was not effective for leprosy prevention (World Health Organization 1988a).

The reasons for the variation in protection of and reaction to BCG was deemed to be due to different strains of *M. leprae* in the BCG vaccine trials and to different epidemiological situations that created variable levels of exposure to the organism. Interference from prevalent environmental mycobacteria or immunogenetics were also noted (Fine 1982). However, as the genetic relationship between the BCG vaccine strains is now known, the success of these and future trials can be better evaluated (Curtiss et al. 2001). Mukherjee reported that while BCG vaccines containing killed purified *M. leprae* from armadillo tissues led to long-lasting cell-mediated immunity in mice and guinea pigs, it would be prohibitively expensive to breed armadillos and create the vaccines (Mukherjee 1989). An alternative strategy would be to develop a vaccine to prevent nerve damage that would stop bacteria from entering the nerve cells or arrest *M. leprae*–related nerve destruction.

An effective vaccine would also protect against strains that are susceptible to and resistant to drugs. The BCG derived from *M. bovis* is the most studied. The protection rate for this vaccine varies. A meta-analysis of the role of the BCG vaccine in the prevention of leprosy considered the results of seven experimental and nineteen observational studies (Setia et al. 2006). The latter showed an overall protective effect of 26 percent and the former 61 percent. The research team noted that "the issue now is not whether BCG is effective" but rather "What is the best way to use BCG to protect against leprosy? Who should be vaccinated, when should they be vaccinated and how often?" (Smith 2004, 8). As the BCG vaccine has variable, unpredictable, and incomplete protection for both leprosy and tuberculosis, new vaccines need to be developed to address the almost static global prevalence of leprosy (Duthie, Gillis, and Reed 2011).

Fortunately, new knowledge about the *M. leprae* genome and related bioinformatic processing has changed the way potential antigens are being identified and new vaccines are being explored. Now that both the leprosy and tuberculosis genome have been sequenced, researchers are in a position to identify new molecules that have vaccine potential (Cole et al. 1998, 2001). This knowledge has also made it easier to develop a vaccine, but adding protection for leprosy to a new tuberculosis vaccine would be helpful (Rodrigues and Lockwood 2011, 469). On October 12, 2017, the Infectious Disease Research Institute and the American Leprosy Missions issued a press release that described the clinical trials of the first vaccine developed specifically for leprosy (Lepvax; see Duthie et al. 2018).

Comorbidities, Prognosis, and Cause of Death for People with Leprosy

Predicting the probable course of the disease and the outcome for any person with leprosy will be affected by the type of leprosy they have. Lepromatous leprosy yields a poorer prognosis (Bryceson and Pfaltzgraff 1990). However, while the treatment given is very important, ethnicity (some groups are more resistant; see Silva Nery et al. 2019), age, sex, pregnancy, malnutrition, and intercurrent infection are additional variables that will affect the disease's progression.

Leprosy and HIV (human immunodeficiency virus) can co-occur, but leprosy has not been noted to be more common in people with HIV (Scollard et al. 2006). However, HIV can result in increased susceptibility to leprosy, as has been seen in some areas where leprosy is endemic. HIV leads to a decline in T-lymphocyte function, which lowers resistance to infections, but leprosy may also make people more susceptible to HIV because of compromised immunity. Depression of the immune system could also change the type of leprosy a person experiences, which can lead their infection to downgrade to lepromatous leprosy (Turk and Rees 1988). Antiretroviral treatment for HIV may lead to activated subclinical *M. leprae* infection in some people and the exacerbation of existing leprosy lesions (Sarno et al. 2008; Talhari et al. 2010; Lockwood and Lambert 2011). It is possible that there may be more people with HIV and leprosy in endemic areas for HIV than are known (Scollard et al. 2006). People co-infected with HIV and leprosy also show a greater likelihood of peripheral nerve damage (particularly of the motor nerves) than patients with leprosy alone (Xavier et al 2018).

A number of other conditions may co-occur with leprosy, including endocrine dysfunction (leading to osteopenia [an abnormal decrease in mineralized bone], osteoporosis, and diabetes), musculoskeletal complaints such as joint disease and changes in the connective tissues of the tendons or ligaments that attach to bones, and respiratory disease such as sinusitis, and oral diseases such as dental caries and periodontal disease (Almeida et al. 2017; Jacob Raja et al. 2016). For example, if the testicles are affected (endocrine dysfunction) and a man has low testosterone levels, bone density can decrease and osteopenia and osteoporosis can develop (Saporta and Yuksel 1994; Ishikawa et al. 2000). While musculoskeletal changes and complaints can present challenges for people with the infection, people with underlying leprosy where leprosy is not endemic may present primarily with musculoskeletal complaints that could delay diagnosis and treatment for leprosy (Chauhan et al. 2010). The facial sinuses (upper respiratory tract) become inflamed in lepromatous leprosy due to invasion of the leprosy bacillus (Kiris et al. 2007; Suzuki et al. 2013), and diabetes is increasingly being associated with leprosy, especially for people with lepromatous leprosy (Saraya et al. 2012). Finally, periodontal disease develops more readily in people with leprosy because of low resistance to *Porphyromonas gingivalis,* one of the common organisms that causes this gum disease (Ohyama et al. 2010).

Tuberculosis was and is a common cause of death in people with leprosy. As far back as the nineteenth century, Hansen and Looft listed tuberculosis as the key cause of death in people with tuberculoid leprosy and the second most common cause of death for people with lepromatous leprosy (Hansen and Looft 1895). People experiencing leprosy reactions can be especially weakened and more susceptible to tuberculosis (Mitsuda and Ogawa 1937). While reports of co-infection have been fairly rare in recent times (0.02 "cases" per 100,000 people; Rawson et al. 2014) and genetic studies have found no shared susceptibility genes (Wang et al. 2015), in the 1930s, Mitsuda and Ogawa found that people with leprosy who had died from other causes also had tuberculosis. Nephritis was the second most common cause of death and septicemia was third. Involvement of the heart, stomach, intestines, urinary bladder, ovaries, and testes was also associated. Kidney failure can be a major cause of death for people with leprosy today too; kidney disease has a high prevalence in people with leprosy (Grange and Lethaby 2004). However, a more recent study of causes of death in people with active, non-cured leprosy in China revealed that suicide was the most common cause (16 percent); the second and third leading causes were related to the cardiovascular system and failure of the body organs (Shen et al. 2011).

Thus, not only do people with leprosy experience the effects of this infection, other health-related conditions may develop.

Methods of Diagnosing Leprosy in the Past

In comparison with how leprosy is diagnosed and treated today, physicians and other healers in the past drew upon a huge range of options, often based on cultures of disease in particular times and places. This section considers some of these options through a series of questions.

Who Diagnosed Leprosy in the Past and What Methods Did They Use?

A famous English physician once remarked to his students: "Gentlemen, there are three things of importance in medicine; the first is diagnosis, the second is diagnosis, and the third is diagnosis" (Lie 1938, 215). Modern diagnosis has a scientific basis and includes a variety of methods, but in the past the situation was very different.

Some parts of the world have more information than others about diagnosis in the past. China is one example. We know that Chu Chen-heng (AD 1280–1367) used a "Five Deaths" classification for leprosy based on signs and symptoms: skin numbness and loss of sensation, loss of limb strength, and nasal bridge flattening. This suggests that some cultures looked for a range of clinical features for diagnosis. However, in twelfth- and thirteenth-century England, diagnostic tests described for the medieval world would not have identified leprosy accurately. Some tests such as pricking the hands and feet may identify lack of sensation, but they did not appear until the late thirteenth century, and the fact that medieval medical texts described diagnosis in relation to the four humors hindered correct attributions of disease (Satchell 1998, 48).

The first known leprosy "diagnosis" in Europe was in Siena, Italy, in AD 1250 (Demaitre 2007, 37). However, during the medieval period, lay people such as village elders, members of the clergy, patients in leprosaria, members of parish councils, magistrates, civil officers, and juries often diagnosed leprosy (Browne 1975a, 23). Later, physicians became more prominent in making judgements, but tensions existed between lay and medical practitioners. As Magilton (2008b, 14–15) notes, "determining whether or not an individual was leprous was not a medical monopoly." One question that perhaps will never be answered is whether all people with leprosy came into contact with the diagnosticians. Probably many poor people did not because physicians usually attended the rich (Magilton 2008c). Opportunities for diagnosis for those living in urban communities may also have been greater.

People in the medieval period who diagnosed leprosy were probably not knowledgeable enough to be able to differentiate between the conditions classified under the word *lepra,* a Greek word meaning "scaling" or a term used for various skin diseases characterized by crusts and scales on the skin (Dols 1979; Andersen 1969). Although it is likely that lay healers and physicians could have recognized lepromatous leprosy (Browne 1975a), not all regions of the world had access to accurate descriptions of the disease. Thus, we cannot be certain that historical diagnoses of leprosy were correct. As medical literature became available to more people and literacy increased, the signs and symptoms listed for leprosy could be used to guide people in diagnoses, but fewer people may have been diagnosed because diagnostic criteria also became stricter and more focused (see Rawcliffe 2006, 158–168). "Invisible" clinical features were likely often missed or hidden by clothing. Therefore, many people with leprosy may have gone undiagnosed because they were not visibly affected or diagnosis was not accurate enough to detect them.

However, some authors argue that people in medieval England really did know how to diagnose leprosy but that diagnosis probably improved when physicians and barber-surgeons became more involved (Satchell 1998). When physicians became the diagnosticians is debated, but this milestone is documented in parts of Europe, showing that the transition had occurred by the early years of AD 1300 (McVaugh 1993, 222). By the following century, some of the larger towns in those countries had physicians who regularly examined people suspected of having leprosy. Nevertheless, many early diagnosticians had little training and described not what they saw but what they thought they ought to see.

What Were the Main Methods of Diagnosis?

Observation of a range of signs and symptoms formed the basis for a diagnosis. Noting skin lesions and facial features and assessment of the blood

Table 3.1. Eight of the most convincing signs documented for all forms of leprosy by twelve authors dating from the first–second centuries AD to the sixteenth–seventeenth centuries AD

Signs of leprosy	Number of authors reporting
Hair, eyebrows, and/or lashes thin, fall out	11
Fetid breath, body, or sweat	10
Hoarse voice	9
Lips thicken, turn livid	9
Anesthesia of extremities, etc.	8
Extremities swell, fissure, or drop	7
Nodules on forehead, chin, or cheeks	7
Nasal cartilage corrodes	5

Adapted from Demaitre (2007), Table 7.2.

and urine were the main methods used. By the fourteenth and fifteenth centuries in Europe a set of protocols and checklists for leprosy diagnosis had been developed by a variety of "schools," according to the form and stage of development of the disease. Many signs were needed to diagnose the infection, including the person's behavior, and these lengthy checklists were developed in an attempt to make more accurate diagnoses. Table 3.1 shows the top eight signs (of a list of thirty) that were most likely associated with leprosy. These eight signs were mentioned by at least five of twelve authors from antiquity whose writings on diagnosis range in date from the first to the second centuries AD to the sixteenth to seventeenth centuries AD (Demaitre 2007, 220). One type of medical judgement about a diagnosis of leprosy was termed the *Iudicium leprosorum,* a process Guillaume des Innocens first described in his book *Examen des éléphantiques ou lépreux* (Lyon, France, 1595), which was based on the writings of Arabic, French, Greek, and Latin authors (Demaitre 2007, 35). In this process, several diagnosticians assessed a person suspected to have leprosy and each contributed to the diagnostic procedure. The diagnosticians recorded the patient's history, asked specific questions, and examined them through sight, touch and smell, for example by looking at their urine or taking a sample of blood.

Facial Features

Some authors have stressed that diagnosis emphasized the face. This is seen in the writings of John of Gaddesden (AD 1280–1361), physician to King Edward II of England, who said that a diagnosis could not be made until the face was deformed (Clay 1909). More recently, this focus on the face has

Figure 3.4. Image of person with leprosy in Norway by Johan Ludvig Losting (1810–1876). Credit: Bergen Museum.

been reiterated by artistic depictions of people with leprosy that concentrate on facial disfiguration (Figure 3.4; and see Rawcliffe 2006). However, because the face may not be affected in all people with leprosy, the people most diagnosed would likely be those with lepromatous leprosy. By the sixteenth century in Europe, physicians and other diagnosticians were considering the whole body rather than just the face (Demaitre 2007, 229). Diagnostic tests were not as important as examination of the body as a whole, but loss of sensation was thought to be an early and primary sign of leprosy.

Skin Lesions

Many historical descriptions of the diagnosis of leprosy refer to skin lesions. For example, in Chinese texts, people suspected of having leprosy were observed through a charcoal flame and were diagnosed if they had a nodular face (Skinsnes 1964). In seventeenth-century Finland, diagnosis was also based on skin lesions of the face and other areas of the body, along with characteristic changes to the voice (Murenius 1908). In Europe, cancer and morphea, a rare skin condition (localized scleroderma), were linked with leprosy more than any other disease, but many skin diseases became associated, for example vitiligo, impetigo, and psoriasis (Demaitre 2007). People who lived in poverty-stricken urban environments in the medieval period were prone to developing skin conditions and leprosy was likely to be diagnosed frequently because it was often confused with other skin conditions (Brody 1974, 21).

Urine and Blood

In medieval England and elsewhere, in addition to a general examination of someone with suspected leprosy, physicians used many tests to see if there was humor imbalance. They tended to focus on the urine and blood (Rawcliffe 2006, 64–72).

For example, in late medieval Wales, if a raven's egg placed in a suspect's blood hardened when the blood had cooled, then the person had leprosy (Cule 1970). If a blood sample from a person was put in water and the blood stayed on the surface, the person was diagnosed with leprosy (Peniarth MS 27 Part II, p. 13; cited in Cule 1970). Finally, in a translation of a fourteenth-century English clinical description of leprosy, if salt sprinkled into the blood dissolved, the person was diagnosed with leprosy (Major 1949). Some methods appear equally nonsensical, for example testing the blood for grittiness or clotting as an indication of leprosy, but on reflection they might have some basis in scientific evidence (Ell 1984; Demaitre 1985). In advanced lepromatous leprosy there is increased platelet "stickiness" in the blood that might lead to clotting (platelets are the red blood cells that help stop bleeding). In the thirteenth century, Theodoric of Cervia noted that the blood of people with leprosy was greasy, thin or thick, and viscous (Campbell and Colton 1960). Gilbertus Anglicus, an English physician (ca. 1180–ca. 1250) trained at the Schola Medica Salernitana, a medieval medical school in Italy, suggested that the blood of a person with leprosy contained fine sand particles in his 1510 *Compendium of Medicine* (*Compendium Medicinae*). Eight of the thirty authors Demaitre (2007, 220) documented also described "clotty, sandy, ashy blood" in people with leprosy.

Blood and urine diagnosis can also be observed in historical artwork. An early sixteenth-century woodcut by Hans von Gersdorff (*Feldtbuch der Wundtartzney*) showing a person with skin lesions is a good example that shows how an image can be interpreted in different ways. The woodcut

Figure 3.5. Leprosy being diagnosed by a committee. Woodcut from Hans von Gersdorff, *Field Book of Wound Treatment* (Strasbourg. John Schott, 1517). Credit: National Library of Medicine, Bethesda, Maryland, USA.

shows several diagnosticians examining a person (Figure 3.5): one person is making a uroscopy examination; a surgeon is palpating the head while a possible second doctor looks on; and a barber is washing a cloth soaked in the patient's blood to see the residue. Salt and vinegar in two small bowls are ready for other tests of blood and urine. Martin makes a similar but different interpretation (1921, 452). He observes that a cloth is stretched over the barber's "dish" and that there are two small bowls likely of (recommended) brass. These are for a process in which blood is run into one brass bowl on top of salt; "if the blood does not decompose it signifies that the case is leprosy" (Martin 1921, 452). Fresh well water was put into the other brass bowl; if the blood added did not mix with the water, then the person was considered to have leprosy. This was repeated with vinegar; if the combination of vinegar and water bubbled, the person was diagnosed with leprosy. In addition, if fine granules such as sand were retained when the blood was strained through a linen cloth, a diagnosis of leprosy followed. Ober's interpretation was that the illustration showed skin lesions (nodules) of leprosy on the person, examination by a physician, another physician holding a urine flask while an older man ("possibly one of the town's elders") looked on, and a servant on the left who was rinsing a concave object in a basin that was "perhaps the patient's wig" (Ober 1983, 179). Attempting to make sense of historical artwork showing medical procedures is not an easy task, especially in relation to specific diseases.

Sensation Loss

There is debate about how frequently sensation loss (or anesthesia) was described as a sign of leprosy in historical sources. There are references to this diagnostic criterion in early Chinese sources (Skinsnes 1964). Ko Hung's *Prescriptions for Emergencies* (AD 281–361) says that the first symptom of leprosy (*lai ping*) is numbness of the skin or a sensation of worms creeping. Demaitre's (2007, 220) table of signs and symptoms described as being used to diagnose leprosy in the past shows that at least eight authors mentioned sensation loss as a sign of leprosy. Gilbertus Anglicus (AD 1180–1250), an English physician, was one of the first medieval doctors in Europe to mention anesthesia as a diagnostic sign (Andersen 1969). In the fourteenth century, John of Gaddesden also described sensation loss in skin lesions that affected only one limb, possibly suggesting tuberculoid leprosy.

Was Diagnosis Accurate?

There is scarce information about what cohorts of society were actually diagnosed with leprosy in the past. Who was diagnosed? The rich? The poor? The old? The young? People in rural areas? People in urban areas? Some authors suggest that people in rural areas were diagnosed less often (Lynnerup and Boldsen 2012), but were people with leprosy in rural areas more accepted into society? There is certainly bioarchaeological evidence that both sexes and all ages were buried in medieval leprosarium cemeteries and that people with leprosy were buried in both urban and rural cemeteries. In addition, there is little evidence that rich people with leprosy were buried in non-leprosarium contexts (see chapter 5). Does this mean that rich people were more likely to be segregated in leprosaria?

Did diagnosticians diagnose correctly? If diagnosis today is a challenge for the medical profession, even in Western developed countries, and leprosy is known as the great imitator (Breslow 1968), how successful was diagnosis in the past? Diagnosis was not without its problems, although Satchell (1998, 52) has noted that "sources other than medical texts suggest that diagnosis was not entirely random in England between 1100 and 1250." Furthermore, Vilhelm Møller-Christensen's work on skeletons from the medieval leprosarium at Naestved, Denmark, identified that a majority of the individuals buried there had leprosy-related bone changes, which is used as support for accurate diagnosis (Richards 1977; Demaitre 1985). However, none of the sources provide conclusive evidence that diagnosis was always correct, although Richards (1977, 98) argues that if leprosy is clearly described in historical texts and shown in artwork and there is evidence of skeletons buried in leprosaria cemeteries with bone changes of leprosy, then accurate diagnosis was practiced. Supporting his claim, Richards cites the three-stage description of leprosy by the Frenchman, Bernard de Gordon (thirteenth–fourteenth centuries AD):

> Dusky redness of the face, harsh voice, thinning and loss of hair, widespread scabs and boils and nodules.
> Enlarged eyebrows; swollen and obstructed nostrils; nasal-sounding voice; lumps on earlobes, face, and elsewhere; pins-and-needles sensation; and muscle wasting.
> Destruction of the nasal tissues, difficult breathing, rough and hoarse voice, thin lips, disfigured face, mutilated hands and feet.

However, Guy de Chauliac, the author of *Inventarium sive Collectorium Partis Chirurgicalis Medicinae* (first published in 1363), described the clinical features as certain and uncertain (Richards 1990). He also described three recommended stages of diagnosis: suspicion, strong suspicion, and certain diagnosis (the person should be sent to hospital), and said that anyone who was declared to be free of the disease could receive a certificate of proof. He listed the certain signs and symptoms as thickening of the eyebrows and loss of eyebrow hair, disfiguration and obstruction of the nostrils, scars around the eyes and ears, and a harsh nasal voice. These signs were still used for diagnosis on the Åland Islands in Finland in the seventeenth century. Nevertheless, Richards (1977, 102) claimed that the best pre-scientific description of leprosy appeared in 1816 in Norway (that of Pastor Welhaven of Bergen), which made it possible to make better diagnoses. In Europe, after around 1300, many authors differentiated ambiguous signs and symptoms from those of leprosy at the same time that doctors were becoming the diagnosticians. Did the decline in leprosy after 1300 correlate with doctors making more accurate and careful diagnoses because they were concerned with the social consequences of their decisions (Demaitre 1985)? If that was the case, were many diagnoses in Europe before that time inaccurate?

However, physicians included other diseases under the word "lepra" after the fourteenth century and diagnosed only a small percentage of people who were brought to them with leprosy, even though many more had the infection (Demaitre 2007). One such disease was venereal syphilis (VS), and some suggest that there may have been confusion between leprosy and syphilis diagnosis (Brody 1974). This is because of the medieval idea that leprosy was transmitted sexually and because venereal syphilis is also transmitted by sexual intercourse. In addition, in both diseases, skin lesions appear, the nasal structures can collapse, and there can be swelling of the lower legs. Even in clinical contexts there has been confusion about differentiating the two infections (see Doyle 1953; Dupnik et al. 2012). Clearly misdiagnosis must have occurred. In addition, definitions of "lepra" were often imprecise and varied and could change according to different sociocultural contexts (Rawcliffe 2006, 158–168).

One bioarchaeological study concluded that leprosy was not confused with venereal syphilis because skeletons buried at some medieval leprosy hospitals do not have evidence of venereal syphilis (Crane-Kramer 2002). However, more recent data indicate that three individuals with treponemal disease (VS) were buried in the late medieval leprosarium cemetery associated with St James and St Mary Magdalene in Chichester, Sussex, England, suggesting either that people with diseases besides leprosy were admitted to the leprosarium or that these people were misdiagnosed as having leprosy (Lee and Boylston 2008a). However, in a different region of the world in the tenth century, the work of the Arabic physician Ali ibn al-Abbas al-Majusi shows no confusion between leprosy and venereal syphilis (Dols 1979). His medical text mentions treating people who had not lost extremities with mercury; al-Majusi inferred that these individuals had venereal syphilis, not leprosy. This suggests that the difference between leprosy and venereal syphilis was understood, although mercury was used to treat both venereal syphilis and leprosy (Porter 1997, 175).

Ell (1988, 502) believes that "diagnostic accuracy was very high, at least for these northern European regions." He emphasizes that "nearly all the patients dying in medieval leprosaria had lepromatous leprosy. This argues that medieval diagnosis of leprosy was conservative, tending to select out the most extreme cases" (Ell 1986, 301). He bases this on data from the skeletons buried in the leprosarium cemetery at late medieval Naestved, Denmark, that Richards (1977) and Demaitre (1985) referred to. Magilton (2008c, 10) also suggests that it would be "unsurprising to find a high percentage of skeletons in a leper hospital because those buried there would be the ones with the most advanced disease."

Schmitz-Cliever (1973b) also considered that diagnosis was accurate because 38 of the 41 (92.6 percent) skeletons excavated from the medieval leprosarium in Melaten, Germany, had bone changes characteristic of leprosy. A high proportion of skeletons with leprous-related bone changes (68.5 percent of 54 excavated) has also been reported at the leprosy hospital site of St Mary Magdalen, Winchester, England (Roffey and Tucker 2012). This situation was also seen at the St James and St Mary Magdalene cemetery in Chichester for the early part of the cemetery that was used when the site served the leprosarium (Magilton, Lee, and Boylston 2008). Magilton (2008a, 113) concluded that "the apparent absence of lepers from normal cemeteries of the Middle Ages suggests that the disease could be diagnosed in its early stages when no skeletal changes had taken place" and emphasizes that disfigurement was the most important diagnostic feature. While facial features were important for diagnosis, most excavated skeletons with the lesions that are characteristic of lepromatous leprosy are located in cemeteries that were not associated with leprosy hospitals. This evidence does not support Magilton's view.

A number of studies of skeletons with leprosy-related bone changes have suggested that one of the reasons for the preponderance of such skeletons in non-leprosaria contexts may have been that diagnosis was not accurate (e.g., for Britain; see Roberts 2002). In addition, some medieval leprosaria cemeteries have a low proportion of buried people with skeletal changes of leprosy. Thus, the statement that "nearly all the patients dying in medieval leprosaria had lepromatous leprosy" because they were correctly diagnosed with lepromatous leprosy (Ell 1986, 301) cannot be regarded as universally true. People buried in leprosaria with no evidence of leprosy in their skeletons could have been highly resistant to the bacterium, they could have had tuberculoid leprosy and thus not necessarily have shown any bone changes, they may have died before the bone changes occurred, or they may have been misdiagnosed. These scenarios also argue against the idea that medieval diagnosis of leprosy was conservative and focused on the individuals with the most extreme expression of the infection.

Perhaps of relevance too is the French surgeon Guy de Chauliac's fourteenth-century *Inventarium sive Collectorium Partis Chirurgicalis*

Medicinae, which advised caution about attributing a leprosy diagnosis (Rawcliffe 2006, 157). Physicians would have been very careful who they diagnosed with leprosy because such a diagnosis could have had considerable negative implications (Demaitre 1985). While there may be reasons why a person might not want a diagnosis of leprosy, at times people might actually have wanted that diagnosis, perhaps to ensure that they would have access to water, food, and shelter in a leprosarium. Some people in medieval Islamic society intentionally disfigured themselves to appear to have leprosy in order to encourage others to give alms. This involved covering themselves with a mixture of the leaves of the indigo plant, basil, cubeb (*Piper cubeba*, or tailed pepper), and copperas, or ferrous sulfide (Bosworth 1976). Faking leprosy to acquire a diagnosis is certainly evident in Europe, and some people even disputed negative diagnoses (Demaitre 2007).

Specific Treatments of the Past

There are references to many treatments aimed at people with leprosy, but few relate to prevention. However, it was considered that there was no cure for leprosy, for example in medieval Europe (Demaitre 2007, 244–249). Evidence for past treatment of leprosy can be categorized as bathing regimes, specific dietary practices, herbal and mineral remedies, the use of blood, surgery, and hospital admission. Many of these treatments aimed to restore balance to the humors. Like diagnostic methods, some seemed logical for the time, but others were incomprehensible. For example, alchemist's gold, earth from an anthill, perfumed water in which the Christ child had been washed, the blood of a turtle from the Cape Verde Islands, and the blood of an infant were all used at some time in the past to treat leprosy (Brody 1974).

Bathing and Other Treatments for the Skin

A rather logical type of treatment for leprosy was bathing, which was presumably aimed at cleansing the skin lesions of leprosy. Numerous authors have indicated its widespread use to treat leprosy. However, there would have been many diseases whose clinical features incorporated skin lesions and bathing was often recommended for treatment of skin diseases. As the cause of the infection is a bacterium, bathing would not rid a person of the organism, but it might have soothed skin lesions and ulcers that developed on the hands, legs and feet. Corrosive ointments were also used to target lesions associated with leprosy (Rawcliffe 2006, 219, 224, 238). There is no structural evidence of baths used to treat people with the infection, but one is described in an inscription dated to the mid-sixth century AD at Scythopolis (modern Beth Shean, Israel) (Avi-Yonah 1963).

Greek and Arabic authors describe bathing as a treatment for people with leprosy, and in the Near East bathing in sulfurous water was recommended (Dols 1979; Mitchell 1993). This is noted for the city of Al-Hammah in southern Tunisia, where water leaving the city formed a sulfurous lake called the "Lake of the Lepers" around which many people with leprosy lived (L'Africain 1956). Several remedies involving bathing and the application of remedies to the skin lesions of people with leprosy are described for nineteenth- and twentieth-century southern India (Buckingham 2002). For example, mercury and arsenic ointments and poultices for sores were used, as was gurjon oil that was rubbed into the skin (from the *Dipterocarpus turbinatus* tree, which is native to the Andaman and Nicobar Islands). Physical cleansing of the body was also very important in Hindu tradition and bathing was practiced in India for most forms of leprosy. In Europe, specifically in England, bathing was commonly resorted to. Leprosaria were often located near springs or wells (Rawcliffe 2006). For example, at Harbledown leprosy hospital near Canterbury, Kent, the building was located near a spring that had apparent healing properties. People with leprosy also took baths that included juniper in eighteenth-century Iceland and Norway (Richards 1977, 74). In the seventeenth and eighteenth centuries, volcanic mud baths were also popular. The Arnamagnæanic manuscript, a fourteenth-century Danish medical book, recommended that ants and their anthills should be placed in a bath used for leprosy treatment (Ehlers 1895). This relates to the idea that "like cures like": one of the symptoms of leprosy is a feeling of ants crawling under the skin.

Diet and Rebalancing the Humors

A proper diet and lifestyle were advocated for the prevention of leprosy in medieval Europe. Diet was intricately bound up with balancing the humors (Mitchell 2000b). It is unknown how many people with leprosy had access to a well-balanced diet, but many leprosaria in medieval England had farms and gardens and such a diet was potentially possible (Rawcliffe 2006, 327–328), at least for those who were patients. It could be argued that people who lived in leprosaria were advantaged over those without leprosy. The latter may have been poor and may have had little prospect of a good healthy lifestyle. Some medieval European authors recommended specific components of diets for people with leprosy, for example snake meat, blood (including that of the hare—*Lepus*), mouse and pigeon dung, toads and frogs, and turtles (Demaitre 2007, e.g., 257, 266). Other "remedies" included refraining from eating fish (Hutchinson 1906), the medieval Islamic practice of drinking wine in which vipers had been placed (Dols 1979), a sulfur elixir in wine in China in the sixth and seventh centuries AD (Skinsnes and Chang 1985), heads of scorpions in medieval Damascus (Walker 1934), and the flesh of dead infants in China (Skinsnes 1964).

Laxatives and purgatives also featured in medieval England, along with bloodletting (Rawcliffe 2006, 212–213, 232–236). Bloodletting to rid the body of contaminated blood was also used during the Ming Dynasty in China (AD 1368–1643); Hsueh Li Chai, an Imperial physician, described the practice in his leprosy chapter in *Lei Yang Chi Yao* (Confidential Text Regarding Malign Ulcers; Skinsnes and Chang 1985). A "special" laxative was used in China to cause diarrhea and thus rid the body of leprosy "germs." For women, a similar treatment was used to drive germs to the uterus and create the opportunity to "sell" leprosy via sexual intercourse (Skinsnes 1964). Bloodletting featured prominently, and in seventeenth- and eighteenth-century Iceland, Norway, and Sweden, leeches and cupping were used to let blood in addition to standard bloodletting (Richards 1977). There are conflicting opinions about the value of bloodletting and its effect on leprosy. However, loss of blood lowers the body's stores of iron (Kluger 1978). If bacteria need iron to grow and reproduce, reducing the availability of iron may help the body cope with an infection (see also Stuart-Macadam 1992), but because the leprosy bacterium is slow growing this may have had little effect (Mitchell 2000b).

"Surgery"

Specific surgical treatment is rarely mentioned, but there is written and artistic evidence to suggest that cautery was used on people with leprosy. In this process, hot irons were applied to skin lesions at specific points of the body (see Figure 3.6; Rawcliffe 2006, 232–239 on medieval England; Demaitre 2007 on medieval Europe; and Dols 1979 for Greek and Arabic medicine). There is some evidence of amputation of the ends of the bones of the lower limbs (tibiae and fibulae) in skeletons from two medieval leprosaria sites in England (Roffey and Tucker 2012; Lee and Boylston 2008b; Figure 3.7) and in the skeleton of a person with leprosy buried in a nineteenth-century cemetery in London that was not associated with a leprosarium (Walker 2009). These rare examples may indicate surgical treatment for feet damaged by leprosy (as is seen today).

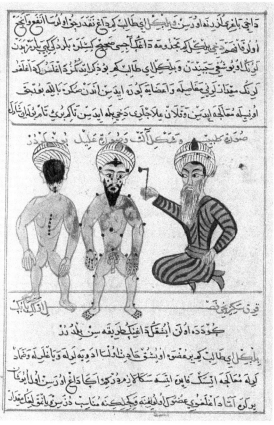

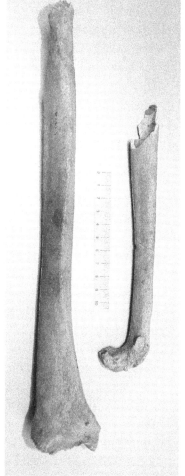

Above: Figure 3.6. Cauterization points for treating people with leprosy. From *Chirurgia imperiale,* ca. 1300 AD. Credit: Bibliothèque Nationale de France, Paris.

Right: Figure 3.7. Evidence of amputation of the foot of a person who was buried in the medieval leprosarium of St James and St Mary Magdalene, Chichester, England (Skeleton 137a [right bone in the picture, compared to a normal looking bone on the left]). Source: Magilton et al. (2008, figure 19.8). Reproduced by permission of Chichester District Council/ Donald Ortner/Saul Crawshaw/John Magilton and the Council for British Archaeology from CBA Research Report 158.

Blood Use and Other Treatments

While bloodletting was used in an effort to re-balance the humors in people with leprosy, there is evidence that blood was used in other ways to treat the disease. For example, in southern India, preparations of iron and iodine were used to improve blood quality in people with leprosy in the nineteenth and twentieth centuries (Buckingham 2002), and a fourteenth-century Danish medical book in the Arnamagnæanic Manuscript Collection documents smearing the blood of a male goat over the body (Ehlers 1895). In medieval Norway, an oily fluid (spermaceti) present in the heads of sperm whales was used (Larson 1917). Mercury was also used in some parts of medieval Europe (Richards 1977; Demaitre 2007). The use of mercury suggests again that there may have been confusion about what disease was being treated because mercury was also used to treat syphilis in medieval Europe (Porter 1997, 175; see also Rasmussen et al. 2008). In southern India in the nineteenth century, carbolic acid treatment was described. This is also called phenol, one of the oldest antiseptic substances; it is a derivative of benzene, a poison. People with leprosy were placed in wicker cone-shaped structures covered with an air-proof material; the base of the cone was sealed with sandbags. The person with leprosy sat inside and inhaled vaporized carbolic acid through a tube from a kettle outside the structure (Johnston 1874, 1875).

Herb and Mineral Therapy

Many herbs were recommended for the treatment of leprosy in the past. In the medieval period in Europe these included alder, betony, borage, bugle, dock, elecampane, garlic, honeysuckle, juniper, nettle, rue, sage, scabious, and violet. Culpeper's 1653 herbal (1805) reveals that over 50 percent of these herbs were also used for skin and lung disorders. Very few were recommended particularly for leprosy, but fumitory and scabious are mentioned specifically for this disease. Sometimes herbs were mixed with animal parts and other products such as urine, feces, minerals, or food, and many were used to make ointments for the skin (Roberts 1987; Rawcliffe 2006, e.g., 213–226). Section 595 of the 1861 Welsh document "Meddygon Myddfai," which includes the writings of physicians of Myddfai in Wales, describes this sixteenth-century treatment for leprosy from the Welsh Leech Book: "Take a toadstool, or if this is not obtainable a ground fungus that is called boletus, and the red alder leaves, and purified butter, and boil together in sheep's milk, and straining through new linen, and anointing frequently with this, you shall be healed with God's help" (quoted in Cule 1970, 50). The Italian surgeon Theodoric Borbognoni (1205–1298) described a popular ointment called *saracenic* that was used for leprosy; it was usually mixed with mercury, fats, and oils and rubbed into the skin (Campbell and Colton 1960). Again, the link with mercury suggests confusion with syphilis. Yu Chang (1585–1664) also mentioned "drunken goddess powder" in his *Yi Men Fa Lu* (Standard Methods in Medicine); it was made with seven herbs plus mercury (Skinsnes and Chang 1985). Many herbalists in China were convinced that they could cure the disease, but there was one type of leprosy that was incurable (chicken-claw leprosy), which was described as a loss of sensation and curled fingers.

Perhaps the main herbal remedy for leprosy that has been written about for hundreds of years was made from the seeds and the oil from the seeds of chaulmoogra trees, or *Taraktogenos kurzii* (later called *Hydnocarpus kurzii*), *Hydnocarpus wightiana,* and *Hydnocarpus anthelmintica* trees. There are about forty species of *Hydnocarpus,* mostly found in Southeast Asia. Their fruits are the size of an orange and they have hard, angular, marble-sized seeds (Norton 1998). The oil was administered by intramuscular or subcutaneous injection, orally (the oil), and topically (the pounded seeds mixed with ghee, or clarified butter) (Buckingham 2002). Subcutaneous twice-weekly injections for ten weeks seemed to be the best method of administration (Cochrane 1947). Intramuscular injection was very painful and caused leprosy reactions and fever. The oil is purported to activate lipids that destroyed "foreign" leprosy-related lipids. Another explanation for how the oil worked was that irritation caused by the injections attracted phagocytes (cells that kill microbes) to the leprosy bacilli (Norton 1998). The oil has been used for a long time for leprosy treatment in Ayurvedic medicine in India and for other skin diseases (Parascandola 2003), mostly from the seeds of *Hydnocarpus pentandrus* (Sahoo et al. 2014). As early as the fourth–sixth centuries AD, the Indian *Sushruta Samhita* described treatment with chaulmoogra oil (Lowe 1947; Dharmendra 1947), and Dharmendra (1947) notes that the *Sushruta Samhita* also describes the use of seeds of *Hydnocarpus laurifolia* to prevent leprosy. In his assessment of the use of the oil for the treatment of leprosy, Oommen (2002, 203) noted that "the 'law of signature[s]' states . . . that if a disease is endemic in any particular geographic location, nature itself provides a cure in the same geographic location. . . . . The oil of hydnocarpus has both wound healing and anti-leprosy properties and could be considered as a useful adjunct (not as an alternative) in the management of leprosy." This seems a logical assumption. Relevant analogies are dock leaves (which contain oxalic acid) to alleviate nettle stings, both of which grow in the same locations in both the UK and the United States, and the use of local honey to treat pollen allergies (Saarinen, Jantunen, and Haahtela 2011).

However, the effectiveness of chaulmoogra treatment is reported as variable. Mouat, a British surgeon who was professor of materia medica at the Bengal Medical College from 1841 to 1853, noted that the indigenous population of India used the seeds of the chaulmoogra tree for treating skin diseases, including leprosy. Mouat did some experiments that involved dressing leprosy-related ulcers with the oil and giving a tablet of the seed three times daily. He found that the ulcers healed rapidly (Mouat 1854). In 1879, Cottle also described administering chaulmoogra oil and an ointment of the oil mixed with lard to affected skin. He recommended that the oil be given in small doses after food, noting that constipation and sickness were side effects. However, he reported that it had a "high reputation as a remedy for scrofula, skin diseases, and leprosy" (1879, 969). Chaulmoogra oil became increasingly important from the 1870s onward and it was used until the 1940s, when dapsone was introduced (Rastogi and Rastogi 1984; Drabik and Drabik 1992; Norton 1994; Buckingham 2002).

The first comprehensive chemical analysis of the oil was done at the Wellcome Chemical Research Laboratories in London in the early twentieth century. Clinical tests on humans followed with varying results. Chaulmoogra oil contains substances that have high bactericidal activity in vitro (Morrow, Walker, and Miller 1922). The activity is in the fatty acids. It is specific against the acid-fast groups of bacteria like *M. leprae* and is inactive toward other bacteria. Twenty-one patients with leprosy at a San Francisco hospital in California were treated intramuscularly weekly with ethyl esters of the fatty acids of the chaulmoogra oil (Morrow, Walker, and Miller 1922). No patient became bacteriologically negative during treatment. However, another study in Hawaii concluded that younger patients responded better than older patients. In a much earlier study on Culion Island in the Philippines, injections of the ethyl esters were administered weekly to 4,000 people with leprosy. After 18 months of treatment, 56 percent of patients had improved, 36 percent had not, 6 percent had worsened, and 2 percent had died. A large majority (92 percent) saw no worsening of the infection. The improvement rate was better for those who had had the disease for a shorter time (Callender 1925). A recent study of the anti-leprotic properties of chaulmoogra oil concluded that one of the mechanisms for controlling leprosy was host lipases (enzymes) that destroy foreign lipids (fats), which include the cell wall of *M. leprae* (Sahoo et al. 2014).

The herb was also used historically for leprosy treatment in China, where it was known as *ta feng tzu* (seeds) or *ta feng yao* (oil) (Skinsnes and Chang 1985). Hydnocarpus was mentioned during the Sung Dynasty (AD 960–1279), when it was often used as a general balm for all kinds of skin lesions (Skinsnes 1964). By the Ming Dynasty (AD 1368–1644), it was used specifically for treating leprosy. For example, in the seventeenth century, Yu Chang mentioned in *Yi Men Fa Lu* that chaulmoogra oil was used to kill leprosy "worms." Hsueh Li-Chai's sixth-century text *Lei Yang Chi Yao* (Confidential Text Regarding Malignant Ulcers) noted that *ta feng tzu* was used for all kinds of sores, and in *Pen Ts'ao Ching Shu* (Simplified Herbal Text), Miao His-yung said it was used for both leprosy and ringworm. In China, chaulmoogra oil was much more widely used by the late eighteenth and early nineteenth centuries (Skinsnes and Chang 1985). There, as in India, it was used for both leprosy and other diseases, suggesting that diagnosticians may have confused leprosy with other skin diseases.

Was Segregation Practiced and Was It Strictly Enforced?

The most recent interpretations of the evidence suggest that not all people with leprosy were segregated in the past (Rawcliffe 2006, 230; Demaitre 2007, 241). Beliefs about whether people with leprosy were isolated from their communities originally derived from descriptions in the Bible about "lepers" being outcast. However, the Bible does not specifically describe leprosy as it is known today because the Hebrew word *tsar'ath* (also *zara-ath, tzaraat, zara'at,* and *sāra'at* in various sources) was mistranslated as leprosy and was subsequently associated with uncleanliness (Lie 1938). This mistranslation has led to direct links made between leprosy and isolation from the rest of society, but in recent years this has been reevaluated. Clearly, people with some type of affliction are described in the Bible as being segregated, but there is no convincing evidence that the illness was leprosy. In the Bible, "cleansing the leper" involved cleansing that sought to rid the person of sin. Cleansing had three stages: outside the camp (separating), waiting within the camp (marginal and not clean enough to enter his/her own house), and cleansing through sacrifice at the door of the meeting tent, an act that completely purified the person so they could rejoin the community of worshippers (Lewis 1987). While this may refer to people with leprosy, it likely included people with other complaints that may have been skin diseases (e.g., eczema) or diseases that included skin lesions (e.g., syphilis or tuberculosis).

Although several hundred years of historical evidence in Europe described that people with leprosy should be segregated, generalizations about whether people with leprosy were segregated should be avoided because the reasons for segregation, if practiced, varied in different times and places (Richards 1977). For example, although both positive and negative attitudes toward leprosaria were expressed in conjunction with English leprosaria founded from AD 1100 to 1250, little hard evidence exists that people with leprosy were persecuted (Satchell 1998, v). This also applied to segregation. By the late twelfth century, the church expected that people with leprosy would be separated from society (ibid., 29), but this did not necessarily mean segregation because the practice often depended on context. However, Mitchell (2000b) maintains that segregation occurred widely in the crusader states (eastern Mediterranean) with leprosaria being located outside towns. By law, higher-status people with leprosy had to join a specific military order, the Order of St Lazarus, which was established for those with leprosy (Mitchell 1993). However, leprosy was more tolerated by the thirteenth century. People without leprosy were also admitted to the order and leprosaria were built inside city walls. It is possible that the King of Jerusalem's (Baldwin IV) leprosy increased tolerance of the disease because he was not segregated (Mitchell 2000a).

Nevertheless, extensive evidence shows why segregation was practiced and why particular institutions and places were highlighted as suitable, for example leprosaria and islands. There is also archaeological evidence for leprosy hospitals and for skeletons with leprosy that have been excavated from their associated cemeteries.

Much has been written about leprosaria, although there are few large-scale global studies. Researchers prefer to concentrate on individual leprosy hospitals (e.g., Magilton, Lee, and Boylston 2008) or on leprosaria in specific countries (e.g., Satchell 1998) or in regions of a country (e.g., Hart 1989). Clearly not all people with leprosy were segregated in the past, for many reasons, and people with other diseases might have been either misdiagnosed and segregated incorrectly or deliberately placed in leprosaria, even though they had another disease. Sometimes the very sick were refused admission because of the potential expense of caring for them and because they could not raise money for the hospital (Magilton 2008c, 19). Thus, the profile of people in each leprosy hospital would have been varied and would likely have changed over time. Leprosaria were also classed as places of sanctuary for the sick and poor, including those with leprosy, where access to food, water, and shelter were guaranteed. As the number of new leprosy hospitals declined after the fourteenth century, for example in England, those that survived changed their function to admitting people with a range of health problems (Gilchrist 1995). For example, the St. Jørgens Hospital in Bergen, Norway, one of the oldest hospitals in Scandinavia, which was established in the early fifteenth century for people with leprosy (Leprosy Museum 2003), changed its policy in the early sixteenth century to admitting people who did not have leprosy.

While people were segregated into leprosaria, the prevailing view, at least for medieval England, is that both people with leprosy and leprosaria were well integrated within towns (Rawcliffe 2013, 318). At the same time that leprosaria were founded, many general hospitals were also being opened, illustrating the increase in benefactors who wanted to contribute to the spiritual well-being of their community. Virtually all of these hospitals were religious institutions that encouraged patients to repent of their sins so their soul could be cleansed (Magilton 2008b). Although there is little specific evidence for medical and surgical care as we understand it today, people in leprosaria did receive some general overall care (Rawcliffe 2006, 322–337).

Conclusions

This chapter has shown how a diagnosis of leprosy is made today, including by observing signs and symptoms. It has illustrated that diagnostic tests have progressed in recent years with the advent of molecular methods. They have also contributed to detecting drug-resistant *M. leprae* strains and bacterial strain-specific markers for assessing exposure to *M. leprae* and tracing transmission patterns. This work will contribute to eliminating leprosy, but difficulty accessing health care facilities for a diagnosis remains. Treatment of leprosy has focused on drug therapy, but a holistic approach to managing leprosy in a person is needed that includes both social and medical perspectives. Clearly, drug therapy has been important for reducing leprosy globally since the 1940s, when dapsone was first used for treatment. Free multidrug therapy is now available for all who need it. However, access to the antibiotics used for leprosy can be a challenge for people for many reasons, leading to no treatment. This is being addressed but could take time. Effective vaccines for preventing leprosy need development, especially so because the rate of success in using the BCG vaccine is variable. Molecular research is helping the world rise to this challenge. Drugs and vaccines are well and good but using nonmedical therapy is equally important, particularly the prevention and treatment of damage to the hands and feet due to loss of sensation.

In the past, diagnosis could be done by inexperienced people who sometimes used methods that seem nonsensical to us now. In retrospect, it can seem that sheer chance mostly seemed to determine whether a person was diagnosed with leprosy, misdiagnosed, or not diagnosed at all. However, logical tests are described, such as looking at urine and blood and focusing on the skin lesions. Treatments included bathing, caring for skin lesions, cautery of body parts, and herbal remedies. The most widespread "treatment" described in the historical literature was segregation to leprosy hospitals in marginal areas, such as outside towns or on islands. It is unclear how many people in the past were segregated in this way compared to the number who were more readily accepted within their communities.

4

The Bioarchaeology of Leprosy

> The only way of obtaining definite evidence of the occurrence of early leprosy is to make use of the fact that the lepromatous type of leprosy causes characteristic and permanent changes in the bones.
>
> Møller-Christensen (1967, 295)

This chapter considers how leprosy affects bones and teeth. Some skeletal changes (or lesions or alterations—damage or abnormal change in the tissues of the body caused by disease or trauma)—have been identified and described as specifically caused by leprosy (pathognomonic; direct effects) and nonspecific to leprosy, meaning that they may be related to leprosy but could be caused by other conditions (not pathognomonic; indirect effects). Clinical literature is the main source of information about diagnostic criteria for leprosy in bioarchaeology, assuming that the lesions have not changed throughout the time that leprosy has afflicted humans. However, it is important to determine whether the characteristics of the bone changes identified in clinical literature have been influenced by modern treatments such as drug therapy. Thus, considerations must be given to the strengths and weaknesses of using clinical data as a basis for recording and interpreting evidence of disease in skeletal remains (Mays 2012). The bone changes described in clinical literature may not be applicable to paleopathology (the area of bioarchaeology that focuses on disease) and bioarchaeologists may not be able to make specific diagnoses. For example, the expression of bone lesions due to disease may have been changed, reduced, or enhanced, or they may not be present if a person was treated with antibiotics. Thus, it is important to use information from medicine before the antibiotic era for diagnosing leprosy.

The diseases that do produce lesions in the skeleton are seen in only a small proportion of people and the proportion that is afflicted depends on many factors. In the case of leprosy, whether the bones are affected relies primarily on where on the immune spectrum the person is located (are they resistant to *M. leprae* or not?) and on factors such as nutritional status.

Immune system strength is particularly relevant to identifying leprosy in skeletal remains. What is seen archaeologically will be only the tip of the iceberg of the presence of leprosy in the past. We should also be aware that the early stage bone changes in both patients and archaeological skeletons are often not recognized and that the lesions that are recognized in the latter are generally chronic in nature. That is because the people concerned had had the infection for a long enough period for bone damage to occur. The lesions described and illustrated in clinical and bioarchaeological literature are usually related to the most nonresistant form of the disease (lepromatous leprosy). In addition, extensive and perhaps severe bone changes cannot develop overnight. There can be a long time frame from the early bone changes to very recognizable lesions, and people in the past may have died before those late-stage lesions developed. Recognizing early stage lesions in the bones of archaeological skeletons would be very helpful for diagnosis in future work. However, those early lesions may not be identified in radiographs (X-rays) of people with leprosy today as being related to leprosy and thus would not be within the bioarchaeological criteria for diagnosis. Finally, a person with subclinical *M. leprae* infection and with no bone changes will likely be undetectable archaeologically unless biomolecular methods are used successfully, such as ancient DNA analysis.

A person with tuberculoid (high-resistance) leprosy may not develop bone changes, while one with lepromatous (low-resistance) leprosy is more likely to. Leprosy affects the skeleton in only about 3–5 percent of people with the disease (Paterson and Rad 1961), but the bone changes will vary in their expression according to the type of leprosy experienced and may also develop and change if a person moves along the immune spectrum. However, much higher rates of skeletal involvement have been found in some studies, for example 15 percent of 150 patients with leprosy in Hawaii (Chamberlain, Wayson, and Garland 1931), 29 percent of 505 in the Carville Leprosarium in Louisiana (Faget and Mayoral 1944), and 54 percent of 894 in the Hay Ling Chau Leprosarium in Hong Kong (Paterson and Rad 1961). These data indicate that frequency rates of bone changes may vary considerably depending on the group of people and/or the context under study.

It is fortunate that medical doctors such as Vilhelm Møller-Christensen (1903–1988) and leprologist Johs Andersen (1923–2005; 1969), both from Denmark, took an interest in leprosy and the related bone changes (Bennike 2002, 2012). Møller-Christensen noticed that there was relatively little information about how leprosy affected the skeleton when he was trying to diagnose the disease in skeletal remains from a Danish cemetery. He studied the skeletons of people who had died with leprosy in the medieval period in Denmark and patients with bone changes of leprosy in Brazil, Thailand, and Singapore (Bennike 2012, 362). In 1974, he published a comparative study of patients with leprosy in Bangkok and ancient skeletons from Denmark that showed evidence of leprosy. Paleopathologists worldwide now use his research to attempt to diagnose leprosy (see especially Møller-Christensen 1953a–c, 1961, 1978), and other scholars have used his work to extend the diagnostic criteria. Likewise, Andersen, a doctor who worked with people with leprosy in Africa and India, contributed much to the paleopathology of leprosy after he was inspired and encouraged by Møller-Christensen. Andersen also contributed his knowledge of the impact of leprosy on living people to paleopathological interpretations.

While recognizing leprosy in human remains has depended on understanding how it affects the skeleton, it is perhaps the use of nonspecific bone changes that have proved most challenging when making diagnoses because other diseases can lead to the same or similar bone changes. The bone changes of leprosy I identify and describe in the following sections are based on the work of Abraham (1908); Andersen (1969); Andersen and Manchester (1987, 1988, 1992); Andersen, Manchester, and Ali (1992); Andersen, Manchester, and Roberts (1994); Barnetson (1951); Carpintero-Benétez and García-Frasquet (1998); Carpintero-Benétez, Logroño, and Collantes-Estevez (1996); Chamberlain, Wayson, and Garland (1931); Cooney and Crosby (1944); Dyer (1921); Dyer and Hopkins (1911); Enna, Jacobsen, and Rausch (1971); Erickson and Johansen (1948); Esguerra-Gómez and Acosta (1948); Faget and Mayoral (1944); Harbitz (1910); Harris and Brand (1966); Hirschberg and Biehler (1909); Karat, Karat, and Foster (1968); Kulkarni and Mehta (1983); Lechat (1961); Lechat and Chardome (1955); Leloir (1886); Manchester (1989, 2014) and other editions in the period 1989–2014; Marks and Grossetete (1988); Michman and Sager (1957); Miller (1913); Møller-Christensen (1953a, 1953c, 1961, 1974, 1978); Møller-Christensen et al. (1952); Murdock and Hutter (1932); Ortner (2002, 2008); Paterson and Rad (1961); Paterson and Job (1964); Reichart (1976); Rothschild and Rothschild (2001); Sane et al. (1985); and Thappa et al. (1992).

The observations these authors have made are based on both clinical (primarily from radiographs/"X-rays") and bioarchaeological data from skeletal remains from archaeological sites. Radiographs have led to greater understanding of some of the bone changes identified in living patients but may not show more subtle changes that can be seen only on the bones themselves. Many have noted that the skeletal changes of lepromatous

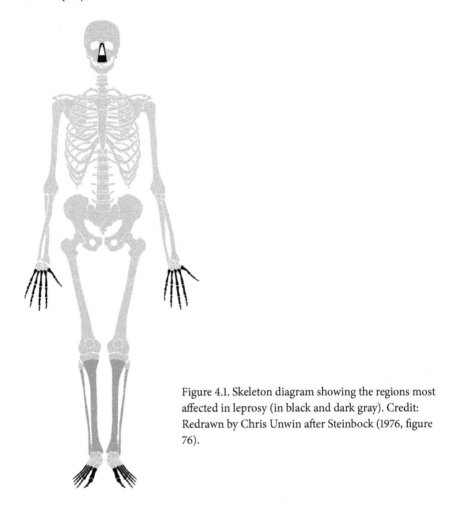

Figure 4.1. Skeleton diagram showing the regions most affected in leprosy (in black and dark gray). Credit: Redrawn by Chris Unwin after Steinbock (1976, figure 76).

leprosy are the same or very similar in both living people and archaeological skeletons. As Andersen (1969, 101) has said, "As far as the low resistance type of leprosy is concerned, the clinical picture of leprosy in Denmark from the middle of the 13th century till the middle of the 16th century is identical with the picture of leprosy from the middle of the 20th century in North East India." This suggests that if the bone changes described in clinical data are observed in skeletal remains, a diagnosis of leprosy can be made. However, whether this is the same disease that is reported in historical texts is debatable, and while some authors of clinical data specify their criteria for selecting the individuals they studied, some do not. For example, Andersen used strict criteria to select patients for study in the Purulia Leprosy Home and at a hospital in Western Bengal (Andersen 1969,

92–93). Patients 1) had to be in an abacillary condition in order to exclude the possible influence of drugs on the development of bone damage (prior to the introduction of sulfone therapy for leprosy in 1958); and 2) had to have been diagnosed by a reputable leprologist.

M. leprae can affect the bones of the face, hands, and feet (Figure 4.1). There is as yet no discussion about whether leprosy caused by *M. leprae* and *M. lepromatosis* produces the same bone changes, and until the ancient DNA of *M. lepromatosis* has been positively identified in a skeleton with bone changes consistent with leprosy, discussions of similarities and potential differences cannot proceed. In the following discussion, I specifically refer to skeletal data published from three medieval leprosaria, two from England (Chichester: Magilton, Lee, and Boylston 2008; Winchester: Roffey and Tucker 2012) and one from Denmark (Naestved: Møller-Christensen 1978). These are the best archaeological skeletal leprosy datasets currently available, but not all these publications have provided complete sets of comparative data and there are occasional discrepancies in the data in Møller-Christensen's many publications on Naestved, Denmark (1953c, 1961, 1969, 1978).

DIRECT EFFECTS OF *M. LEPRAE*

Leprosy affecting the facial bones was originally termed *facies leprosa* or Bergen Syndrome, the latter because bone changes had been identified in Norwegian patients with leprosy in Bergen. Vilhelm Møller-Christensen was the first person to describe these changes in skeletal remains (Møller-Christensen, Bakke et al. 1952). Johs Andersen (Bennike 2002) and Keith Manchester, a medical doctor (Roberts 2012), later named the *facies leprosa* changes as rhinomaxillary syndrome. The rhinomaxillary changes observed in skeletal remains are considered secondary to infection of the soft tissues in the mouth and nose. Because associated soft-tissue changes in leprosy are implicit in the term *facies leprosa*, that term should not be used in bioarchaeology (Andersen and Manchester 1992, 122).

The radiological picture for leprosy includes nasal bone absorption. The first part of the face affected is the anterior nasal spine, followed by the anterior or front part of the maxilla, the alveolar margins, and the nasal bones (Paterson and Job 1964). Paterson and Job suggested that these changes were partly due to nonspecific osteitis (inflammation of the outer [cortical] layer of the bone, potentially caused by a range of disease causing organisms) secondary to ulceration of the soft tissues and partly due to the spread of lepromatous leprosy lesions in the nasal mucosa to the bones of the face. Leprosy-related soft-tissue ulcers can, if not treated, affect the underlying bone. This leads to an inflammatory response within the periosteum (the membrane covering the bone surface) and subsequent new bone formation and/or destruction. Møller-Christensen (1974) categorized the bone changes of rhinomaxillary syndrome into Type 1 (involvement of the anterior nasal spine) and Type 2 (involvement of the alveolar process of the maxilla and loosening or loss of the incisors).

Lepromatous leprosy or near-lepromatous leprosy are the only types that have this bone change. Lepromatous leprosy is diagnosed if all the facial bone changes occur together (Manchester 2002; Andersen and Manchester 1992). Andersen's (1969) study of patients with leprosy in India found that 40 of 56 patients were affected by rhinomaxillary syndrome, but it had been noted in other studies (e.g., Chamberlain, Wayson, and Garland 1931; Murdock and Hutter 1932) that bone changes characteristic of leprosy in the maxillae or the anterior nasal spines of people with leprosy could not be identified (Møller-Christensen, Bakke et al. 1952). This may have been because the patients studied did not have lepromatous leprosy. The distribution of lesions in rhinomaxillary syndrome reflects the cooler temperature of exposed mucous membranes and skin (see Brand 1959). Lower temperatures encourage the invasion of *M. leprae* and leprosy-related lesions can develop in areas where the temperature is less than $33.8°$ C ($92.8°$ F; Scheepers 1998). Leprosy bacilli flourish in low temperatures, as seen in the soft palate (back of the roof of the mouth), uvula, and pharynx in patients with lepromatous leprosy (Reichart 1974). Patients in Reichart's study ranged in age from 17 to 74 years, had had leprosy for an average of 21.4 years, and had been treated for an average of 5.3 years. Ten people had changes to their soft palate, including the uvula, related to the infection. These areas are constantly exposed to a cooling air stream and are likely to experience a temperature lower than $37°$ C ($98.6°$ F), thus potentially favoring the survival of *M. leprae*. Rendall, McDougall, and Wilks (1976) also recorded temperature in 100 patients with leprosy. The incisor region was colder than the molar region, the region of the upper incisors was colder than the region of the lower incisors, and the upper molar regions were colder than the lower molar regions. Even when a person breathed with their mouth closed for ten minutes, the temperature around the upper teeth was cooler. Leprosy bacilli could need a low temperature to survive because a low temperature is good for bacilli growth or because the low temperature depresses cell-mediated immunity (Purtilo et al. 1974; McDougall et al. 1975). The relationship between oral temperature and oral lesions in leprosy has also revealed that the anterior palate is frequently involved (in 75.7 percent of patients in one study; Scheepers 1998); that region has a temperature of 27.4° C (81.3° F).

The development of lesions in the soft tissue and bones of the face also depends on many factors beyond temperature: "the type of leprosy and its course, the time of onset of the disease, the duration of the disease, the time of onset of therapy, the type and duration of treatment, and racial and geographical factors" (Reichart 1976, 397; see also Rodrigues et al. 2017 for a recent study of how leprosy affects the oral cavity).

Andersen and Manchester (1992) describe the progression of rhinomaxillary syndrome:

The nasal airways are blocked (*M. leprae* enter the airways and cause swelling of the mucous membrane and a nasal discharge).
The olfactory nerve is involved, resulting in loss of the sense of smell.
The loosening of the chondro-osseous junction of the nasal bones causes the bridge of the nose to dip.
The nasal septum perforates and a saddle-nose deformity develops, but there is preservation of the nasal bones.
Pus-forming bacteria infiltrate the mucous membrane and submucosal tissues and the nasal mucosa ulcerate and atrophy.
The nasal conchae/turbinate bones within the nose absorb.
Lepromatous papules may occur on the lips.
Nodules may appear on the palate and tongue.
The palate perforates.
The central and lateral incisors loosen and eventually may fall out.

Absorption, Remodeling, and Possible Total Loss of the Anterior Nasal Column

In living patients, rhinomaxillary skeletal changes are detected using radiography or by palpation of the nasal spine (Møller-Christensen, Bakke, et al. 1952; Figure 4.2). Recording of skeletons is based on Møller-Christensen (1965, 1978), Andersen and Manchester (1992) and Manchester (2014).

Bone-forming and destructive lesions are seen: there is surface pitting and progressive resorption of the anterior nasal column with initial loss of cortical bone and medullary cavity exposure. Although Møller-Christensen grades the changes, it should be noted that applying any grading of bone changes in disease in archaeological skeletons risks introducing

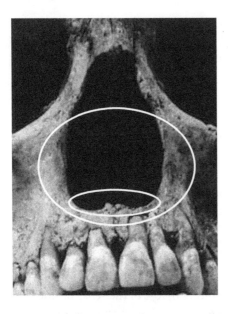

Figure 4.2. Loss of the anterior nasal spine and remodeling of the nasal aperture from a skeleton from Naestved, Denmark. Photo by Charlotte Roberts.

intraobserver and interobserver error in recording, thus making comparison of data challenging (see also Boldsen 2001); the progress of bone changes is

Well-defined reduction of the anterior nasal column;
Advanced atrophy of the nasal spine accompanied by a small remaining anterior nasal column;
Complete absence of the anterior nasal column, the original base of which is covered with cortical bone.

In archaeological studies of leprosy, a high frequency of anterior nasal column involvement has been found, for example in 85 of 123 complete skeletons (69.1 percent) and 85 of 99 definitely diagnosed with lepromatous leprosy or tuberculoid leprosy (85.8 percent) at Naestved, Denmark, and in 42 of 72 observable skeletons with other facial changes of leprosy (58 percent) at Chichester, Sussex, England (Lee and Manchester 2008). Michman and Sager (1957) associated changes in the anterior nasal column and the alveolar process of the maxilla with increased duration of lepromatous leprosy. Other studies link anterior nasal column changes to inflammatory damage in the nasal cavity (Møller-Christensen 1978). A dry climate has been correlated with damage to the anterior nasal column too. Marks and Grossetete (1988) hypothesized that nasal secretions may dry out and damage the nasal mucosa when they are removed, ultimately leading to greater absorption of the anterior nasal column. Studies of this hypothesis are needed. This study also found a low correlation between changes to the anterior nasal column and changes to the alveolar process of the maxilla, suggesting that they are likely unrelated and separate processes. However, at Naestved, 60 percent (74/123) of individuals exhibited changes in both the anterior nasal column and the alveolar process of the maxilla. In addition, of the 99 skeletons that had both rhinomaxillary syndrome and hand and foot bone damage, 74 (75 percent) had both nasal and alveolar changes (Møller-Christensen 1961).

Absorption, Recession, and Remodeling of the Alveolar Process of the Maxilla

This bone change starts in the alveolar process of the maxilla and the absorption extends to the incisors and possibly beyond them to other teeth. In this process, the tooth roots are exposed and may be lost (Figure 4.3). Møller-Christensen (1978) described changes to the alveolar process of the maxilla as always associated with inflammatory changes of the oral surface of the palate; 38 of 72 individuals were observed with these changes at Chichester (52.8 percent; Lee and Manchester 2008). The rate of loss of anterior maxillary alveolar bone height has also been measured in patients with leprosy to explore its extent; Møller-Christensen first did this work in the 1950s. In later work, Subramaniam et al. (1983) measured the distance between the apical foramen, the crest of the alveolar bone, and the cementoenamel junction on both sides of the maxillary central incisors twice in living patients (in 1978 and again in 1982) to see if alveolar bone had been lost over the course of four years. They calculated alveolar bone height on the four sides of the two maxillary central incisors and calculated the percentage of alveolar bone loss for each patient as 100 minus the mean alveolar bone height. For patients with lepromatous leprosy, the average bone loss was 29.5 percent. For patients with tuberculoid leprosy, it was 21.8 percent, but this was not significantly different from the bone loss for patients with borderline leprosy (22.2 percent). Seven individuals also had a net *gain* in bone height. In a follow-up study after ten years, mean alveolar bone loss was significantly greater again in people with lepromatous leprosy than for people with borderline leprosy but not for people with tuberculoid leprosy (Subramaniam et al. 1994; see also Kasai et al. 2018). More recently, Boldsen (2001) described alveolar bone loss extending from maxillary canine to maxillary canine as one of the features for diagnosing leprosy.

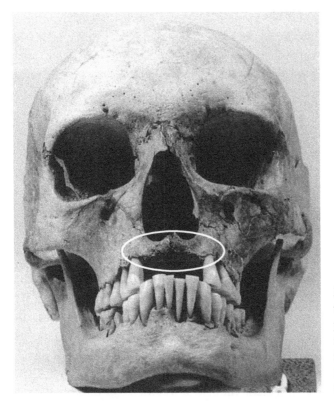

Figure 4.3. Absorption of the alveolar process of the maxilla from a skeleton from Naestved, Denmark. Photo by Charlotte Roberts.

Møller-Christensen (1965), Andersen and Manchester (1992), and Manchester (2014) have described the progressive changes to the alveolar process of the maxilla.

- Smooth recession of the prosthion upward; slight exposure of the central incisor tooth root.
- Extension of atrophy involving the sockets of the incisor teeth; extension of the prosthion recession; exposure of the roots of all the incisors, which may become loose. Resorption of the tooth sockets that first affects the anterior walls and then the posterior and superior walls.
- Severe atrophy and the loss of at least one of the incisors. Anterior resorption of the alveolar process of the maxilla that creates a crescentic profile around the midline of the maxilla. Extension of the absorption that may go beyond the lateral incisors. Minor inflammatory pitting on bone surfaces but retention of cortical bone thickness, perhaps due to endosteal deposition.

Absorption, recession, and remodeling of the nasal aperture margins. Bilateral symmetrical absorption and remodeling of the lateral and inferior margins from sharp to blunt in the inferior third to half of the nasal aperture. Surface pitting at the inferior border and exposure of the medullary cavity in the early stages; later capping of the medullary cavity with cortical bone. Smooth remodeling and no surface pitting of the lateral margin; retention of the cortex covering. Absorption of the alveolar process of the maxilla, likely attributable to damage to the autonomic nasopalatine nerve in this cooler region, leading to dysfunction in the blood vessels and a decreased blood supply.

Inflammatory Changes to the Palatine Process of the Maxilla: Nasal Surface

The changes on the nasal surface of the palatine process are concentrated toward the midline and extend along the length of the bone, but the main alterations are seen in the central area (Figure 4.4) The changes are consistent with nonspecific infection of the periosteum and underlying cortex/outer layer of the bone. Møller-Christensen (1961) found this bone change in 115 of 123 complete skeletons with leprosy at Naestved (90.6 percent),

Figure 4.4. Porosity due to inflammation on the nasal surface of the palate from Skeleton XL58 from York Minster, England. Photo by Jean Brown.

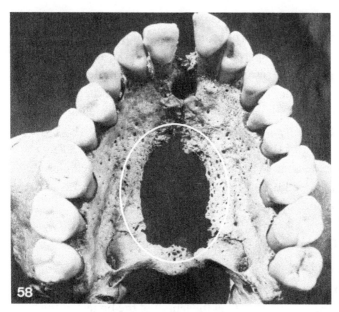

Figure 4.5. Perforation of the bone of the hard palate of Skeleton 262, Naestved, Denmark. Source: Møller-Christensen (1978, figure 58).

and Lee and Manchester (2008) observed it in 58 of 72 individuals with leprosy at Chichester (80.6 percent). Møller-Christensen (1965), Andersen and Manchester (1992), and Manchester (2014) have described the manifestations of this process:

> Slight inflammatory pitting consisting of small closely spaced pits, usually near the transverse palatine suture or on the lateral nasal crest;
> Formation or destruction of new bone; later, erosive lesions that may be related to leprous granulomata (small areas of inflammation);
> General superficial inflammation;
> Possible perforation of the palate (Figure 4.5).

Inflammatory Changes to the Palatine Process of the Maxilla: Oral Surface

This bone change (Figure 4.6) is not necessarily associated with changes to the nasal surface of the palate, but at Chichester both surfaces were affected in 38 of 72 individuals with a leprosy diagnosis (52.8 percent; Lee and Manchester 2008).

Møller-Christensen (1965), Andersen and Manchester (1992), and Manchester (2014) describe the bone changes:

Slight inflammatory pitting;
New bone formation;
Possible perforation of the palate;
Related pitting and/or new bone formation on the turbinate bones and nasal septum.

Inflammatory pitting of both sides of the septum may lead to its perforation and eventual total absorption in the late stages of rhinomaxillary syndrome. However, one should remember that there are many causes of septum perforation, including trauma, nose picking, cancer, rheumatological problems, and topical nasal drug use. Bolek et al. (2017) added exposure to a heavy metal (nickel) to this list. Coarse pitting affects the turbinate bones (nasal conchae), which may be absorbed and lost. This is all likely caused by the infective processes. The frequencies of these changes in the turbinate bones and the septum are rarely noted in clinical or bioarchaeological studies (but see Ghosh et al. 1951; Kasai et al. 2018). However, in the individuals diagnosed with leprosy at Chichester, the turbinate bones were affected

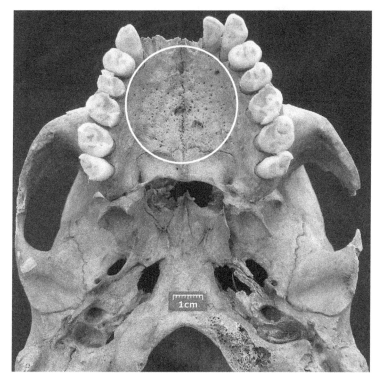

Figure 4.6. Porosity and clear destructive lesions of the oral surface of the palate, Skeleton 101, skull 5, Kaldus, Poland. Photo by Tomasz Kozlowski.

in 32 of the 72 individuals with preserved and observable turbinate bones and the nasal septum was affected in 18 individuals (Lee and Manchester 2008). However, in Ghosh, Dharmendra, and Dey's (1951) study of people with lepromatous leprosy, the septum was more commonly infiltrated with leprosy bacteria than the turbinate bones.

A potential facial bone change in leprosy that has not been identified in archaeological skeletons to date is caused by paralysis (palsy) of the motor nerve that can occur in the face because of damage to the seventh cranial, or facial, nerve. Sensory loss as a result of involvement of the middle and lower branches of the trigeminal nerve (fifth cranial nerve) can also develop (Dharmendra 1978). Upper facial palsy may also be a sequel to involvement of the zygomatic branch of the seventh cranial nerve, leading to weakening and motor paralysis of the orbicularis oculi muscle and eyelid weakness. This occurs alongside lagophthalmos (where the person cannot bring their eyelids together), which ultimately leads to corneal ulcers. When the buccal and mandibular branches of the facial nerve are affected, palsy in the lower face may also occur (see Figures 2.16 and 2.17). In a bioarchaeological context, paralysis of these nerves may be revealed in asymmetrical facial bones due to lack of muscle-induced remodeling. In addition, there could be abnormalities or differences in the shape of the orbits, especially if only one side is affected.

Grading the Bone Changes of the Alveolar Process of the Maxilla, the Anterior Nasal Spine, and the Palatine Process of the Maxilla

Møller-Christensen's (1978) grading of the bone changes of the alveolar process of the maxilla, the anterior nasal spine, and the palatine process of the maxilla may be considered subjective; scholars have likely recorded these features inconsistently. In bioarchaeology, this has been a matter for debate because grading systems have been suggested for a number of pathological conditions in an attempt to quantify bone changes and make suggestions about the severity of pathological lesions. However, systems like these can introduce inconsistencies in the recording of lesions. What one scholar might think are slight bone changes, another might think are moderate. This ultimately leads to data that cannot be reliably compared. While the grading systems described for leprosy-related bone changes are assumed to reflect the severity and duration of the processes involved, the grades for each feature do not necessarily reflect duration (Subramaniam et al. 1983). Recording presence or absence may be a better way of recording, or, if a grading system is being used, at least introducing tests for intraobserver and interobserver error to ensure consistency in recording between different researchers.

Ortner's (2002) work is relevant to assessing the duration of bone lesions in skeletons with leprosy. Both active and healed pitting and/or new bone formation can be seen on bones of the skeleton. The former represents a reaction to an active disease at the time of death and the latter represents chronicity (Wood et al. 1992). While Ortner noted that active absorption of bone is rarely identified in archaeological skulls or in the bones of the hands and feet in individuals with leprosy, he described some evidence. He suggested that active destruction may occur at an early stage of the infection and that later remodeling makes that earlier destruction invisible to bioarchaeologists, or that it could be absent because the disease is slow in development. Slow destructive remodeling is very characteristic of a chronic disease such as leprosy.

Roffey and Tucker (2012) reported that 23 of 31 individuals at Winchester with facial bones preserved had rhinomaxillary syndrome (74.2 percent), and Lee and Manchester (2008) noted that 72 individuals at Chichester had some facial bone changes, 37 of whom (51 percent) had definite rhinomaxillary syndrome.

Differential Diagnoses for Rhinomaxillary Syndrome

In palaeopathology, differential diagnoses (other possible disease diagnoses) must be considered for any pathologically related bone changes, in addition to the possibility that they may represent normal variation for the population being studied. However, Møller-Christensen suggested that *facies leprosa* "seems to represent specific changes in advanced cases of lepromatous leprosy" (1965, 604). Waaler (1953), who examined four "ancient specimens with leprosy" from the collections of the Armed Forces Institute of Pathology in Bethesda, Maryland, and three patients from Norway, found that damage to the alveolar process of the maxilla and anterior nasal spine in one patient was identical to the evidence from medieval Naestved but that the change in the anterior nasal spine was not related to inflammation of the bone. However, he found inflammation of the underlying connective tissue and involvement of small nerves of the mucous membrane. He suggested that these changes were the result of nerve involvement and were either the same process as that seen in the fingers and toes or were caused by atrophy of bone caused by inflammation of adjacent soft tissues. Waaler did not consider the changes to be specific to leprosy "but rather a characteristic symptom of the disease" (ibid, 607). Michman and Sager

(1957) believed that mastication microtrauma was an alternative contributor to atrophy of the alveolar bone. Brand (1959) believed that inflammatory congestion in the nose caused blockage with consequent mouth breathing and a lower oral temperature. Because mouth breathing cools the mouth in the area of the alveolar process of the maxilla, leprosy bacilli will thrive, likely predisposing the bones in this area to further destruction. Mouth breathing can also lead to the formation of plaque (see Reichart, Ananatasan, and Reznik 1976), as Baker and Bolhofner (2014) and Roffey and Tucker (2012) have noted in their work on ancient leprosy, but of course the presence of calculus (calcified plaque) on the teeth of skeletons without leprosy-related bone changes does not mean that the living person had leprosy. Calculus deposits on teeth of archaeological skeletons from prehistory to more recent periods are extremely common across the globe (e.g., see Roberts and Cox 2003). Plaque develops as a result of carbohydrates in the diet and poor oral hygiene, usually in an acidic oral environment (Hillson 1996, 254–260).

Facial bone lesions that could be mistaken for leprosy have been documented. Although some studies consider loss of the anterior nasal spine to be pathognomonic of leprosy, Cook (2002, 81) feels that they "overstate the specificity of this lesion." While Møller-Christensen (1965) considered destruction of the anterior nasal spine and alveolar process of the maxilla to be pathognomonic of leprosy, Ortner (2003, 269) notes that syphilitic and tuberculous infections, neoplastic disease (tumors), and leishmaniasis (a parasitic infection) may all cause the same changes. The anterior nasal spine also protrudes more in "European" skulls than in other groups, such as those from Africa (Møller-Christensen 1978) and therefore its loss in leprosy may be noticed more readily in them. "Nasal guttering," for example in skulls from Africa, may further simulate the leprosy-related changes in the alveolar process of the maxilla. Normal variation in the facial bones in different ancestral groups, specifically "guttering" of the inferior part of the anterior nasal (or pyriform) aperture, may look like leprosy-related remodeling in this area (Richard Wright, pers. comm., August 2013). Detailed studies of the morphology of the pyriform aperture in skulls reveal that different climatic zones can also influence morphological features, including the width of nasal apertures, which are large in *Homo neanderthalensis* (Jaskulska 2014). Finally, pitting, especially on the oral surface of the maxilla, may be considered a normal variant for some populations, while ritual ablation (removal) of the anterior maxillary teeth (see Tayles 1996) and antemortem tooth loss from dental disease could lead to alveolar resorption. Thus, it is important to look for other lesions beyond rhinomaxillary syndrome when diagnosing leprosy. Ortner (2003, 260) suggests that "the *facies leprosa* syndrome, in combination with atrophy and truncation of the fingers and toes, would appear to be almost pathognomonic for leprosy." However, it is necessary to account for the variation in expression of the features of leprosy that affect the pyriform aperture and for potentially culturally induced alterations, depending on the part of the world that is the focus of archaeological studies.

Rhinomaxillary Syndrome and Other Associated Skeletal Changes

When Møller-Christensen (1974, 432) considered the bone changes of rhinomaxillary syndrome and associated changes in the postcranial skeletons of people buried in the medieval leprosarium cemetery at Naestved, he reported that 100 percent of the skeletons with hand and/or foot bone changes attributable to leprosy had inflammation of the nasal cavity and that 85 percent of 115 skeletons with rhinomaxillary syndrome had "typical deformities of the hands and/or feet." Only 2 percent had rhinomaxillary syndrome alone. Bearing in mind that there are some discrepancies in different publications by Møller-Christensen, in his 1961 work he recorded that while 132 of the 185 skeletons (71.3 percent) he observed for facial, hand, and foot bone damage attributable to leprosy had facial and postcranial changes, only 49 (26.5 percent) had hand and/or foot bone changes and only 4 individuals (2.2 percent) had just facial bone damage. This corresponds well with Roffey and Tucker's (2012) study at medieval Winchester, where only four individuals with rhinomaxillary syndrome had no postcranial changes and six had postcranial changes but no rhinomaxillary syndrome. While skeletons with rhinomaxillary syndrome often have associated postcranial bone changes that most likely represent lepromatous leprosy, in tuberculoid leprosy this is normally not the case.

The other direct effect of leprosy to consider in diagnosis is leprous-related osteomyelitis, or infection of the bone and its marrow. This usually occurs in the bones of the extremities, is only seen in lepromatous leprosy, and is the result of the spread of *M. leprae* through the bloodstream to medullary cavities (Resnick and Niwayama 1995a, 2326). Another possible direct effect of *M. leprae* is leprogenic odontodysplasia, a condition characterized by constriction and shortening of the roots of the permanent maxillary incisors (Figure 4.7) This condition is believed to be caused by leprosy

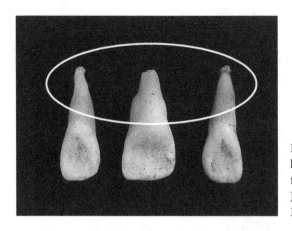

Figure 4.7. Teeth showing leprogenic odontodysplasia from a skeleton from St. Jørgen's, Odense, Denmark. Photo by Vitor Matos.

bacilli invading the dental tissues after the tooth crowns have formed, leading to malformation of the roots. In addition, when *M. leprae* have been present in the pulp cavity for a long time, granulomatous tissue and leprosy bacilli cause necrosis (death) of the pulp and a pinkish-red discoloration of the tooth crown or pink spots, often on the upper incisors (Asano 1958; Itakura 1940; Sano 1958). This has never been reported in the bioarchaeological record.

Leprogenic odontodysplasia was first described in skeletons buried at medieval Naestved (Danielsen 1970). It has not been documented in living people with leprosy (Reichart 1976) and has rarely been reported since 1970 in archaeological human teeth. However, dysplastic incisor and canine tooth roots in a child with possible congenital syphilis from Mississippi have been described (Cook 2002). It was suggested that both treponemal disease and trauma could lead to similar changes in tooth roots, including pressure erosion of soft tissue masses on the tooth roots. Other possible associations include dentinogenesis imperfecta, osteopetrosis, hypoparathyroidism, and chromosomal abnormalities. While all of these conditions affect the development of the roots of permanent teeth, they usually affect all the tooth roots, unlike in leprosy (Danielsen 1970). However, other leprosy-related bone changes would be expected in order to confirm a diagnosis.

These dental alterations are assumed to be developmentally induced and to correspond to the areas of the embryological premaxilla (soft tissue and bone) that envelops the developing incisor teeth. In leprosy, this area is infiltrated by bacteria. The germs (cells) of the developing maxillary incisors are under the anterior part of the base of the nasal cavity, where heavy and early infiltration of bacilli in lepromatous leprosy is seen (Cochrane and Davey 1964). This region also has a low temperature that is exacerbated by mouth breathing when there is nasal congestion; this can influence bacillary growth (Job, Karat, and Karat 1966). Bacteria thus enter the mouth via the bloodstream into the gums and the teeth (Garrington and Crump 1968). Danielsen (1970, 18) suggested that the "dental abnormalities appear to be caused by and be specific to low-resistance leprosy in childhood." There is of course a close anatomical relationship between the nasal mucosa and the odontogenic cells of the maxillary permanent incisors. Danielsen found leprogenic odontodysplasia occurred only in people with marked rhinomaxillary syndrome, but rhinomaxillary syndrome could occur in children without leprogenic odontodysplasia. Teeth extracted from patients with leprosy may also contain pulp that harbors *M. leprae,* most commonly in the maxillary incisor pulp cavities (Scollard and Skinsnes 1999). This has implications for selecting biological samples for ancient DNA analysis, where teeth may be an obvious choice for sampling.

Leprogenic odontodysplasia is very rare in archaeological contexts. There was no evidence of this condition in the several hundred skeletons buried at medieval Chichester (Ogden and Lee 2008), but recent data has been published from elsewhere:

- In the central incisors of a child with rhinomaxillary syndrome who was 11–12 years old and lived in medieval Sigtuna, Sweden (Kjellstrom 2012);
- In the right central incisor of a person 13–19 years old with rhinomaxillary syndrome and damage to their foot bones, buried in a leprosarium cemetery in Denmark in the sixteenth or seventeenth century AD (Matos 2009; Matos and Santos 2013b);
- In the central incisors and possibly one lateral incisor of a child 8–10 years old with some porosity of the nasal aperture margins from medieval Winchester, England (Roffey and Tucker 2012).

It is unclear why leprogenic odontodysplasia is not more commonly seen, but in living people it may be absent because the maxilla is not routinely X-rayed. Unless the affected teeth are lost because of recession of alveolar bone and the condition is noted by a person who recognizes it, leprogenic odontodysplasia would not be observed. The absence of teeth in an archaeological context, especially postmortem loss of single-rooted teeth like the incisors and canines, antemortem loss because of both shortened roots and loss of alveolar bone, and the fact that few juvenile skeletons have been identified with leprosy (Lewis 2002) may explain why the condition is not noted more in archaeological skeletons (Roberts 1986).

Indirect Effects of *M. leprae*: The Peripheral Nerves

The indirect effects of *M. leprae* involve neuropathy of the sensory, motor, and autonomic nerves and the consequent effects on other soft tissues and the bones of the skeleton. When different types of leprosy-related changes occur together or stress points alter, for example through the foot, they may aggravate each other. For example, contracted ("bent") toes due to posterior tibial nerve paralysis can worsen an ulcer on the sole of the foot because the architecture and consequent functioning of the foot change (Karat, Karat, and Foster 1968; Barnetson 1951). Thus, while I have divided nerve involvement into that related to the sensory, motor, and autonomic nerves, there is often a synergistic relationship between the types of nerve damage and their subsequent effects. Naturally, this relationship may also alter through the person's life as changes to the soft tissues and bones develop.

Sensory Nerve Involvement

Leloir (1886) was one of the first people who wrote about bone changes in leprosy, noting that the ulceration seen on feet and hands was associated with a decrease in bone size of the digits. Harbitz (1910, 147) called the changes "lepra mutilans" when hand and foot deformities were present as an indirect result of chronic leprous neuritis (inflammation of nerves) and the accompanying loss of sensation (also described in Heiberg 1886). He described the phalanges as "sloughing off" (absorption), leading to the deformations that develop in the hands and feet. In the early part of the twentieth century, Herrick and Earhart (1911, 811) considered the differential diagnoses for the bone changes Harbitz and Heiberg had described: "These trophic bone lesions, while not absolutely pathognomonic of the disease, are very characteristic of it. Any condition which might present a similar bone lesion could be readily differentiated by the previous history of injury or disease."

The belief that fingers and toes affected by leprosy "drop off" is incorrect. The soft tissues of the affected fingers or toes contract around what is left of the bone after it is absorbed. Karat, Karat, and Foster (1968, 149) explain that "it is the tendency of the soft tissue to contract in the absence of bony support that has given an air of mystery to the process of shortening of the limbs in leprosy." The bone changes in the hands and feet in leprosy are the result of muscular imbalance and repeated trauma to soft tissues that have lost sensation (Paterson and Job 1964). Sensory loss usually occurs before motor and autonomic nerve involvement (Enna, Jacobsen, and Rausch

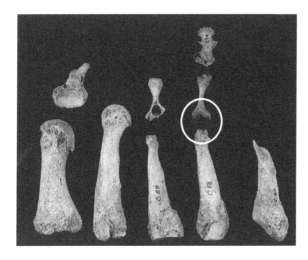

Figure 4.8. Cup-and-peg deformity of foot bones in Skeleton 88, St James and St Mary Magdalene, Chichester, England. Photo by Don Ortner.

1971), and while anesthesia is a late development in lepromatous leprosy, it can develop early in tuberculoid leprosy (Andersen, Manchester, and Roberts 1994). Loss of sensation occurs on the ulnar side of the hand and fibular side of the foot first because of damage to the fine cutaneous nerve fibers, although the sensation loss does not usually go beyond the elbow or knee. Ulcers can develop on the hands and feet at maximum stress points. The person may also lose their sense of deep pain. Because the proprioceptive nerve is affected, the person loses spatial sensation in their joints, meaning they lose their ability to sense the relative position of their body parts and the strength of effort being employed in movement. This may lead to subluxed and dislocated joints and changes in the normal architecture of the hands and feet. The proximal ends of the phalanges may become cupped and the distal ends may become "pegs" (this is known as cup-and-peg deformity; see Figure 4.8).

As the disease progresses, a secondary bacterial infection may spread to bones and joints, causing osteitis, osteomyelitis, and septic arthritis. Some joints may fuse together, restricting normal movement. Bone changes in the hands will start at the ends of the distal phalanges as a result of soft tissue damage. These changes will not go beyond the metacarpophalangeal joints, except possibly in the thumb. Bone alterations in the feet will typically start in the proximal phalangeal shafts or the metatarsals. All of the foot bones and joints may be affected. While the bone damage initially develops as a result of loss of sensation in the feet and hands and subsequent trauma to and ulceration of the skin, sensory loss alone does not cause bone changes seen (Paterson and Job 1964). For example, if a nerve is cut but the affected limb is protected from trauma, the skin will become thin and the bones will become osteoporotic (lose mass), but they will not collapse or be absorbed.

The Hands

Paterson and Job (1964) reported that fingertip absorption was the most common radiological finding in leprosarium patients. This was supported by their findings of erosion of the distal phalangeal tufts associated with loss of the normal pulp of the underside of the fingertips along with tight, smooth skin in old healed areas of ulceration and infection. They found that when fingers affected by damage to the motor nerve contract early in the infection such that the fingertips are not subjected to trauma, there is no evidence of erosion. However, where there is just right-angled contraction of the distal interphalangeal joints, the tips of the fingers or toes may be subject to trauma and the soft tissues and bone may gradually absorb. In a person with leprosy who has lost sensation in a hand but has no paralysis, ulcers tend to occur on the palm and at the tips of the fingers. A person with contracted fingers due to motor nerve involvement and paralysis may develop ulcers at the contracted interphalangeal joints (Andersen, Manchester, and Roberts 1994). Not all of the fingers will be equally affected; the index and long fingers may be affected first because they are used more. As these fingers shorten due to bone absorption, the other fingers will become affected and will subsequently be damaged. In the hand, the carpal bones are not involved as frequently as other bones, but there have been fewer studies of carpal involvement in people with leprosy.

Bone alterations in the hand may also include destructive lesions in finger bones due to acute inflammation that initially presents as swelling of the soft tissues overlying the affected fingers. Another change is that cysts, or fluid-filled sacs, may cause the bones to look like a honeycomb. Radiographs of the hands of people with leprosy have shown radiolucency (bones that are almost transparent), honeycombing (clustered cystic air spaces), atrophy (wasting away) due to osteoporosis, localized cystic areas in the metacarpals and phalanges, and enlargement of the shafts of the metacarpals (Murdock and Hutter 1932). These changes have been likened to the changes caused by dactylitis in tuberculosis.

Carpal disintegration, or loss of the normal structure and appearance of the carpal (wrist) bones, can also occur in people with leprosy. Although Kulkarni and Mehta (1983) suggest that carpal disintegration is not seen much because there is no repeated trauma due to weight bearing, as is seen in the feet, Sane et al. (1985) report that it can result from compressive and shearing forces transmitted across the wrist of someone whose hand is neuropathic due to leprosy. These compressive and shearing forces are superimposed on the carpal bones, which may already be affected by weakening of the ligaments after infection or osteoporosis or fractures. However, Sane and colleagues eventually concluded that carpal disintegration was rare in leprosy. This conclusion seems to be confirmed by the lack of evidence in palaeopathology, although poor preservation and/or incomplete retrieval of carpal bones during excavation may explain some of the lack of evidence.

Sane and colleagues (1985) examined three groups of patients at the Bandorawalla Leprosy Hospital in Pune, India, to document carpal disintegration:

Individuals with a history of repeated subclinical trauma to the wrist joint (caused by sport or heavy manual labor) who did not respond to treatment;

Individuals with stresses and strains across the wrist predominantly due to bearing weight using their upper extremities as support (e.g., people with serious foot ulcers or people who had had an amputation below the knee and had to use their arms to support themselves);

High-risk groups (e.g., people who played sports and people who did heavy manual labor who already had tarsal disintegration).

They identified three stages of carpal changes. The first was a history of a sprained wrist with pain that limited daily activities. In this stage, there was no radiological abnormality apart from possible osteoporosis and cystic cavities. In the second stage, the person experienced no pain but there was swelling and early deformity of the carpus, possible ulcers on the anterior aspect of the wrist, callosities (hardened areas of the skin), and crackling (crepitus) of the wrist when it was moved. At this stage, radiology revealed changes in the normal anatomy of the carpus: the capitate, lunate, and scaphoid bones were destroyed due to neuropathy and osteoporosis, osteomyelitis, and carpal collapse were visible radiologically. In the third stage, the person continued to experience no pain but the wrists were deformed. The wrist crackled when moved and the function of the hand had deteriorated. Radiology revealed destruction of the carpus and it was no longer possible to identify individual carpal bones. The distal ulna and the radius could be destroyed later as a result of bearing weight on the arms. Sane et al.

identified the relevant factors in leprosy that may affect wrist stability and disrupt intercarpal and radiocarpal joints and their associated soft tissues: osteoporosis, cysts in the carpal bones, contracted fingers leading to carpal instability, and, following ulceration of the wrists, infection that may spread to the carpus, including joint surfaces (septic arthritis).

The Feet

The bone changes of the feet due to sensory loss are often preceded by thickened and hardened skin that cracks and enables infection to enter the foot and cause ulcers (Price 1959). Sensory loss in the feet leads to damage due to pressure on specific areas of the undersurface of the foot when a person is walking. The areas affected by pressure may change as the foot becomes progressively deformed. The alterations are caused by uneven distribution of weight in a foot that is functionally unbalanced and/or greatly deformed. In a static foot, the body weight is distributed evenly over the two soles. Each heel takes 25 percent of the total body weight and the lateral metatarsal heads and the two sesamoid bones of each foot take another 25 percent. This weight distribution changes when a person walks (Andersen 1961; Yawalkar 2009, Figure 158).

Ulcers in leprosy occur most prominently in the forefoot (Price 1959; Cross 1972; Yawalkar 2009). An ulcer under a metatarsal head can lead to foot drop and subsequently to problems with the Achilles tendon. Ulcers are also commonly located under the first metatarsophalangeal joint (great toe) and its proximal phalanx and under the base of the fifth metatarsal in the midfoot. Ulcers in the latter case are associated with paralysis of the lateral popliteal nerve. Ulcers of the tips of the toes can occur when there are contracted or bent toes and foot drop (Barnetson 1951) or if the person wears unsuitable footwear (Price 1964a, 1964b). Ulcers in the forefoot may expose the metatarsophalangeal joints and adjacent bones, and subsequent secondary infection can extend along the metatarsals and cause pathological fractures (Price 1961). The heads of the metatarsals may also fracture (Barnetson 1951). Ulcers on the lateral side of the midfoot can potentially affect the cuboid bone through the fourth metatarsal joint and the calcaneocuboid joint. In the former, the ulcer may spread to all other tarsometatarsal joints except the first. In the latter case, the ulcer spreads across the foot and invades the talonavicular and anterior talocalcaneal joints. Ulcers are rare on the insteps of the feet unless the foot arch has collapsed (Srinivasan and Dharmendra 1978d). The midfoot and hindfoot can also be affected on the lateral side of the sole, under the fifth metatarsal, under the heel, and under the lateral tuberosity of the calcaneum. The posterior calcaneal tuberosity can be subsequently avulsed (separated) and the calcaneum thus fractured. Ulcers under the tarsus are less frequent (Hirschberg and Biehler 1909). Heel ulcers usually occur in the center of the heel, but only when there is foot drop in the later stages of foot damage. However, clinical studies report heels as the least common places for ulcers in leprosy (e.g., see Hasan 1974; Price 1959; Cross 1972; see also Paterson and Job's 1964 radiological description of calcaneal osteitis secondary to a deep penetrating ulcer of the heel). Heel ulcers also do not heal as well when they are treated as ulcers on the rest of the foot do (Srinivasan 1976).

While ulcers do occur in people with feet deprived of sensation, not all anesthetic feet ulcerate. Changes to the bones of the feet in a person with leprosy may also occur with no history of ulceration (Miller 1913; Hasan 1974). Ulcers can also overlay the lateral malleolus of the fibula, the heel of the palm, and the point of the elbow. This can result from a person sitting on the ground cross-legged and raising themselves up on the palms of their hands (Srinivasan and Dharmendra 1978c). Identification of bone changes associated with ulceration in these regions has not been reported in archaeological skeletons. Many studies of ulceration patterns in people with leprosy relate them to activity patterns. For example, Price (1959) found a higher frequency of ulcers under the metatarsophalangeal joints in women and suggested that they were related to the fact that the women carried loads on their heads. Cross (1972) found a higher rate of ulcers in men in two groups of people with leprosy in Melanesia. One group lived in the bush and were nomads and the other group were coastal village dwellers who fished and kept gardens. Frequency rates were no different between bush and coastal people, but more of the men's feet were affected than the women's (29.1 percent versus 21.4 percent). In women, 20 percent of their ulcers were in the middle tarsus region, likely the result of carrying heavy loads.

Factors that can affect whether ulcers in leprosy develop include congenital and acquired deformities of the foot, the extent of loss of sensation, the characteristics of the gait of the person, and the pattern of weight bearing and of motor nerve–related paralysis (Srinivasan and Dharmendra 1978c). In the second half of the twentieth century, attempts to treat ulcers included putting plaster casts on affected limbs and prescribing bed rest (Price 1960; Ross 1960; Pring and Casiebanca 1982; Kaplan and Belber 1988). However, plaster casting was not popular because of the inconvenience it caused, because of the associated stigma, and because osteoporosis can develop because of immobilization for lengthy periods.

As infection progresses into the foot due to overlying ulcers, ligaments and tendons that link the tarsals of the foot together may also become infected. The arch of the foot is normally maintained by the shape of the bones of the tarsus and its associated ligaments. The medial arch is most affected in leprosy. Its loss can lead to damage to the talus, the medial cuneiform bone, and the base of the first metatarsal. The middle and lateral cuneiform bones and the bases of the second and third metatarsals may also become involved. It is less common for the lateral arches to be affected, but when they are, it leads to damage to the talus and the calcaneum. When the transverse arch is lost, ulcers can develop under the second, third, and fourth metatarsophalangeal joints (Andersen, Manchester, and Roberts 1994). If the longitudinal arch collapses, ulcers can occur in the midfoot in the region of the tarsometatarsal joints.

When the arches collapse, a person will use the tendons to force muscles to work in order to prevent strain on the ligaments. The plantar aponeurosis (thick tissue on the sole of the foot) links the calcaneum to the metatarsal heads and bases of the phalanges. The long plantar ligament links the calcaneum to the cuboid bone and the second through fourth metatarsals. If these structures are infected, the underlying ligaments may partially detach, the foot arch may collapse, and the tarsal joints can fracture due to the pressure of bearing weight. The resulting new bony prominences in the sole of the foot are then subject to friction and pressure and new ulcers will develop (Price 1964a). If the long plantar ligament collapses, the skin overlying the tuberosity of the cuboid bone and the anterior tuberosity of the calcaneum can develop secondary ulcers. The plantar aponeurosis and the ligaments under the tarsal joints are responsible for the shape of the medial arch. Weakness or detachment of a major part of the posterior attachment of the plantar aponeurosis leads to collapse of the medial arch and a heel ulcer. Secondary ulcers also develop at the new bony prominences over the navicular bone tuberosity, the medial cuneiform bone, and the base of the first metatarsal following arch collapse. If there is total foot drop, a central ulcer can develop.

Tarsal disintegration in leprosy is better documented in the literature than carpal disintegration. Both are the result of the joints of the bones becoming abnormal as a result of infection with leprosy. Tarsal disintegration occurs in the neuropathic foot of people with leprosy and is caused by insensitivity, muscle paralysis, altered muscle pull, repeated plantar ulcers, and secondary pyogenic infection (Kulkarni and Mehta 1983). Tarsal disintegration is more common in people with tuberculoid leprosy. The high levels of bone stress in leprosy may be the result of a reduction in the mechanical strength of bones due to osteoporosis (Patil and Jacob 2000; see also Paterson and Job 1964).

When joints deprived of a nerve supply are subject to repeated stress or even normal use, disorganization of the bones, their joints, and supporting soft tissues can result (Harris and Brand 1966). Tarsal disintegration occurs when there is "vigorous activity" (ibid., 5) and loss of pain sensation. In leprosy, damage to the tibial and common peroneal nerves leads to anesthesia of the foot and movement can proceed without pain. If the posture is disturbed as the tibia transmits the forces of the body weight to the talus, then to the calcaneum, and then to the anterior foot bones, these forces will change, ligaments may rupture, and minor fractures can occur. Kulkarni and Mehta (1983) have described four cumulative stages of tarsal disintegration in people with leprosy:

Swelling in the talonavicular region, loss of sensation and loss of a protective reflex; neuropathic joints;
Muscle paralysis, flat foot, fixed or mobile deformities, loose ligaments;
Instability with involvement of the tarsals, loss of normal anatomical configuration, fractures, absorption of bone;
Marked instability, grossly deformed foot, severe radiological changes.

The primary changes are to the body of the talus, the posterior talocalcaneal joint and the calcaneum, the neck (head) of the talus, the talonavicular joint, the navicular bone, the naviculocunieform joint, the anterior calcaneum, the calcaneocuboid joint, the cuboid bone, and the cuboid-metatarsal joints. The secondary changes are general joint degeneration and total disintegration of the foot. Lee and Manchester (2008) described tarsal disintegration in fifteen individuals from medieval Chichester, but this bone change is not commonly reported in archaeological contexts.

Septic Bone Change in Leprosy

Septic bone changes in leprosy can be nonpyogenic or pyogenic (Andersen, Manchester, and Roberts 1994). Nonpyogenic septic changes are the result of direct bone damage caused by *M. leprae* infiltrating the bones via the blood vessels. This leads to osteitis in the cortex and osteomyelitis in the medullary cavity of bones. Paterson and Job (1964) describe these changes as secondary to infection of the distal end of the bone of a digit (a finger or toe) and to plantar ulcers of the feet. In X-rays, the trabecular structure and later the cortex in the affected area become hazy and the bone looks like it has been destroyed. Changes can also occur as a result of granulomatous lesions, or lepromas (nodules containing bacilli), which are the direct result of the leprosy bacterium. These may lead to erosive lesions on the outer cortical surface of long bones. Multiple granulomas can cause irregular erosions and marginal new bone formation in the metatarsals, the metacarpals, and the phalanges. Lesions in the medullary cavities of the bones of the metacarpophalangeal and interphalangeal joints consist of foci or cysts in the subchondral zones (on the bone under the joint cartilage). A cup-and-peg joint deformity may result. A medullary cavity leproma in the carpus may cause "ballooning" of the bone, with or without cortical thinning, much like the dactylitis that is caused by tuberculosis (Andersen, Manchester, and Roberts 1994).

Pyogenic lesions of bones and joints in leprosy occur when environmental pathogens enter the bone, followed by deep-tissue anesthesia and ulcers (Andersen, Manchester, and Roberts 1994). A bone abscess or cyst may develop in deeper areas of the bone, leading to osteomyelitis. Because of inflammation, the small blood vessels (arterioles) are dilated, which promotes osteoclasts (bone-destroying cells), resorption of sequestra (dead bone), and involucrum (new bone) formation. Nutrient foramina (holes where arteries enter the bone) may also be enlarged. The joint surfaces of bones can be destroyed and their surfaces become irregular, and the bones of the joint may fuse. Arthritis of bones in people with leprosy may be due to mechanical causes as a result of relaxation of tendon and joint capsules caused by nerve damage. The joints become abnormally mobile or loose and damage to the joint surfaces occurs (Murdock and Hutter 1932). However, increasing age is also strongly associated with joint degeneration.

Differential Diagnosis of Bone Changes Due to Damage to the Sensory Nerves

Loss of sensation, ulcers of the feet or hands, and absorption of the toe and finger bones may develop as a result of other diseases. These include diabetes (a group of diseases that affect how the body uses blood sugar), the late stages of syphilitic infection (an infectious disease that can affect the nervous system), and spina bifida (abnormal development of the spine and spinal cord that leaves a gap in the spine) that can lead to the formation of ulcers (e.g., see Chaudhury 1970). In living people, other signs and symptoms would help in a differential diagnosis, but in an archaeological skeleton the bone damage indirectly caused by ulcers may be more difficult to assign to leprosy. Paterson and Job (1964) also describe nerve inflammation, congenital indifference to pain, and traumatically damaged Charcot (or neuropathic) joints as conditions that could lead to collapse and disintegration of the tarsals. They also document the absorption of the distal phalanges in diseases where there is a reduced blood supply: skin diseases such as psoriasis where a cup-and-peg deformity can develop, "clubbing" of the fingers (enlarged finger ends) in hypertrophic pulmonary osteoarthropathy caused by various lung diseases, frostbite, Raynaud's disease (which affects the blood vessels, leading to poor circulation to fingers and toes), acro-osteolysis, scleroderma ("hard" skin), syringomyelia (a spinal cord problem), and tabes dorsalis (spinal column degeneration associated with syphilis). The bone changes related to sensory nerve damage are therefore not pathognomonic for leprosy.

Motor Nerve Involvement

M. leprae affects the motor nerves in a variety of ways that can lead to loss of function. These include:

- Paralysis of muscles, muscle groups, and tendons; loss of synergistic action and balance; subluxation (partial dislocation) and dislocation at the interphalangeal joints (hyperflexion);
- Hyperextension of the metacarpophalangeal and metatarsophalangeal joints; hyperflexion of the interphalangeal joints (contracted fingers due to involvement of the ulnar nerve or contracted toes and foot deformities due to involvement of the posterior tibial nerve);
- Drop foot due to damage to the lateral popliteal nerve; absence of dorsiflexion (the ability to raise the foot upward);
- Loss of the longitudinal and transverse arches and subsequent ulcers of the plantar surfaces of the feet;
- Facial nerve damage and paralysis of the lower eyelids leading to lagophthalmos and possible eye infection.

The potential bone changes as a result of motor nerve damage are

Deformities leading to trauma and subsequent ulcers of the hands and feet when there is sensation loss; destruction of soft tissue and underlying bone;

Infiltration of pyogenic bacteria causing septic arthritis of joints, and osteitis and osteomyelitis of the bones of the hands and feet and limb deformity that influence the site and mode of progression (Andersen, Manchester, and Roberts 1994);

Subluxation of joints following damage to ligaments and joint capsules;

Inflammatory change especially around joints (caused by soft tissue damage); septic arthritis of the metacarpophalangeal and interphalangeal joints secondary to ulceration caused by contracted fingers;

Palmar grooving at the distal ends of the proximal phalanges due to pressure-induced absorption of the proximal ends of the midphalanges from flexion contracture of fingers (also called the "ditch sign"; Enna, Jacobsen, and Rausch 1971; Paterson and Job 1964) (see Figures 4.9 and 4.10);

Carpal "squeezing" (or dislocation);

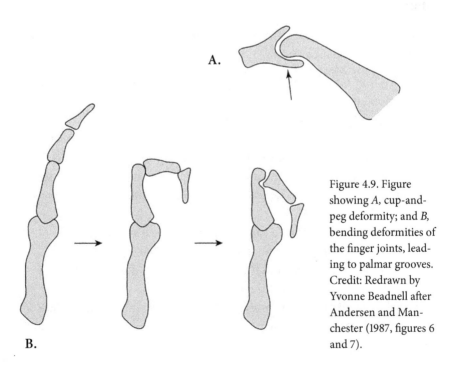

Figure 4.9. Figure showing A, cup-and-peg deformity; and B, bending deformities of the finger joints, leading to palmar grooves. Credit: Redrawn by Yvonne Beadnell after Andersen and Manchester (1987, figures 6 and 7).

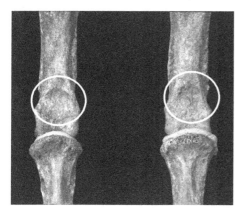

Figure 4.10. Hand phalanges showing palmar grooves as a result of flexion deformity, or "bending," of the finger joints in a skeleton from St James and St Mary Magdalene, Chichester, England. Source: Magilton et al. (2008, figure 14.8). Reproduced with permission of Donald Ortner and of Catrina Appleby of the Council for British Archaeology.

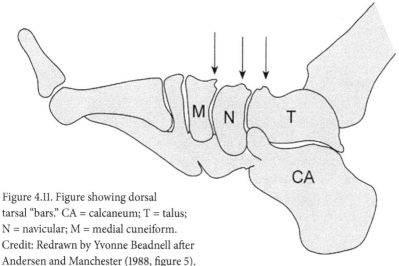

Figure 4.11. Figure showing dorsal tarsal "bars." CA = calcaneum; T = talus; N = navicular; M = medial cuneiform. Credit: Redrawn by Yvonne Beadnell after Andersen and Manchester (1988, figure 5).

Figure 4.12. Bone formation on the dorsal surfaces of the tarsals ("bars") in Skeleton 159, Cannington, Somerset, England. Credit: Used with permission from Keith Manchester.

Dorsal tarsal exostosis (new bone formation) due to collapse of the longitudinal arch (indicating a "dropped" foot) when the motor part of the posterior tibial nerve is affected (Figure 4.11 and 4.12); loss of function and proprioception; development of "bent" (contracted) toes;

Navicular "squeezing" (dislocation) and flattening of the navicular bone.

The contracted fingers in people with leprosy are the result of disruption of the function of the long extensor and flexor and intrinsic hand muscles caused by ulnar nerve paralysis (Andersen and Manchester 1987, 77; see also Murdock and Hutter 1932; Dyer and Hopkins 1911). Dyer and Hopkins observed that the tendons and flexor muscles shrink more than the extensors and thus the fingers cannot fully extend. The metacarpophalangeal joints are extended or hyperextended and the interphalangeal joints are hyperflexed. As hyperflexion is constantly present, partial dislocation (bones of joints that are not totally in normal anatomical line with each other) may occur at those interphalangeal joints. People may continue to try to use their hands and damage the dorsal surfaces of the interphalangeal joints, leading to ulceration, bacterial invasion, and pyogenic arthritis.

Until 1987, motor nerve damage in leprosy and subsequent bone changes had not been reported in palaeopathology or at least had not been recognized and recorded as such (Andersen and Manchester 1987). "Grooves" on the palm side of the distal ends of the proximal phalanges of the hand are now noted in archaeological skeletons that exhibit other bone lesions of leprosy, likely indicating long-standing pressure from the base of the middle phalanx. If there is lateral deviation at the interphalangeal joints with hyperflexion, the groove may not be equally distributed across the phalanx (Andersen and Manchester 1987). The corresponding proximal end of the middle phalanx may also be broad and flattened and may have a palmar "lip" or may be cupped. Andersen and Brandsma (1984) noted this bone damage in X-rays of people with leprosy. Lee and Manchester (2008) recorded evidence of palmar grooves for 51 individuals from medieval Chichester; the majority (43) of that group exhibited other bone changes of leprosy. Clinically, contracted toes are seen nearly as often as contracted fingers, but they are less obvious and the patient is less aware of them (Srinivasan and Dharmendra 1978a). This abnormality most commonly affects the outer (more lateral) toes and may lead to ulcers on the metatarsal heads.

While dorsal tarsal exostoses have been noted in people with leprosy, they cannot be viewed as pathognomonic for leprosy. However, they may be helpful for a more rounded diagnosis in bioarchaeology. The navicular bone is the mainstay of the longitudinal arch, and stresses caused by a flat (dropped) foot can lead to plantar (pertaining to the sole) disruption of the navicular bone in leprosy (Andersen and Manchester 1988). This can be associated with damage to the nerve endings responsible for proprioception, the loss of the ability to detect changes to body position and movement, and deep-rooted pain (Hirschberg and Biehler 1909). Thus, the dorsal ligaments of the talonavicular, cuneiform-navicular and cuboid-navicular joints can be stressed and strained (see Figures 4.11 and 4.12). This leads to ridges of new bone on the superior (dorsal) surfaces of the tarsals at the sites of attachment of the ligaments. The ligaments of the navicular and talus bones are most affected; the cuboid bone is affected less frequently. Fusion of the tarsal bones may also occur. Lee and Manchester (2008) have done the most systematic study of this bone change in the skeletons from medieval Chichester; they found that 52 individuals had tarsal bars.

Differential Diagnosis of Bone Changes Due to Damage to the Motor Nerves

Bone changes as a result of contracted fingers and toes and dorsal tarsal exostoses are not pathognomonic for leprosy, although their presence may contribute to a diagnosis of leprosy in an archaeological skeleton. Essentially, any condition that causes contracted fingers could lead to palmar grooves. One example is paralysis of hands caused by a stroke. Any condition that causes a collapse of the foot arches may also cause tarsal exostoses. In addition, many infections can lead to septic arthritis, osteitis, osteomyelitis of the bones, and these infections can lead to a reaction (porosity, new bone formation) on the surface of the bone that reflects inflammation of the periosteum.

Autonomic Neuropathy

The autonomic nervous system has two parts, sympathetic and parasympathetic (Wilson 1990, 267), and damage to it leads to loss of arteriole control and harmony between bone-forming and bone-destroying cells (osteoblasts and osteoclasts, respectively). Damage to the sympathetic parts of the system leads to permanent dilatation of the arterioles and peripheral hyperemia (too much blood in a part of the body), which is significant for the progression of septic bone changes (Andersen, Manchester, and Roberts

1994). The potential bone changes resulting from autonomic nerve damage in leprosy are

> Concentric diaphyseal remodeling of the shafts of the metacarpals, metatarsals, and proximal and mid-phalanges, creating an hourglass appearance; the proximal phalanges are affected first;
>
> Cortical bone absorption externally and formation of bone internally due to remodeling that leads to a reduced and eventually lost medullary cavity; this may also be caused by anesthesia and loss of deep sensation in the palms and soles (Figure 4.13);
>
> Possible subsequent shaft fractures within the bone loss area and absorption and loss of the distal fragment, a process known as acroosteolysis (Andersen, Manchester, and Ali 1992);
>
> Remodeling, most often seen in the distal third of the shafts of the metatarsals and metacarpals. Concentric and "knife-edge" (mediolateral) remodeling only in the metatarsals; the fifth metatarsal is most affected (Hirschberg and Biehler 1909). This has also been unfortunately called the "licked candy stick" appearance (Enna, Jacobsen, and Rausch 1971, 301; Faget and Mayoral 1944, 6; and see Figure 4.14). The dorsoplantar diameters of the metatarsal shafts remain intact. Possible loss of the metatarsal heads and cup-and-peg deformity at the distal ends, shortened phalanges, and possible destruction of the mid-phalanges. Remodeling in the phalangeal shafts that starts at the mid-diaphysis.

Diaphyseal remodeling leads to "progressive loss of the diameter of the metatarsals, metacarpals and proximal phalanges, associated with maintenance or increase in the thickness of the cortical bone at the site" (Andersen, Manchester, and Ali 1992, 211). Radiological changes include erosion of the subperiosteal layers of the cortical bone and thickening of its inner layers and the formation of new bone on the medullary side (Paterson and Rad 1961; Paterson and Job 1964). While the specific cause and disease related processes responsible for remodeling in autonomic neuropathy "are unknown" (Andersen, Manchester, and Ali 1992, 214), neurovascular dysfunction following autonomic neuropathy in leprosy may be at play. The process could involve arteriolar dilatation due to autonomic nerve damage leading to hyperemia in the extracortical area of the bone and then high oxygen tension in the blood. That would stimulate osteoclast activity and cause resorption of the cortex (Andersen and Manchester 1992).

Enlarged nutrient foramina in bones may be a consequence of dilatation

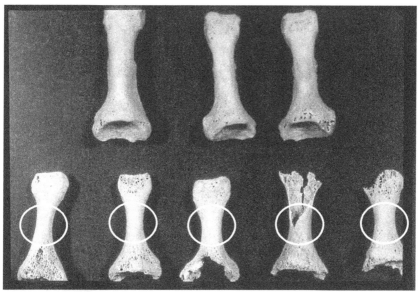

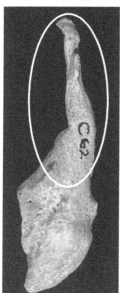

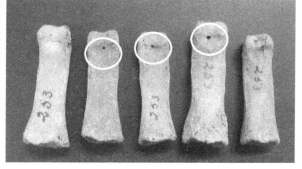

Above: Figure 4.13. Diaphyseal remodeling of foot phalanges, Skeleton 382, Stonar, Kent, England. Photo by Jean Brown.

Left: Figure 4.14. Mediolateral sharp-edged remodeling of a metatarsal in Skeleton 62 from St James and St Mary Magdalene, Chichester, England. Photo by Jean Brown.

Below: Figure 4.15. Enlarged nutrient foramina, Skeleton 253, Lødingen, Norway. Photo by Charlotte Roberts.

of arterioles (Figure 4.15). Murdock and Hutter (1932, 603) surveyed over 1,000 X-rays and reported that 140 patients with leprosy had enlarged nutrient foramina. They suggested that this bone change was the result of "atrophic responses to pathologic leprous vascular supply" (1932, 603; see also Faget and Mayoral 1944). However, holes in nutrient vessels with a diameter of less than 1 millimeter may be normal. Paterson and Job (1964) also noted erosion and death of bone and suggested that it was caused by problems with blood circulation caused by leprosy-related inflammation of the blood vessel. The endothelium (the inner lining of the blood vessels) can also be infected with *M. leprae,* leading to blockage and necrosis (Fite 1941).

In the foot, the bone changes of leprosy caused by autonomic nerve problems are usually bilateral, but in the hand they can be unilateral. In normal circumstances, if the bone loss is the result of disuse atrophy, it would be expected that the cortex would be thinned and osteoporotic and the medullary cavity would be increased (Andersen and Manchester 1987). However, it is possible that disuse atrophy caused by another condition alongside leprosy could cause the changes in the bones of the feet and hands. Lee and Manchester (2008) recorded 34 individuals at medieval Chichester with concentric remodeling of the foot phalanges.

Differential Diagnosis of Autonomic Nerve Bone Changes

Concentric absorption of the metatarsals has been noted in other conditions that involve loss of sensation and nonspecific changes of soft tissues, for example diabetes, inflammation of the nerves unrelated to leprosy, congenital indifference to pain, and syringomyelia (a cyst in the spinal cord; Paterson and Job 1964). However, similar alterations have been found in people with no disease involving the nervous system and in people with no detectable sensory loss, reminding us that these bone changes are not pathognomonic for leprosy.

The Relative Frequency of Foot and Hand Bone Alterations in Leprosy: Past and Present

More foot than hand damage has been recognized in living people with leprosy. Hirschberg and Biehler (1909) note that the toes are affected in people with leprosy earlier than the hands. This was also seen in a radiographic study of patients in an outpatient clinic in Canada and in patients at Carville, Louisiana (Rothschild and Rothschild 2001). Bone absorption was the most common bone change. In his discussion of the atrophy of hand and foot bones in leprosy, Barnetson (1951, 302–303) explained that the "trauma of weight bearing is more constant and more intense than the trauma the fingers are subject to." This is borne out by clinical studies. Andersen (1969, 114) found more bone changes to the feet than the hands in a study in India. Because more foot damage was seen in the living than in archaeological human remains, he claimed that it was because the patients in his study did "more walking on unprotected feet than the inmates of the St Jørgen's Gård [medieval Naestved, Denmark]."

In general, more damage to foot than hand bones has also been observed in skeletal remains from archaeological sites. For example, only four of thirty-four individuals with preserved hand bones from the medieval Winchester leprosy hospital cemetery exhibited damage to those bones (Roffey and Tucker 2012). However, more skeletons from medieval Chichester had hand bone changes that may be attributed to leprosy, including new bone formation mainly on the metacarpals (20 individuals) and palmar grooves (51 individuals, 43 of whom had bone changes of leprosy elsewhere in the skeleton) (Lee and Magilton 2008b). The lack of damage to the bones of the hand may indicate that affected individuals were aware of their sensory loss and were protecting their hands from injury, although we do not know if people with leprosy in the distant past protected their hands and feet from further damage after they had experienced trauma-related ulceration. Nevertheless, 87 of 123 people with leprosy at Naestved (70.7 percent) had foot damage and 64.2 percent (79/123) had hand damage (Møller-Christensen 1961), indicating that hand damage at least at this site was fairly common.

In other studies that have focused on smaller groups of archaeological skeletons from medieval leprosy hospital contexts, changes to the bones of the hand are not common. Overall, the reason for the lower frequency of changes to hand bones, both clinically and archaeologically, may be that it is more common for people with leprosy to notice the effect of trauma on the palms of the hands than on the soles of the feet and to implement a mitigation strategy to prevent hand damage.

OTHER ASSOCIATED BONE CHANGES

A number of other bone changes have been identified in skeletal remains when leprosy has been diagnosed, both in archaeological skeletons and in living people with leprosy. This section highlights some of these observations.

Osteopenia and Osteoporosis

Osteopenia is a precursor of osteoporosis and is characterized by a decrease in bone amount without a risk of fracture (Brickley and Ives 2008, 151). Osteoporosis is characterized by a decrease in bone amount with consequent weakening of the affected bone and a risk of fracture. Osteoporosis may be seen in people with leprosy. Ishikawa and colleagues (1999) found that acute inflammation due to ulceration can lead to osteoporosis in the bones of the hands and feet and may occur transiently in leprosy reactions. It is particularly seen in men with leprosy because the testes can be affected. In this study, bone mineral density levels were significantly correlated with bioavailable testosterone in all three regions of the skeleton that are often affected by osteoporosis (the radius, the lumbar vertebrae, and the neck of the femur). Ishikawa and colleagues concluded that bioavailable testosterone may be one of the main factors that affects bone mineral density in male leprosy patients.

Osteoporosis has also been associated with leprosy in archaeological populations. Lee and Boylston (2008b) identified sixty-five individuals with the condition at medieval Chichester. In addition, twelve of twenty-nine males (41 percent) had osteopenia. At medieval Naestved, 36 percent of 123 skeletons with leprosy had osteoporosis of the metatarsal heads and subarticular areas of the phalanges (Møller-Christensen 1961). Localized osteoporosis can also occur in people with lepromatous leprosy in the form of destructive bone lesions, and be seen underneath active localized skin lesions. When osteoporosis occurs in hand and foot bones, pathological fractures may be the result. This is another factor that contributes to shortening of the fingers and toes after fractures heal. Finally, osteoporosis can develop in limbs as a result of disuse. However, diagnosing osteoporosis in archaeological remains can be challenging because of postmortem changes to bones that can make them appear to be osteoporotic. Bioarchaeologists may mistake these postmortem changes that lead to loss of bone mass and/or thinning of the bone cortex for osteoporosis. Furthermore, suggesting that osteoporosis is the result of leprosy in archaeological skeletal remains is complicated by the fact that it cannot be known whether osteoporosis or leprosy occurred first. In addition, osteoporosis may simply be an indicator of increasing age.

Periosteal Lesions on the Long and Short Tubular Bones

Periosteal lesions of the bones of the forearms, hands, and feet have been observed in people with leprosy and in the skeletons of people from archaeological contexts (Figures 4.16, 4.17, 4.18, 4.19). When these lesions are in a limited area, they may be the result of an overlying ulcer (see Boel and Ortner 2013). However, the suggestion of an ulcer is much less commonly reported in archaeological skeletons than diffuse periosteal lesions. Periosteal changes on the tibia and fibula are reported most often in relation to

Above: Figure 4.16. Bone changes on the tibia suggesting an overlying leprosy-related ulcer in Skeleton 62, St James and St Mary Magdalene, Chichester, England. Photo by Jean Brown.

Left: Figure 4.17. Periosteal new bone formation on the lower leg bones (tibia and fibula). Photo by Charlotte Roberts.

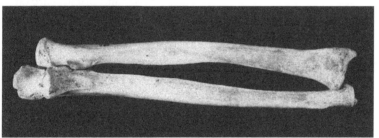

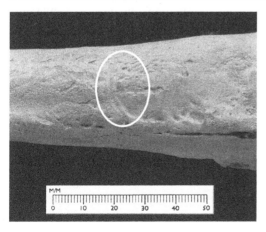

Above: Figure 4.18. Periosteal new bone formation on the forearm bones from a skeleton with leprosy from St James and St Mary Magdalene, Chichester, England. Photo by Mary Lewis.

Left: Figure 4.19. Imprint of blood vessels in new bone formation. Photo by Jeff Veitch.

leprosy. These are caused by inflammation of the periosteum, most often seen on the surfaces of the lateral tibia and medial fibula, corresponding to the attachment of the interosseous membrane between the two bones. The interosseous membrane is likely important in the spread of nonspecific inflammation due to leprosy from the foot up the leg. Periosteal lesions are more commonly seen on the metatarsals and foot phalanges than in the metacarpals and hand phalanges (Andersen 1969). Additionally, patients with leprosy usually have swollen legs due to infection of the feet that causes secondary phlebitis (inflammation of the veins) or lymphangitis (inflammation of the lymphatic system). Chronic venous stasis (slow blood flow) also develops as the valves of the veins become involved (Price 1961), which may also contribute to periosteal lesions seen on the bones of the lower leg (Jopling and McDougall 1988).

Some authors suggest that periosteal lesions of the lower leg bones develop in leprosy as a result of ascending infection from infected feet and that the less common forearm periosteal lesions are the result of hand infection, all originally initiated by a loss of sensation in the hands and feet and the development of ulcers (e.g., Srinivasan and Dharmendra 1978d). However, others have shown that periosteal lesions (and osteitis) are often linked to leprosy reactions (Paterson and Job 1964; Enna, Jacobsen, and Rausch 1971). Paterson and Job (1964, 436) described the radiological signs of periosteal lesions in living people with leprosy as follows: "Usually in leprosy these outer layers [of new bone] are reabsorbed, but in some cases they consolidate into new bone on the outer aspect of the cortex." Periosteal new bone formation on the tibiae and fibulae can be very common; this condition was observed in nearly half (20) of 56 patients in one leprosy hospital (Andersen 1969). In 17 people, the bone changes were bilateral, and in 34 feet of the 20 patients the sole of the foot was ulcerated (three people also had loss of sensation but no ulcers). While Dyer suggested as early as 1921 that leprosy-related ulcers of the skin were not necessary for periosteal reaction to occur, periosteal lesions of the lower leg bones have also been observed in patients with leprosy who have midfoot plantar ulcers and no damage to the bones of the feet (Manchester 2002). Andersen was of the opinion that the inflammatory lesions of the lower leg bones were sterile, suggesting that the periosteal reaction response was actually to local toxins from bacterial infection of the feet. The advantage of observing leprosy-related tissue changes in living patients is that the ulcers can be observed in relation to the presence or absence of periosteal lesions of the lower leg bones if the lesions can be identified on X-rays. Andersen (1969,

116) concluded that "both the osteoarcheological picture and the radiological findings indicate that these lesions are characteristic of leprosy, but that they are not specific lesions." This is a point to recall when diagnosing leprosy in skeletal remains. Periosteal lesions of lower leg bones can occur in many conditions affecting the skeleton (Weston 2008, 2009), including tuberculous and treponemal infection, scurvy, and trauma. In Weston's macroscopic and radiographic study of bones in pathology museums deriving from people with documented medical histories, periosteal lesions were not seen to be diagnostically unique for *any* pathological condition.

At medieval Chichester, notwithstanding the relatively lower frequencies of periosteal lesions on forearms (15 individuals, 9 of which were definitely diagnosed with leprosy) and on the metacarpals and phalanges (25 individuals, 19 diagnosed with leprosy), there was a relatively high frequency of alterations to the lower leg bones (Lee and Manchester 2008; see also Lewis, Roberts, and Manchester 1995a, 1995b). The tibiae and fibulae of skeletons excavated from Naestved and Winchester also had high rates of periosteal lesions. At Chichester, 75 individuals of 384 excavated had bone changes of leprosy and another 61 had nonspecific changes that may have been related to leprosy. In the adult population, 128 individuals had periosteal changes on their lower leg bones (46 percent). Ninety-two of those 128 had bilateral changes. Sixty of those 92 (65 percent) also had more specific bone changes of leprosy. At Naestved, 472 individuals were represented, including 123 complete skeletons. Overall, 329 skeletons (70 percent) were diagnosed with definite leprosy. Ninety-one of the 123 complete skeletons had periosteal lesions on the bones of the lower legs alongside inflammatory lesions in the nasal cavity (73.8 percent). Møller-Christensen (1953c, 1961) also noted grooves on the lateral side of affected tibiae, something many have thought are the imprints of blood vessels on new bone formation (Figure 4.19). At Winchester, 46 of 54 skeletons (85 percent) had bone changes of leprosy. The facial, hand, and lower leg bones of 22 individuals were preserved; of those, 20 (90.1 percent) were affected in all three regions. In one skeleton, no hand bones were affected and in another no lower leg bones were affected. Of those whose lower leg and hand bones were preserved (36), 31 people were affected in both regions (86 percent). Thus, there was an association between periosteal lesions on lower leg bones and other bone changes of leprosy in the skeletal remains from these two leprosaria cemeteries, suggesting that these changes are likely associated with leprosy. However, the presence of periosteal changes on the lower leg bones alone cannot be considered indicative of the infection.

168 · Leprosy

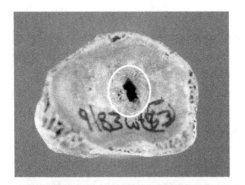

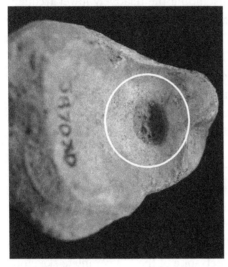

Above: Figure 4.20. Developmental defect of the proximal articular surface of a proximal first foot phalanx. Photo by Jean Brown.

Left: Figure 4.21. Osteochondritis dissecans of the distal tibial joint surface. Photo by Charlotte Roberts.

Joint Defects

In Møller-Christensen's 1961 report on the skeletons from medieval Naestved, he described centrally located depressions on the first metatarsal head and/or on the proximal articular surface of the first proximal foot phalanx. Andersen (1969, 177) called this "osteoarthropathia centralis metatarsophalangea prima." Due to an inability to see them adequately using X-rays, these lesions have not been studied in living people with leprosy. They were recorded in the bones of 50 of 91 individuals with leprosy at Naestved who had completely preserved feet (55 percent). Thirty-two individuals (8.3 percent) of the total skeletons from medieval Chichester have also been identified with what Lee and Manchester (2008) call "osteochondritis dissecans." This appears to be very similar to the osteoarthropathia centralis metatarsophalangea prima Andersen described, although he did not suggest links with leprosy. Osteochondritis dissecans is a traumatically induced condition often found in juvenile individuals where part of the joint surface fragments and/or separates (Resnick, Goergen, and Niwayama 1995). In contrast to Andersen, Møller-Christensen considered osteoarthropathia centralis metatarsophalangea prima to be "a characteristic symptom of leprosy" (Møller-Christensen 1961, 42), likening it to osteochondritis dissecans. Lesions such as these are also found quite frequently in skeletons without leprosy in archaeological contexts and are more likely to represent a developmental defect in the joint surface (Figure 4.20; see Roberts and Manchester 2010, 121). Larger defects may represent osteochondritis dissecans as a result of trauma (such as an osteochondral fracture; see Figure 4.21) or an inherited predisposition (see Resnick, Goergen, and Niwayama 1995, 2611).

Porosity of the Orbits

Another bone change that has been described in skeletons with leprosy is cribra orbitalia, or porosity of the orbits (Figure 4.22). It was present in over two-thirds of 99 skeletons with leprosy buried at Naestved (69.7 percent). The highest frequency was found in individuals 1 to 14 years old (72 percent of 25). Around 50 percent of male and nearly 60 percent of female skulls studied were affected. Bennike et al. (2005) report cribra orbitalia in nonadult skeletons at Naestved. This bone change was highest in skeletons of individuals 14.5 to 20 years old. Møller-Christensen (1965), who labeled the bilateral and symmetrical lesions *ursura orbitae,* found it in skulls with bone changes of leprosy 134 percent more frequently than in those without. He suggested that "leprosy or one of its complications is responsible for this difference, and this indicates that *ursura orbitae* has some pathologic significance" (ibid., 610). Lewis (2008) also records cribra orbitalia in 35 of 62 (56 percent) children's skulls examined from Chichester (none with

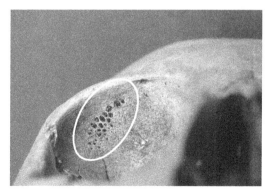

Figure 4.22. Porosity in the eye sockets. Photo by Charlotte Roberts.

bone changes of leprosy) and in 29 percent of 298 adults affected with leprosy (9.7 percent) (see also Lee and Boylston 2008b). There was a lower frequency in the leprosarium part of the site than in the almshouse part (Lee and Magilton 2008a). Møller-Christensen (1978) proposed that cribra orbitalia might be correlated with leprosy because the disease can affect the lacrimal glands and the eyes, but he also suggested that in people who do not have leprosy, malnutrition or poor hygiene might be the cause. He mentioned the viral infections of mumps and smallpox as possible causes but deemed this connection less likely. He indicated that the lesions might be the result of "pressure of an enlarged lacrimal gland" (ibid., 116). In the twenty-first century, chronic inflammation of the eye (see Ortner 2003, 2006, 2008), iron deficiency anemia (Stuart-Macadam 1985), infection in general as an adaptive response to increased pathogen load (e.g., parasitic intestinal infection leading to hemorrhage and iron deficiency anemia; see Stuart-Macadam 1992), scurvy and rickets (Brickley and Ives 2008), and vitamin B_{12} deficiency (Walker et al. 2009) have been among the many suggested causes of cribra orbitalia. Ortner's (2006, 203) opinion corroborates Møller-Christensen's thought that cribra orbitalia found in association with leprosy in skeletal remains probably represents a "vascular response to chronic infection of the eye that is known to occur in some cases of leprosy and leads to blindness." Ortner also notes that while the change is "reactive bone formation" (ibid., 267), cribra orbitalia is actually characterized by holes, or lytic lesions, and new bone formation is more likely the result of scurvy in this area of the skeleton (see Brickley and Ives 2008). On balance, the evidence suggests that the lesions may well represent an adaptive response to infection; this conclusion follows the work of Stuart-Macadam. However, cribra orbitalia is not a specific indicator of leprosy because skeletons without leprosy have cribra orbitalia in many contexts from different time periods and regions of the world, unless of course these individuals all had leprosy but had not developed related bone changes at the time of death.

Respiratory Involvement

People with leprosy today can develop associated upper and lower respiratory tract disease that may affect the nose, the sinuses, the nasopharynx, the larynx, the trachea, the bronchi, and lung tissue (Kaur et al. 1979). Because the ears are connected to the respiratory system via the Eustachian tube, ear infections can also develop, the ears perhaps acting as a reservoir of the infection.

Involvement of the paranasal sinuses can lead to *M. leprae* migrating into the nose and beyond (Hauhnar, Mann, et al. 1992, 390). As the nose is infiltrated with bacteria in a high proportion of patients with lepromatous leprosy, there may be a highly infectious discharge from the nasal mucosa and the sinuses may become affected (Barton 1979). Barton studied the paranasal sinuses of sixteen patients with lepromatous leprosy and found that all had abnormalities; the most frequent observation was mucosal thickening of the maxillary sinuses. These changes were due to *M. leprae*'s infiltration of the sinus mucosa. The maxillary sinus is often most affected of all the facial sinuses (Soni 1988, 1989; Hauhnar, Kaur, et al. 1992), mostly in its antero-inferior area, where maximum stagnation of secretions containing *M. leprae* bacilli occurs (Soni 1989; see also Gupta, Tiwari, and Singh 2004). However, other sinuses may be affected. The ethmoidal, maxillary, frontal, and sphenoid sinuses were all affected in one study (Kiris et al. 2007).

Archaeologically, there have been very few studies of skeletons with leprosy for evidence of sinusitis. Boocock, Roberts, and Manchester (1995a, 1995b) have provided the only detailed studies of skeletons from a leprosarium. They found that 54.9 percent (73) of 133 individuals from medieval Chichester had bone alterations in one or both maxillary sinuses, although there was no difference in frequency between those who had bone changes of leprosy and those who did not. However, it should be remembered that sinusitis can be caused by a variety of factors, not just bacterial infections such as leprosy (e.g., see review and clinical references in Roberts 2007). Thus, sinusitis is not pathognomonic for leprosy.

While Roberts, Lucy, and Manchester (1994) have suggested that periosteal lesions on rib surfaces are caused by pulmonary infection, most frequently tuberculosis, there has been little bioarchaeological focus on the co-occurrence of bone changes of leprosy and rib lesions. However, five of seven people with leprosy had rib lesions from Area A at medieval Chichester (the earliest, leprosarium phase), suggesting the co-occurrence of leprosy and lung infection in these individuals (Lee and Boylston 2008a). Another study that focused on a leprosy hospital cemetery from the nineteenth century in St. Eustatius in the Caribbean Netherlands found lesions that Gilmore (2008) thought could be related to leprosy. The damage was on the posterior aspect of the hyoid bone, the thyroid cartilage, and the third through sixth cervical vertebrae of a skeleton. The hyoid bone is situated adjacent to the epiglottis and opposite the cervical spine. Its posterior surface is normally smooth, but in this skeleton there were destructive lesions and woven (immature) bone formation on all the bones and the

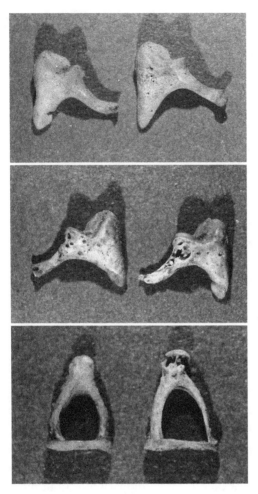

Figure 4.23. Destructive lesions of the bones of the middle ear of a skeleton with leprosy from St James and St Mary Magdalene, Chichester, England. Photo by Tjasse Bruintjes.

thyroid cartilage, likely due to the spread of inflammation, perhaps caused by leprosy. As leprosy may involve the epiglottis and vocal chords, this skeletal involvement could indicate that the person had a hoarse voice, which can be a characteristic sign in someone with leprosy. In fact, most lesions in the respiratory tract in leprosy occur in the larynx (Tze-Chun and Ju-Shi 1984). The epiglottis can also become thickened, nodular, and ulcerated (Soni 1992).

Hearing problems can also occur in people with leprosy (Awasthi et al. 1990). Infection of the ear bones has been observed in skeletons from medieval Chichester. One hundred and thirty-six ear bones were extracted from 97 temporal bones of 89 individuals from that site. Just over 50 percent of the bones had erosive lesions (Figure 4.23), probably due to middle ear infection as a consequence of leprosy. Bruintjes (1990, 632) suggested that "the results tentatively point towards a more than average middle ear involvement in the individuals with osseous rhinomaxillary changes due to lepromatous leprosy." In leprosy, the thickening of the nasal mucosa, which narrows the pharynx, can affect normal drainage of the middle ear. In an archaeological study of the skeletons from Chichester, Dalby (1993) found that in X-rays of the mastoid processes of the skull, the mean size of mastoid air cells was no different in people with leprosy and people who did not have leprosy (the size of these cells can indicate a mastoid infection that is often due to middle ear infection). This may be because the mastoid air cells develop in childhood and thus probably before the person contracted leprosy.

Less Commonly Noted Bone Changes

Other bone changes in skeletal remains showing signs of leprosy are even less commonly reported in bioarchaeology but are worth exploring in the future. People with leprosy tend to have poorer oral health (e.g., see Reichart, Ananatasan, and Reznik 1976; Matos et al. 2017), and if they need to rely on others to help them with oral hygiene and/or experience contracted fingers, this could lead to inadequate oral hygiene (Rawlani et al. 2011; Ferreira, Gonçalves, et al. 2018). Periodontal disease has been recorded in people with lepromatous leprosy of long duration. In an archaeological context, the dental calculus on the teeth of one person with advanced bone changes of leprosy (Winchester) may have been due to mouth breathing caused by nasal congestion related to leprosy; this probably caused the plaque that eventually developed into calculus (Roffey and Tucker 2012; see also Baker and Bolhofner 2014). One hundred and ninety-one individuals buried at the medieval leprosy hospital at Chichester had dental calculus and 159 had periodontal disease affecting the jaw bone(s) (Ogden and Lee 2008). While there were no data on the association of these conditions with people who had bone changes of leprosy, people buried in Area A, where more evidence of leprosy was found, had more dental disease (Lee and Magilton 2008a). As high rates of dental disease are seen in people with leprosy today, it would be instructive to explore dental disease in individuals with leprosy more closely in archaeological contexts.

The association of trauma with leprosy has also been considered, albeit not very frequently. Judd and Roberts (1998) recorded evidence for fractures in people with and without leprosy at Chichester (see also Judd 2008). The hypothesis of the study was that people with leprosy may become blind

174 · Leprosy

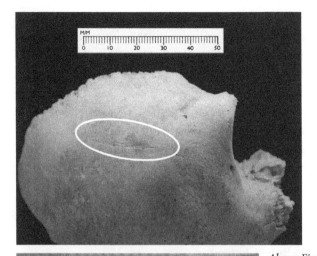

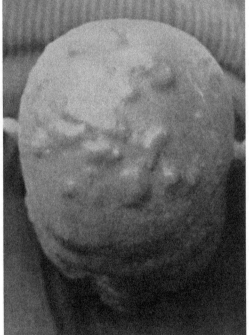

Above: Figure 4.24. Frontal bone groove. Photo by Jeff Veitch.

Left: Figure 4.25. Lepromas on the head of a person with leprosy. Source: Mitsuda (1952, figure 101).

and could develop impairments that lead to clumsiness and loss of proprioception and that because of this, they might be more likely to experience falls and fractures. However, although fractures were seen in skeletons from Chichester, people with leprosy did not seem to have fractures more frequently than those without leprosy. While this archaeological site produced a high frequency of people with fractures compared to those buried in other contemporary urban late medieval sites, the impact of leprosy on fracture rates was not particularly apparent. Only one-third of the long-bone fractures at Chichester were seen in people with bone changes of leprosy (Lee and Magilton 2008a). Grooves in the frontal bone have also been noted in skeletons with leprosy (Figure 4.24) that may be the result of enlarged nerves pressing on the surface of the bone. However, Goode (1993) found that frontal grooves were not significantly associated with leprosy in her study of skeletons with and without evidence of leprosy from medieval Chichester. Indeed, this lesion is usually classed as a normal variant (Saunders and Rainey 2008). Other authors have noted that lepromas (nodules filled with bacteria; Figure 4.25) on the skin of people with leprosy could also cause erosive lesions on the underlying bone (e.g., see Buckley 2008a, 2008b).

The Senses and Leprosy

Diagnosing leprosy in skeletal remains necessitates a consideration of bone changes in specific parts of the skeleton, as for any other disease in paleopathology. Interpreting the effects of those changes in once-living people also requires an appreciation of the associated signs and symptoms that occur with this infection (see Roberts 2017 for an example). While it is normally not possible to identify changes to soft tissue when analyzing skeletal remains, soft-tissue damage in leprosy beyond what has already been described could have caused considerable challenges for people in the past. All of the main five senses can be affected in leprosy—hearing, sight, smell, taste, and touch. However, it should be remembered that what are viewed as "the senses" can vary around the world and through time. Even today there does not seem to be any agreement on the number of senses humans have.

People with leprosy can develop eye infection and ultimately become blind; their hearing can be compromised; sensory nerve damage can lead to a lack of sensation that affects what they may or may not feel when they touch something, especially with their hands and feet, and where ulceration may develop; and there may be congestion in the nose that can damage the normal taste and smell mechanisms. Not only do people with leprosy experience these insults on their senses, but those around them see the person with leprosy; hear their voice, which is often hoarse if the larynx is affected; they may not wish to touch them because of fear and misunderstanding of the disease; and they may be able to sense a distinctive smell if ulcers have become infected. The impact of leprosy on the senses is an area that warrants some research, as is increasingly being done generally in archaeology and history (e.g., Jütte 2005; Woolgar 2006; Graves 2007; Skeates 2010). As Woolgar (2006, 3) states, "An assessment of the practical operation of the senses and their cultural impact is lacking." Woolgar notes that in medieval England, a person who lost even one sense was considered to be evil. This perhaps shows how the effect of leprosy on the five senses may have exacerbated a medieval community's response to the infection.

Diagnosis of Leprosy in Skeletal Remains: What Features Should Be Present?

A clinical study of patients in a leprosarium with all types of leprosy undertaken from 1940 to 1943 found that even in advanced lepromatous leprosy, the patient is "usually free from bone lesions. In our experience bone disease, if it does occur, is limited in extent and of little clinical significance" (Faget and Mayoral 1944, 3). Faget and Mayoral maintained that the bone alterations must occur after a long duration of the disease and that it was more likely that diagnosis would be made in older adults. This was a time when antibiotic treatment was generally unavailable and these conclusions are thus more directly analogous to the past.

It is important to make as accurate a diagnosis of leprosy as is possible for archaeological skeletal remains, but at the same time it is important to remain cautious. Møller-Christensen and Hughes (1962) considered that a cranium with bone lesions of leprosy *might* be an example of someone who had had leprosy and that a diagnosis could not be secure until the long bones had also been examined. If the long bones showed no pathological changes or were not preserved, then a proven diagnosis of leprosy could not be made. Only when a cranium with evidence of *facies leprosa* was accompanied by tibiae and fibulae showing periosteal reaction, bilaterally and symmetrically, could a firm diagnosis of lepromatous leprosy be considered. This is interesting considering the many diagnoses that can be made for periosteal lesions in the lower leg bone. Andersen, Manchester, and Roberts (1994, 21) believed that the bone changes of the hands and feet could "only be considered as highly characteristic of leprosy." They argued that "when a tentative diagnosis of leprosy is made on the basis of bone changes of the extremities alone, it is impossible to arrive at a clinical classification of the disease as with a living patient." They believed that if the bone changes were symmetrical and present in several extremities, it was possible to infer a clinical classification of lepromatous leprosy or borderline leprosy. Individuals with leprosy in the medieval leprosaria cemeteries considered in this chapter exhibited various combinations and frequencies of changes to the bones of the face, hands, feet, and lower legs. In more recent studies that reconsider the changes in lower leg bones, a diagnosis of lepromatous leprosy is usually made if rhinomaxillary syndrome is present, bearing in mind differential diagnostic options. Nevertheless, it is important to remember that leprosy has a broad spectrum of clinical manifestations due to its immune spectrum. It is well to remember Ortner's (2008, 206) observation that "the ability to diagnose leprosy in archaeological human skeletal remains ranges from problematic to highly likely. . . . [It is] the result of great variability in the effect of leprosy on the human skeleton associated with . . . [the] immune response, age of onset . . . [and] diet."

Can people with tuberculoid leprosy be diagnosed in archaeological skeletal remains? Møller-Christensen (1961) suggested that a diagnosis of lepromatous leprosy required the presence of bilateral and symmetrical changes in the bones of the hands and feet and that a diagnosis of tuberculoid leprosy required unilateral changes to the hands and feet. While Lee and Manchester (2008, 208) concur with this suggestion, they emphasized that if bilateral bone changes were encountered, they should be asymmetrical. However, they maintained that skeletal lesions for tuberculoid leprosy were minimal or nonexistent and concluded that "it is therefore difficult to give a firm diagnosis of tuberculoid leprosy in the archaeological record." In more recent research on skeletons from a Danish medieval leprosarium and in a study of clinical records (1947–1985) from a Portuguese leprosarium (Colonia Rovisco Pais, Tocha), Matos (2009; Matos and Santos 2013c) suggested a scheme for differentiating people with lepromatous leprosy from those with tuberculoid leprosy based on bone changes: rhinomaxillary syndrome or rhinomaxillary syndrome and involvement of the hand and foot bones for lepromatous leprosy, and bilateral or unilateral hand and foot bone involvement and no rhinomaxillary syndrome for tuberculoid leprosy.

In related research, Boldsen (2001) created a process that enables researchers to estimate the specificity and sensitivity of lesions of leprosy and calculate leprosy frequency. He used skeletons excavated from a number of Scandinavian sites: St. Jørgensgård in Odense, Denmark (thirteenth–seventeenth centuries AD) and Kvarteret St. Jørgen in Malmö, Sweden (AD 1320–1520), both of which were associated with leprosy hospitals, and

Tirup, which was a rural village parish churchyard in eastern Jutland, Denmark (twelfth–fourteenth centuries AD). St. Jørgensgård revealed evidence for over 1,500 burials. Nine hundred and twenty-four skeletons were excavated and 635 individuals aged 15 years or over were studied. At Kvarteret St. Jørgen, 1,600 graves were excavated and 200 skeletons aged 15 years or over analyzed. At Tirup, 620 burials were excavated and 61 skeletons aged 15 years or older were analyzed.

Boldsen considered bone changes in seven specific areas of the skeleton to be indicative of leprosy: the edge of the nasal aperture, the anterior nasal spine, the alveolar process of the maxilla, the palate, the fibula (subperiosteal exostoses and porotic hyperostosis), and the fifth metatarsal. The aim was to "not determine whether or not any given individual skeleton came from a person who suffered leprosy" but to use "the estimates of sensitivity and specificity for each of the seven symptoms and the estimates [of] point prevalences of leprosy at death . . . to estimate the predictive value of any symptom in each of the three samples by applying [the] formula" (Boldsen 2001, 385). Using this method, Boldsen found that two-thirds of people from the Odense group had bone changes of leprosy, as did 10 percent of the group from Malmö and 25–50 percent of individuals in the Tirup population. Boldsen suggested that because there was a very high correlation coefficient ($r = 0.57$) between changes in the alveolar process of the maxilla and those in the anterior nasal spine, skeletons with both changes likely were those of people who had leprosy in life. This diagnostic method for leprosy in skeletal remains was new at the time for most bioarchaeologists and there are issues to be discussed regarding the skeletal lesions used, particularly the last three on Boldsen's list. However, Boldsen is of the strong opinion that "these estimates are the closest it is possible to come to individual diagnoses," while he notes that they are "probabilistic statements," conceding that "absolute statements about individual disease status are beyond what can be reached by palaeopathological analysis" (ibid.).

In summary, the characteristics and distribution of each bone change thought to be associated with leprosy should be considered before a suggested diagnosis of leprosy is given for archaeological skeletal remains, as should potential differential diagnoses. However, differential preservation of skeletons can compromise diagnosis. In addition, nonspecific bone changes such as palmar grooves and periosteal lesions on the lower leg bones should not be considered to be indicative of leprosy on their own. When leprosy first affects a population, lesions in the skeleton will likely only occur when the disease becomes endemic—that is, as more people become resistant to the infection and live through the acute stages and die when they have chronic bone lesions.

Beyond the Macroscopic: Other Bioarchaeological Methods for Diagnosing Leprosy

Analysis of Ancient DNA and Mycolic Acids

Perhaps the area of palaeopathology that has developed most in the last twenty-five or so years is the ability to detect biomolecules specific to disease in archaeological human remains, specifically the extraction of ancient pathogen DNA (aDNA) fragments particularly from skeletons. This has included aDNA and mycolic acids specific to *M. leprae*. Techniques from molecular biology have been adapted and developed in order to use them to analyze samples from archaeological human remains, ideally to answer archaeological questions that have not been possible using traditional analytical methods (Roberts 2018, 216–223; see also Brown and Brown 2011; Stone 2008; Wilbur and Stone 2012). Interpretation of aDNA data for leprosy has benefited from the sequencing of the modern and ancient genomes of leprosy (Cole, Eiglmeier, et al. 2001; Schuenemann, Singh, et al. 2013) and from the work of Monot et al. (2005, 2009), who have identified strains or subtypes of the bacteria by comparing the single-nucleotide polymorphisms they possess in order to plot the origin and evolution of this infection.

Biomolecular analysis is a destructive analytical method and contamination from DNA from modern environmental microorganisms is a challenge in aDNA analysis (Gilbert et al. 2005). Preservation in different environmental contexts can also be variable (e.g., see Gilbert et al. 2005; Zink and Nerlich 2005). The DNA of ancient pathogens can differentially survive in the diverse environments of the world, between skeletons at one site, and in different bones of the same skeleton buried at the same cemetery site. Because of these factors, it is challenging to predict when aDNA will be preserved in human remains. Some reflections on the quality of data produced from some pathogenic aDNA analyses have been published (e.g., see Gilbert et al. 2004; Bouwman and Brown 2005) and it has been recommended that attention be paid to describing the methods and laboratory facilities used more thoroughly (e.g., see Roberts and Ingham 2008; Cooper and Poinar 2000; Stone 2008; Roberts 2018). It clearly can also be challenging for non-experts to understand the data and published interpretations of those data and appreciate when findings are reliable, a fact that those who do this type of work have increasingly recognized. As Donoghue (2008, 167–168) notes, "There is still much controversy over authenticity of some findings, but there are now sufficient numbers of well-planned and executed studies to give confidence in the reports of ancient DNA from human microbial pathogens." As methods develop and the number of more reliable studies increases, we should be more confident of the authenticity of data produced.

The literature has devoted increasing attention to diagnosis of leprosy in skeletal remains using biomolecular methods (e.g., Taylor et al. 2000; Taylor et al. 2006; and some chapters in Roberts et al. 2002), although not as much as for tuberculosis. However, it should be remembered that even if aDNA of a pathogen can be extracted from a skeleton and identified, it does not mean that the person experienced the infection because they could merely have been a carrier of it and experienced no ill effects (that is, the expression of the disease could have been subclinical). In addition, a positive aDNA result for leprosy for a skeleton cannot be used to prove that nonspecific bone changes were the result of leprosy. Correlation does not mean causation.

Rafi and colleagues (1994a, 1994b) first reported this type of analysis in two articles in 1994, based on a metatarsal dated to AD 600 from Israel. However, the first of these articles (1994a) was not without controversy (Blondiaux 1995; Roberts and Dixon 1995; Stanford 1995; Zias 1995). Many other studies have followed, as outlined in Table 4.1. Skeletons with leprosy have been analyzed from a range of modern-day countries. Most researchers have focused on bone samples from the affected nasal regions and many have found positive results for *M. leprae* aDNA. Analysis of samples taken from other bones, such as the bones of the foot, the lower leg, and the spinal column, has been less successful. These findings support Hershkovitz and Spigelman's (2007, 240) statement that "unless one has the nasal bone or there was lepromatous leprosy, the DNA will be difficult to find." Haas et al. (2000) found that while they were unsuccessful in extracting *M. leprae* aDNA from the hand and foot bones of skeletons with leprosy, bone samples from the rhinomaxillary region were positive for leprosy. Recent research suggests that the turbinate bones of the nose may also preserve *M. leprae* DNA well because their presence in that region of the nose indicates that the bacillus has successfully infected the mucosa (Goulart 2013). Teeth may also be good sources of aDNA, and we should also consider dental calculus as a potential resource for *M. leprae* DNA (Warinner et al. 2014).

Table 4.1. Studies of leprosy in skeletons using ancient biomolecular analysis, in date order by publication

Rafi et al. (1994 a and b)	Israel, 624 AD
Taylor et al. (2000)	Orkney Islands, Scotland, thirteenth–fourteenth centuries AD
Haas et al. (2000, 2002)	Germany, 1400–1800 AD
Donoghue et al. (2001)	Poland, "medieval"; Hungary, tenth, eleventh, fourteenth, and fifteenth centuries AD
Spigelman and Donoghue (2001, 2002)	Israel, 300–600 AD
Haas et al. (2000, 2002)	Germany, 1400–1800 AD; Hungary, tenth century AD
Pálfi et al. (2002)	Hungary, tenth century AD and fourteenth–fifteentth centuries AD
Donoghue et al. (2002)	Israel, 624 AD; Hungary, tenth, eleventh, fourteenth, and fifteenth centuries AD; Poland "medieval"
Strouhal et al. (2002)	Czech Republic, thirteenth–eighteenth centuries AD
Montiel et al. (2003)	Spain, twelfth century AD
Nuorala et al. (2004)	Sweden, tenth–thirteenth centuries AD
Donoghue, Marcsik, et al. (2005a)	Israel, 300–600 AD; Egypt, fourth century AD; Sweden, tenth–thirteenth centuries AD; Hungary, tenth century and fourteenth–sixteenth centuries AD
Donoghue, Erdal, et al. (2005)	Turkey, eighth–tenth centuries AD
Éry et al. (2008)	Hungary, second–ninth centuries AD
Nerlich, Marlow, and Zink (2006)	Germany, 1400–1600 AD
Taylor et al. (2006)	England, tenth–eleventh centuries and thirteenth–sixteenth centuries AD
Likovsky et al. (2006)	Czech Republic, twelfth century AD
Donoghue and Spigelman (2008)	Hungary
Faerman (2008)	Israel, sixteenth–nineteenth centuries AD
Lee (2008)	England, twelfth–sixteenth centuries AD (mycolic acids only)
Taylor et al. (2009)	Uzbekistan, first–fourth centuries AD (and mycolic acids)
Watson et al. (2009, 2010)	England, tenth–eleventh centuries AD; Croatia, eighth–ninth centuries AD; Denmark, thirteenth–sixteenth centuries AD
Monot et al. (2009)	Croatia, eighth–ninth centuries AD; Denmark, thirteenth–sixteenth centuries AD; Egypt, fourth–fifth centuries AD; England, thirteenth–sixteenth centuries AD; Hungary, seventh and tenth centuries AD; Turkey, eighth–ninth centuries AD

(*continued*)

Table 4.1—*Continued*

Pálfi et al. (2010)	Hungary, eleventh century AD (and mycolic acids)
Suzuki, Takigawa, et al. (2010)	Japan, mid-eighteenth–early nineteenth centuries AD
Hadju et al. (2010)	Hungary, 3700–3600 BC (included mycolic acids)
Taylor and Donoghue (2011)	Uzbekistan, first–fourth centuries AD; Czech Republic, ninth century AD; England, tenth–twelfth and thirteenth–sixteenth centuries AD; Hungary, seventh, tenth and eleventh centuries AD; Turkey, eighth–ninth centuries AD
Rubini, Dell'Anno, et al. (2012)	Turkey, eighth–tenth centuries AD; Italy, second–third centuries AD
Lee et al. (2012)	Hungary, seventh century AD (mycolic acids)
Economou et al. (2013a,b)	Sweden, tenth–fourteenth centuries AD
Taylor et al. (2013)	England, ninth–thirteenth centuries AD (and mycolic acids)
Schuenemann et al. (2013)	England, ninth–thirteenth centuries AD; Sweden, tenth–fourteenth centuries AD; Denmark, eleventh–fourteenth centuries AD
Mendum et al. (2014)	England, tenth–twelfth centuries AD
Suzuki et al. (2014)	Japan, fifteenth–eighteenth centuries AD
Donoghue et al. (2015)	Austria, Croatia, Czech Republic, Hungary, Italy, Turkey, and Uzbekistan, sixth–eleventh centuries AD
Gausterer, Stein, and Teschler-Nicola (2015)	Austria, ninth century AD
Inskip et al. (2015)	England, fifth–sixth centuries AD
Molnár et al. (2015)	Hungary, eighth century AD
Roffey et al. (2017)	England, eleventh–twelfth centuries AD
Rubini et al. (2017)	Italy, sixth–eighth centuries AD
Inskip et al. (2017)	England, 885–1015 AD
Köhler et al. (2017)	Hungary, 3780–3650 BC
Schuenemann et al. (2018)	Czech Republic, ninth–tenth centuries AD; Denmark, 1044–1383 AD; England, 415–545 Cal AD and 948–1283 AD; Hungary, seventh–eighth centuries AD; Italy, mid- to late seventh century AD; Sweden, 1032–1115 Cal AD
Krause-Kyora et al. (2018)	Denmark, 1000–1600 AD
Taylor et al. (2018)	Ireland, three sites with date range of 939–1637 AD
Pfrengle et al. (2018)	Norway, fourteenth century AD
Bedić et al. (2019)	Croatia, tenth–eleventh centuries AD
Kerudin (2019); Kerudin et al. (2019)	England, tenth–twelfth centuries AD and fourteenth–seventeenth centuries AD

Along with sequencing of the ancient genome of leprosy, a more recent development in aDNA analysis in leprosy has been exploration of the subtypes (strains) of *M. leprae* (Taylor, Watson, et al. 2006; Taylor et al. 2013; Economou et al. 2013a, 2013b; Schuenemann et al. 2013, 2018) with the goal of exploring the origin, evolution, and spread of leprosy.

Disregarding the potential challenges of analyzing aDNA, its main advantages are that it could provide us with real prevalence rates for disease—that is, if all skeletons were analyzed from a cemetery and all preserved leprosy aDNA. It could also help diagnose disease in people who died before any bone changes occurred or identify people with subclinical leprosy. It can also provide information about the evolution of the strains of *M. leprae*.

Mycobacterium leprae and *Mycobacterium tuberculosis* have robust lipids in their cell envelopes; that is, complex multilayered structures that protect the bacteria from their often-hostile environment. These lipids can also be used as biomarkers for ancient disease (Donoghue et al. 2017). The main lipid classes are of high molecular weight: mycolic acids (70 to 90 carbons) and mycocerosic acids (27 to 34 carbons). Ancient mycolic acids of tuberculosis were first identified in ribs from skeletons from early medieval Addingham, England (Gernaey et al. 2001), and mycocerosic acids were first found in a number of ribs from skeletons from the late Coimbra Identified Skeletal Collection, which is curated in the Museum of Anthropology of Coimbra University, Portugal (Cunha and Wasterlain 2007; Redman et al. 2009). The collection consists of 505 identified skeletons with birth dates from 1826 to 1922 and death dates from 1904 to 1938. In both studies, tuberculosis was confirmed. Both lipid classes can be also found in skeletal remains showing leprosy (see Chapter 5). It is anticipated that further research to isolate and amplify the aDNA of the bacillus and mycolic acids of *M. leprae* will continue on skeletal remains from archaeological sites.

While most work to date has been concentrated on verifying a skeletal diagnosis of leprosy, and increasingly more research is exploring *M. leprae* strains in skeletal remains, it is likely that the future will see even more exciting question-driven research, particularly more research that focuses on identifying genetic variations in the *M. leprae* bacterium that represent different strains associated with geographic locations. However, in my view, biomolecular analysis "for its own sake" should be avoided because it is destructive. Only studies of human remains using biomolecular analysis that answer valid questions or test hypotheses that could not be addressed or tested using any other method should be encouraged. Biomolecular analysis is not a silver bullet for diagnosing disease in human remains. It should be done in conjunction with macroscopic paleopathological evidence and archaeological and historical information so that the data are adequately contextualized. It is pleasing to see more guidance generally for practitioners who work with human remains (museum curators, bioarchaeologists, and commercial archaeologists; see also Roberts 2019). It is also comforting to see some museums requesting to see research designs where destructive methods are to be used, as in aDNA analysis, in addition to projected outcomes and detailed descriptions of sampling methods. For example, in the UK there is guidance on destructive analysis from the Advisory Panel on the Archaeology of Burials in England (2013; see also Squires, Booth, and Roberts 2019). Human remains are a nonrenewable resource that should be safeguarded for future generations of researchers who will be working when methods are even more developed and new ones are realized.

Microscopy

Histological methods of analysis make it possible to study the microscopic structure of the tissues of the body. They were first used on fossil bone at the end of the nineteenth century (Schultz 2001). A very good knowledge of the normal microstructural appearance of bone is needed to be able to recognize abnormal bone due to disease in archaeological human remains, as is an appreciation of the possibility of postmortem changes. Few studies in bioarchaeology have used histological analysis to diagnose leprosy. Based on the extant data, it seems unlikely that a reason for doing more could be supported. Using light microscopy, Schultz and Roberts (2002) studied thin ground sections of tibiae with periosteal reaction in six individuals from medieval Chichester who had been diagnosed macroscopically with leprosy. They concluded that "the probability is very high that the periosteal changes on the surfaces of the tibial shafts . . . were caused by infections secondary to leprosy" (ibid., 96). However, they admitted that the changes may have been caused by other conditions such as treponemal disease. In a more recent contribution, Schutkowski and Fernández-Gill (2010) studied new bone formation on tibiae and fibulae in skeletons from the same archaeological site histologically. They suggest that previously described histological structures (e.g., *polsters* and *greinzstreifen*) are not pathognomonic for leprosy, a fact that Von Hunnius et al. (2006) also noted in their study of treponemal disease in late medieval skeletons from Hull, England.

Imaging

Plain film radiography, a nondestructive method, is the second most common method of analysis used in bioarchaeology. However, there are reports that X-rays can damage aDNA (Immel et al. 2016; Grieshaber et al. 2007), and curators of human remains and researchers who wish to study them should reflect on using aDNA analysis on remains that have already been subjected to radiography. Plain film radiography is not routinely used to diagnose leprosy because the macroscopic bone changes often suffice for diagnosis in paleopathology, but radiographic observations of the bones of living people with leprosy are important for interpreting the paleopathological data. Microradiography (thin sections of tissues, which is destructive), computed tomography (CT), and microcomputed tomography have also been applied to skeletal remains to diagnose disease. Each provides a detailed visualization of the interior of bones, teeth, and soft tissues that cannot be seen externally. Blondiaux and colleagues (1994; see also Blondiaux et al. 2002) have used microradiography in an attempt to differentiate leprosy from treponemal disease in two fifth-century skeletons from France. Using the diagnostic criteria of Coutelier (1971), they used several microscopic methods, including microradiography, to analyze thin sections of tibiae affected by periosteal new bone apposition. They concluded that the microscopic changes were identical in both conditions, corroborating the findings of Schultz and Roberts (2002), Von Hunnius et al. (2006), and Schutkowski and Fernández-Gill (2010).

In a more recent study, Ruffin et al. (2010) used radiography to analyze thin sections of tibiae and fibulae from thirty-three individuals buried in the cemetery associated with a medieval leprosarium in France. Periosteal apposition of new bone was seen on 85 percent of the bones, but new bone was seen on only 45 percent of them radiographically. While this study does not contribute directly to diagnosis of leprosy, it illustrates that subtle new bone formation may not be visualized on X-rays. Thus, clinical diagnostic criteria that are based on X-rays may not always be appropriate for archaeological bones. It seems, therefore, that macroscopic and to a certain extent biomolecular methods are the best way forward for diagnosis of leprosy in human skeletal remains. However, both methods have advantages and disadvantages.

Confounding Factors That Affect Diagnosis of Skeletal Leprosy

The diagnosis of leprosy in archaeological human remains is challenging, not least because all the bone changes could be caused by other pathological conditions, for example those on the lower leg bones. Thus, as Ortner (2003, 45–64) has emphasized, a full and detailed description of the bone lesions is the first step toward a diagnosis, followed by a consideration of the bone changes related to leprosy described in both clinical and bioarchaeological literature. The basic bone changes in the disease are bone formation and bone destruction. The bone that is formed may be unhealed immature woven bone or healed lamellar bone, the former indicating that the disease that caused the bone formation was active at the time of death but did not necessarily cause death (Figures 4.26 and 4.27). The next step is to consider differential diagnoses after documenting the characteristics and distribution pattern of the bone changes. It is important to note that

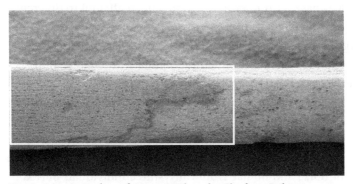

Figure 4.26. Active bone formation. Photo by Charlotte Roberts.

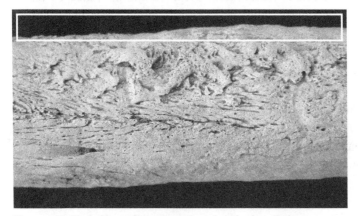

Figure 4.27. Healed bone formation. Photo by Charlotte Roberts.

a skeleton represents the person at the time of their death. It captures the health history of individuals who were buried at a particular time in a specific place. When all lesions are healed (or chronic) in nature, it is impossible to make sense of what disease occurred at what point in time in the person's life if the skeleton has evidence of several diseases. One exception would be when childhood diseases are evident, for example rickets (vitamin D deficiency) or enamel hypoplastic defects, both of which occur in childhood and represent a childhood disease or dietary deficiency. This sort of evidence shows that a person may have been at an elevated risk of early death later in life (Barker 1994; Wood et al. 1992, 349). Another exception would be when there are lesions that were active at the time of death. The process of diagnosing and interpreting leprosy is complex and is confounded by many factors that can make diagnosis even more difficult, as exemplified by Wood et al.'s (1992) seminal paper on the osteological paradox (see also DeWitte and Stojanowksi 2015).

It is important to remember key points from the osteological paradox (Wood et al. 1992, 344) in relation to leprosy:

- The skeletons of only a small proportion of people will be affected in leprosy.
- The soft-tissue alterations in disease do not normally affect the skeleton directly.
- In the early stages of leprosy the nerves and skin are affected, even though some of the soft tissue changes (e.g., lepromas and ulcers) might affect some bones.
- A person could die before bone changes occur (and leprosy is not usually the cause of death for people with leprosy today).
- It is not known how sensitive a lesion may be as an indicator of a disease.

All of these considerations make it difficult to "weigh the countervailing effects of underestimation caused by low sensitivity, and the overestimation caused by selectivity bias" (but see Boldsen 2001). The lesions are also often insensitive to and of low specificity for a disease. This is why considering differential diagnoses is essential. The relevant confounding factors for diagnosing and interpreting the impact of leprosy in the past include preservation bias (see Roberts 2018, chapter 4); the small hand and foot bones and the fragile facial bones needed for diagnosis may not preserve well. It is also not possible to infer accurate frequency rates for leprosy and to know whether a cemetery of skeletons represents the original living population with or without leprosy who contributed to it (see Waldron 1994a). How mobile were people in and out of the community that is represented and did that mobility affect whether people with leprosy are found in the cemetery? Other related factors include not knowing how frail people in the population were (including their immune system strength and where on the leprosy spectrum they might have been), what specific risk factors were present in a population that predisposed people to leprosy, and the overall susceptibility of the population to the infection.

Any group of skeletons may represent an "unknown mixture of people with different levels of frailty and susceptibility to a disease" (Wood et al. 1992, 345). A genetic predisposition to a disease and socioeconomic and environmental risk factors (e.g., high population density) may all contribute to levels of frailty and risk for disease in any one individual. When considering individuals represented by their skeletons at a specific time in their life (their death), it is extremely challenging indeed to disentangle what really led to poor health and death in any one person or population. For example, while it is known that prolonged contact with an infected person is needed to contract leprosy, other factors might affect whether a person contracts the infection. Furthermore, a broad spectrum of the disease may be seen, from lepromatous leprosy to tuberculoid leprosy, and some types of leprosy may or may not produce specific bone changes. Relatively small numbers of archaeological skeletons have been identified with bone changes of leprosy (although they have been identified at many sites) unless the focus of study has been on leprosaria cemetery contexts. As Wood et al. (1992, 345) have said, "Reports on single specimens tell us little about the disease experience of ancient populations." It is when the data are considered from a wide area and over a long period of time that we can begin to understand the impact of a disease such as leprosy on people in the past.

However, it is not possible to estimate *actual* frequency rates for leprosy temporally and geographically. Comparing present disease frequency with that of the past is thus also problematic because frequency rates are calculated differently (see Waldron 1994a, 42). One thing is certain, however: "Disease prevalence and skeletal lesion frequency are two very different things . . . and never must be equated" (Wood et al. 1992, 365). Another factor that Wood et al. (1992, 344) have noted that is particularly pertinent to thinking about frequency rates is the cumulative effect on rates of disease from a burial ground that has been in existence over several hundred years. It is rare that enough dating information is available for a site that would make possible a consideration of disease frequency for those buried over specific time frames within the site (although see Connell et al. 2012 for one notable exception).

A bioarchaeologist usually estimates the frequency of diseases according to the number of individuals affected and the preservation of the relevant parts necessary for diagnosis (e.g., for leprosy, the bones of the face, hands, and feet). The frequency is the overall frequency over the time period of the cemetery and is therefore likely to be less than the true frequency of the disease in the general population because not all skeletons are well preserved, and not all people from the general population will have been buried at the same cemetery site. Thus, while the data that exist for leprosy in archaeological skeletons are a snapshot of the infection's impact on past populations and the tip of the iceberg of the leprosy load, they do provide an idea of which modern-day countries and time periods were affected in the past. However, little bioarchaeological research has been done in some regions; this may explain some of the lack of evidence for leprosy in those places (see Buikstra and Roberts 2012 for an overview of the history of paleopathological study). Diagnosis and interpretation of leprosy in skeletal remains is thus complex, but once data have been collected it is essential that they be placed in context so that the risk factors for the infection within those populations can be identified. Furthermore, while DeWitte and Stojanowski have noted that bioarchaeologists have not really applied the findings of Wood et al. (1992) to many bioarchaeological studies, bioarchaeologists could usefully consider the nuances of the osteological paradox more seriously in any study of the origin, evolution, and history of leprosy (or any disease), but *before the data are collected*. As DeWitte and Stojanowski (2015, 429) have emphasized, "It is crucial for researchers interested in directly assessing heterogeneous frailty and selective mortality to consider these phenomena at the research design stage. We should make decisions about which data to collect and which analytical approaches to take based on the questions at hand before our field or lab work begins, rather than engaging in post hoc decision making about how to analyze data that might not be maximally informative about frailty."

Conclusions

This chapter has explored the direct and indirect bone changes related to leprosy that can be seen in the bones of the skeleton and other possible diagnoses for those changes. Ideally, diagnosing leprosy in skeletons requires a complete, well-preserved skeleton. In particular, the bones of the face, hands, and feet are the key areas to examine for evidence. However, only a few percent of people with untreated leprosy will develop bone lesions and the type of leprosy identified depends on the resistance of their immune system to *M. leprae*. Currently, most skeletons diagnosed in paleopathology will display lesions due to lepromatous leprosy, but as more scholars use the criteria of Matos (2009) for tuberculoid leprosy, we may see more evidence of this high-resistant form. Damage to the peripheral nerves by *M. leprae* is responsible for the (indirect) alterations to the hand and foot bones, while the bone changes of the skull represent the direct effects of *M. leprae* being inhaled into the mouth and nose. People with leprosy may also have comorbidities that can be recognized in the skeleton that are also seen in people with leprosy today, for example osteoporosis, respiratory infections, and dental disease. Lepromatous leprosy may be diagnosed if there is evidence of rhinomaxillary syndrome, or a combination of rhinomaxillary syndrome and damage to the bones of the hands and feet, or bilateral or unilateral damage to hand and foot bones but no rhinomaxillary syndrome for tuberculoid leprosy. Beyond visual or macroscopic diagnosis of skeletal remains, imaging and microscopy have sometimes been used with interesting but not necessarily convincing results. In recent years, biomolecular analysis has been the main advance in analytical methods for studying leprosy, mainly through the extraction and analysis of aDNA. This particular work has benefited from the sequencing of the modern *M. leprae* and *M. lepromatosis* genomes that has provided the opportunity for comparative analyses. The modern data have provided testable hypotheses regarding the origin, evolution, and history of leprosy, and information about strains of the bacteria prevalent in different regions of the world. These data are enabling paleogeneticists, historians, and bioarchaeologists to reevaluate the long history of leprosy in relation to historical accounts of the drivers for movement of people with leprosy across the globe.

5

The Bioarchaeological Evidence of Leprosy

> In medical history when dealing with leprosy, the bioarchaeological data seems to precede the historical sources.
>
> Kjellstrom (2012, 262)

Diagnosis of leprosy through analysis of archaeological human remains is of course a prerequisite for understanding the impact of leprosy on people in the past. This chapter presents the evidence to date from published and unpublished sources. I have also contacted bioarchaeologists around the world in an effort to access unpublished data or find information in outlets local to particular modern-day countries. However, it is likely that resources that have not been located exist and thus there is no certainty that all the evidence is presented here. This is often the nature of archaeological data; some information may never see publication and may remain in gray literature and some may be published in places that are difficult to find.

In order to provide a broad global overview of leprosy in the bioarchaeological record, it is essential to consider the evidence both temporally and geographically. Evidence for leprosy in skeletal remains and preserved bodies from archaeological sites has always attracted attention in bioarchaeology. The US-based Paleopathology Association has made concerted, although undocumented, efforts over the years since its founding in 1973 to coordinate the recording and synthesis of data on leprosy. This has included the establishment in 1981 of the Leprosy Study Group, which was chaired by Keith Manchester (University of Bradford). Aidan Cockburn (USA), Tadeusz Dzierzykray-Rogalski (Poland), David Scollard (then in Hong Kong) and Stanley Rubin (Liverpool) were recruited to the group to cover clinical leprosy, paleopathology, and history. Its aims were:

1. To establish diagnostic criteria,
2. To reevaluate the data in light of new knowledge about diagnosis,
3. To pool the data on leprosy from around the world, and
4. To track the origins and evolution of leprosy and assess why it declined.

Since then, the development of biomolecular analytical techniques have contributed to diagnoses of leprosy in skeletons that have no characteristic bone changes and both the modern and ancient genomes of leprosy have been sequenced (Singh et al. 2015; Cole, Eiglmeier, et al. 2001; Schuenemann et al. 2013). To a certain extent, all the aims of the original Leprosy Study Group have been and are being achieved, but there is much more work to do. Researchers continue to develop diagnostic criteria, hold workshops (e.g., Manchester, Storm, and Ogden 2012), and reevaluate and bring information together, and biomolecular analysis has helped us better understand leprosy's origin. Bioarchaeologists have also developed new areas of research in relation to leprosy. These have included exploring aspects of the identity of people living with leprosy in the past (Roberts 2017; Baker and Bolhofner 2014), tracking the mobility history of people with leprosy using both aDNA and stable isotope analyses (e.g., Inskip, Taylor, and Stewart 2017; Roffey et al. 2017), and establishing the diet of people with leprosy compared to people without the infection (e.g., Brozou et al. 2019). As a result of Aim 1, leprosy diagnoses in some skeletal remains have been reevaluated (Aim 2), and this chapter hopes to fulfill Aim 3. Related to Aim 2, there are increasing discussions about the ethics of excavating, studying, and curating human remains around the world. These concerns include debates about repatriation and/or reburial (Buikstra and Gordon 1981; Buikstra 2006; Scarre 2006; Walker 2008; Roberts 2018, chapter 2; Roberts 2019), and the production of ethics and practice guidance documents for bioarchaeologists.[1] If remains are reburied in the future, for whatever reason, it may not be possible to reevaluate them. Thus, it is essential to make detailed records. Further, "archaeologists in previous generations tended to retain only skulls, discarding the postcranial skeleton" (Manchester 1983, 4), and the postcranial bones that are used to diagnose leprosy may be unavailable for study in regions of the world where paleopathological work will be done in the future. Manchester (1983) also noted that east and southeast Asia and eastern Russia might hold evidence for the earliest data for skeletons with leprosy that have been uncovered so far. Although in recent years more bioarchaeologists have analyzed skeletal remains from sites in these areas (e.g., see Pechenkina 2012), to date there has been very little evidence of leprosy in that region of the world, despite documentary data to the contrary. Biomolecular analyses are increasingly addressing Aim 4 and it is anticipated that much more will be forthcoming. Why leprosy declined continues to be a matter of debate and is still an area for development (e.g., see Crespo, White, and Roberts 2019).

As leprosy is largely absent in New World skeletal remains (see below), it has probably not had the same attention in palaeopathology in the Americas as it has in Europe. However, leprosy has been the topic of regular contributions at the annual meetings of the Paleopathology Association in both the United States and Europe and at the meetings of the American Association of Physical Anthropologists. For example, I organized a specialist workshop on leprosy for the annual meeting of the Paleopathology Association in Salt Lake City, Utah, in 1998. Another workshop on leprosy was hosted at the Paleopathology Association European Meeting in Lille, France, that also enabled participants to see skeletal remains affected by this infection (Manchester, Storm, and Ogden 2012). Finally, I organized a poster symposium at the annual meeting of the American Association of Physical Anthropologists in Calgary, Canada, in 2013. Although the number of new "cases" of leprosy is declining around the world, there appears to be a resurgence of interest in the bioarchaeology of leprosy, especially in biomolecular analyses (see Table 4.1).

The following sections document the bioarchaeological (and in some cases more recent) evidence for leprosy in both the Old and New Worlds.

Leprosy in People Who Did Not Reach Adulthood

Compared to adults, leprosy in nonadult skeletal remains is rarely reported, even though there is an increasing focus among bioarchaeological studies on this demographic group (e.g., see Lewis 2018).

While Lewis (2002, 2008, 2018) has provided very useful overviews of leprosy in past nonadult individuals, historical data for leprosy that affected nonadults is relatively rare. Some of those sources include information from a Chinese book about a child who was infected by the age of three years (Skinsnes 1980: 217 BC) and descriptions of King Baldwin IV of Jerusalem, who was infected as a child (Mitchell 1993: twelfth century AD). Much later, in the sixteenth through eighteenth centuries, descriptions exist of newborn children having been fed purgatives to clear mucus from their mouths and protect them from contracting leprosy (Fildes 1988). In Denmark, children were admitted to leprosaria during this period (Richards 1977). Illustrations of children with leprosy in art are also not common, but examples drawn by Johan Ludvig Losting, an artist from Bergen, Norway, can be seen in Danielssen and Boeck (1847; see also Leprosy Museum 2003). There are multiple possible reasons why the evidence of leprosy in skeletal remains of children is scarce. If children died at a young age, recognizable bone lesions may not have had time to form, but we should also remember that their fragile facial bones and small hand and foot bones may not be preserved in the ground for excavation and analysis. Children may also have developed the high-resistance form of leprosy and thus might not have developed bone lesions by the time they died. In addition, as leprosy can have a long incubation period, children may not have had bone damage by the time they died.

However, a small number of nonadult skeletons (aged up to 20 years) from the Næstved leprosy hospital in Denmark (late medieval period) were identified with lepromatous leprosy (Møller-Christensen 1961). Two individuals who died before the age of 7 had rhinomaxillary syndrome and leprogenic odontodysplasia but no other bone changes. More recent data from Næstved indicates that of 60 fully examined children's skeletons, sixteen had developed bone alterations related to leprosy before the age of 17 years, all but one of whom had the characteristic lepromatous type lesions (Bennike et al. 2005). These children also had a higher rate of "stress" lesions from early childhood, shorter long bones for their dental age, and lower bone mineral content compared to children from a contemporary monastic site in Denmark.[2] All these findings suggest that these individuals experienced interruptions in normal growth, likely as a result of poor living conditions or an inadequate diet or disease.

In Britain and Ireland (modern-day Northern Ireland and Eire), nine children or adolescents with leprosy aged from 12 to 20 years were reported from the St John Timberhill cemetery in Norwich, Norfolk, England (twelfth to fourteenth or fifteenth centuries AD). Five had both rhinomaxillary syndrome and postcranial bone changes (Anderson 1996, 1998, 2009a). At the medieval leprosarium of St James and St Mary Magdalene, Chichester, 104 individuals were excavated who were from perinatal age to 17 years old, but none showed bone changes consistent with leprosy (Lewis 2008). As few children with leprosy-related bone changes have been found at this archaeological site, perhaps children with leprosy were not segregated during the late medieval period, at least in the Chichester area. However, if that were the case, it would be expected that they would be found in non-leprosaria sites. They may also have been buried outside normative burial sites for the period of time and location, a practice that could have been dictated by cultural values about the younger sector of a community. Additional evidence of leprosy and possible leprosy in nonadult skeletons is noted in the following locations in the UK:

Dryburn Bridge, East Lothian, Scotland (prehistoric; a child 6–8 years old; J. Roberts 2007);

Wharram Percy, North Yorkshire, England (AD 950–1100; a child 10 years old; Mays 2007; Taylor et al. 2000);

Scarborough, East Yorkshire, England (after AD 1066; an adolescent; Brothwell 1958);

All Saints, Fishergate, York, Yorkshire, England (eleventh through sixteenth centuries AD; an adolescent; McIntyre 2016; McIntyre and Chamberlain 2009);

Deerness, Orkney Islands, Scotland (thirteenth and fourteenth centuries AD; an adolescent; Taylor et al. 2000);

St Michael le Pole, Dublin, Eire (eleventh century AD; an adolescent; Buckley 2008a, 2008b);

St Mary Magdalen, Winchester (1070–sixteenth century AD, eight adolescents, and one older child with leprogenic odontodysplasia; Roffey and Tucker 2012).

Outside the UK, nonadult skeletons with bone changes of leprosy have been documented in Croatia (Bijelo Brdo), the Czech Republic (Prague), Hungary (Szombathely Ferences Templom), Italy (Martellona), Sweden (Lund and Sigtuna), and Turkey (Kovuklukaya).

Bioarchaeological Evidence of Leprosy in the Old World

Having discussed the scarcity of evidence for leprosy in nonadult skeletal remains, the skeletal evidence is now divided into three broad areas in the Old World: Northern Europe, the Mediterranean and Africa, and Asia (including, for example, islands such as Australia and New Zealand). These regional groupings coarsely reflect similar climatic and environmental factors. The New World is considered as a whole, reflecting the general lack of evidence for leprosy in skeletal remains there. The majority of the evidence presented in this section has been documented through macroscopic recording of skeletons with the bone changes of leprosy. However, in recent years diagnosis of leprosy has increasingly been done with biomolecular analysis of bone samples. Much information presented here has been published, but some has appeared in abstracts for conferences the author has attended and some has been acquired through the goodwill of colleagues who have shared their data before publication.

Globally, evidence for leprosy in human remains is plentiful in some countries, again reflecting varying levels of bioarchaeological work, but perhaps also showing its relative frequency in the past. Leprosy has been found on four of the seven continents of the world. Australasia, South America, and Antarctica have revealed no evidence yet, the latter primarily perhaps because no archaeological human remains have been excavated there (Pearson 2011). This pattern may indicate how leprosy has evolved as a human disease and spread across the globe, but the bioarchaeological evidence is undoubtedly the tip of the iceberg for leprosy in the past, as is the case for other bioarchaeological evidence for disease, for example tuberculosis (Roberts and Buikstra 2003). Indeed, it is only by pooling together historical and bioarchaeological data that the history of leprosy can begin to be explored.

In the Old World, Britain (defined as England, Wales, and Scotland), Denmark, and Hungary have revealed the most evidence. For Britain and Hungary, there is a good range of dates for this evidence. The former has evidence of leprosy probably from around the third to second millennium BC to the nineteenth century AD, and Hungary's evidence dates from the fourth millennium BC to the nineteenth century AD. The evidence that exists pertains to one or a few individuals, depending on the archaeological site studied. It also includes specific bones with leprosy alterations from disarticulated skeletal remains and incorporates individuals represented only by skulls: Austria, China, the Czech Republic, Croatia, Cyprus, Egypt, France, Germany, Greece, Guernsey, Iceland, India, Iran, Ireland, Israel, Italy, Japan, Latvia, Micronesia, Norway, Poland, Portugal, Russia, Spain, Sweden, Thailand, Turkey, Ukraine, and Uzbekistan have evidence (and possibly Canada, Mexico, and the United States). There is currently no bioarchaeological evidence (as far as I am aware) from Belgium, Estonia, Finland, Lithuania, Mongolia, the Netherlands, Romania, Serbia and other former Yugoslavian countries, or Switzerland, but for some of these countries evidence of leprosy exists in historical sources. This disparity might be explained by a lack of bioarchaeological activity in some countries (see information about the history of the study of palaeopathology in Buikstra and Roberts 2012 and Márquez-Grant and Fibiger 2011). In Russia (e.g., Buzhilova 2012: much work on skeletons), the only evidence has been found in a female skull that was recently identified with morphological changes of leprosy. This skull originates from St. Petersburg and is curated in the Rokhlin Collection in the Department of Anthropology of the St. Peter the

Great Museum of Anthropology and Ethnography, Russian Academy of Sciences in St. Petersburgh; it is dated to the nineteenth to twentieth centuries AD (Tukhbatova et al. 2016). Recent unpublished work has achieved a positive DNA result for *M. leprae* of Type 2, subtype F, a strain that is seen in Iran and Turkey today (Pfrengle et al. 2018). However, in contrast to Russia, Latvia (less developed for bioarchaeology) has revealed evidence from Riga that dates to the fifteenth to seventeenth centuries AD. This young man was buried in one of the mass pits at St Gertrude's Church cemetery (Petersone-Gordina, Roberts, and Gerhards 2016). This is the first evidence of leprosy for this region of Europe.

There is generally a paucity of evidence in Asia, the Pacific region, the Middle East, and the New World, even though the highest rates of recent new "cases" have been reported from Brazil (World Health Organization 2019). In contrast, while the majority of skeletal evidence derives from Europe, there is relatively little evidence for leprosy reported in the European region today. Only fifty new "cases" were detected in Europe in 2018. When "cases" are reported, they are suggested to be of people who have either traveled to the country from a high-prevalence country and returned or people who have migrated from a high-prevalence region (e.g., Jensen et al. 1992 for Denmark; Bret et al. 2013 for France). World Health Organization figures for 2018 note that thirty countries of seventy-seven reporting have recorded 1068 "foreign born" new "cases" (World Health Organization 2019).

Figures 5.1 (Europe), 5.2 (global), and 5.3 (Britain and Ireland) show the countries where leprosy in archaeological human remains has been identified and the earliest possible date of such identifications. Note that some dates provided in publications are more precise than others. Often a date range or just a century is given. In the latter case, the start of that century is used. However, absence of evidence is not evidence of absence in bioarchaeology, or indeed in archaeology as a whole. That is because future excavations, analyses, and re-analyses of human remains will undoubtedly change the picture of the distribution of leprosy in the past and its earliest date. The following sections describe and discuss the available data on documented skeletons with leprosy and the context of the burials. The latter is included to explore whether the ways people with leprosy were buried deviated from the norm for their community, thus testing the idea that people with leprosy in the past were all stigmatized and segregated into leprosaria.

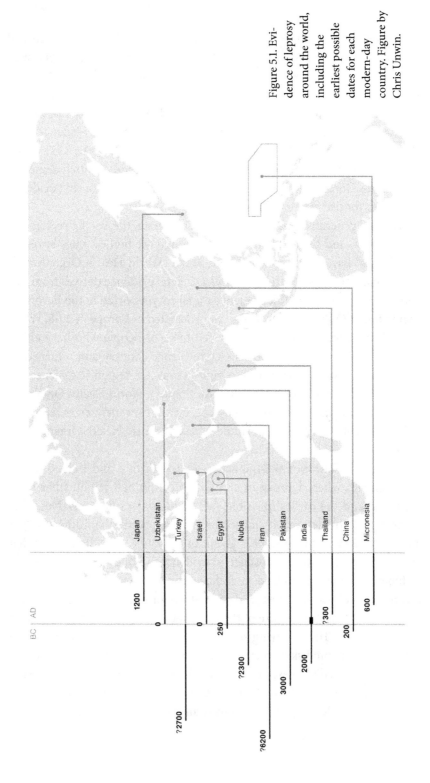

Figure 5.1. Evidence of leprosy around the world, including the earliest possible dates for each modern-day country. Figure by Chris Unwin.

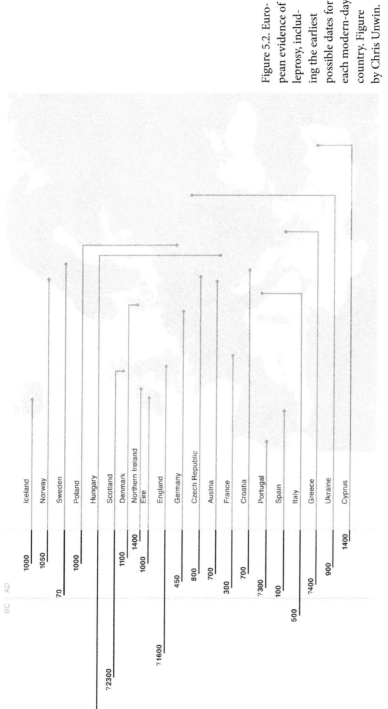

Figure 5.2. European evidence of leprosy, including the earliest possible dates for each modern-day country. Figure by Chris Unwin.

Figure 5.3. Distribution of British and Irish archaeological sites that have revealed evidence of leprosy by period of time. Figure by Chris Unwin.

Northern Europe

Northern Europe has the majority of the evidence for leprosy, in contrast to the lack of evidence there today and the fact that some of the highest frequencies are found in the southern hemisphere. Where there are relevant modern data to report, I have provided them.

Austria

Very little evidence of leprosy in Austria has been described to date. Considering its long history of paleopathological study, this is puzzling

(Teschler-Nicola and Grupe 2012). However, an early medieval skeleton was reported from the rural Slavonian site of Pottenbrunn, west of Vienna (AD 1100; Fabrizii-Reuer and Reuer 2000). A young adult woman (grave 181) had bilateral lesions to the lower leg and foot bones that appear to be related to leprosy. Ancient DNA (aDNA) analysis revealed *M. leprae* DNA Type 3 (Gausterer, Stein, and Teschler-Nicola 2015). She was not buried in any unusual way or in a location that was marginal to the rest of the cemetery. Evidence of leprosy has also been identified at the urban site of Zwölfaxing, just southeast of Vienna (AD 680–830; Maria Teschler-Nicola, pers. comm., August 2008; Donoghue et al. 2006), making the earliest possible evidence the late seventh century AD.

Britain and Ireland

A study of leprosaria in Britain noted that "significant numbers of lepers were suffering Hansen's Disease in England between 1100 and 1250" (Satchell 1998, 66) but that "individuals with signs of Hansen's disease are hardly ever found in any post-Conquest burial contexts apart from the cemeteries of leper-houses" (ibid., 62). The content of this section illustrates clearly that the evidence suggests otherwise. Compared to other parts of the Old World, a lot of evidence has been identified in skeletons from archaeological sites in England but not so in Scotland, Wales, or Ireland (Figure 5.3 and Appendix 3). However, there is not as much evidence as might be expected considering the large amount of documentary data and commentaries therein that indicate that it was a common infectious disease. The lack of evidence from Ireland may be explained by the late development of palaeopathology as a discipline in that country (Buckley 2011). However, the general lack of bioarchaeological evidence could be because the evidence is buried in leprosaria cemeteries, most of which have not been excavated. As will become clear, most skeletons with bone changes of leprosy have been excavated from non-leprosaria contexts. MacArthur (1925, 414 and 420) concluded that while leprosy existed in medieval England, its "importance was exaggerated out of recognition by ignorance" and that "to assert that leprosy in medieval England was a more terrible scourge than plague is nothing short of ludicrous." Further, in 1961, Brothwell suggested that it is difficult to determine whether the literary evidence that leprosy was brought to England in the first century BC is describing the infection or one of the many other conditions referred to by the same term.

However, there may now be earlier evidence than that cited by Newman (1895: 60 BC). Potentially the earliest (prehistoric) evidence of leprosy in skeletal remains in Britain has been described in Scotland from the site of Dryburn Bridge, East Lothian, Scotland (J. Roberts 2007). This skeleton was originally reported by Close-Brooks (1979). McKinley (forthcoming) has also reported "suspicious" rhinomaxillary syndrome on a skull (17632) at Kingsmead Quarry, Horton, Berkshire, southern England, that was redeposited in the base of a ditch; this skull could date anywhere between the late Neolithic (ca. 2000 BC) and the late Anglo-Saxon period (ca. the ninth–tenth centuries AD). Unfortunately, radiocarbon dating was unsuccessful, but McKinley (forthcoming) suggests a Middle Bronze Age date (no later than 1600–1100 BC) based on associated nonburial features. In 1974, Reader reported leprosy in a skeleton from a Roman camp at the rural site of Poundbury, Dorset, also in southern England (fourth century AD). Supporting this find, Brothwell (1958, 287) suggested that leprosy may have been introduced to England from the continent at the time of the Roman invasion in AD 43: "Leprosy was probably brought to England during the Roman invasions." There is also Roman-period evidence from Italy and France, which also were part of the Roman Empire (see below). Although Reader (1974) made an initial diagnosis of leprosy for an older adult from Poundbury, subsequent radiographic analysis suggested that the bone changes had a high possibility of being linked to psoriatic arthritis (Farwell and Molleson 1993, 192) and that there was no suggestion of leprosy (Theya Molleson, pers. comm., July 2008). Reader's report was questioned in the leprosy literature and other diagnoses were suggested (Jonquieres 1977). However, Keith Manchester still considers that this skeleton may show leprosy (pers. comm., August 2016).[3] Unstratified skeletal remains at the Gambier-Parry Lodge Roman cemetery, an urban site at Gloucester in southwest England that dates to the late fourth to early fifth centuries, revealed the upper jaw of a person of juvenile age with rhinomaxillary syndrome (Figure 5.4). However, it is uncertain that the date of the site is correct (Cameron and Roberts 1984; Mary Lewis, pers. comm., August 2008). Some possible Roman evidence has recently been reported in urban London, but this is by no means certain (Walker 2012). While there is sparse evidence for leprosy before the early medieval period in Britain, the evidence described here needs further work to nuance diagnoses.

In England, the early medieval (or Anglo-Saxon) period, which dates from circa 410 to the mid-eleventh century AD, is characterized by rural living (see Roberts and Cox 2003, chapter 4). There is more skeletal evidence at this time, mostly on sites in the south, east, and southwest regions of the country. The earliest reference to leprosy in Anglo-Saxon England

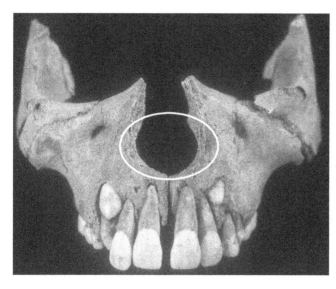

Figure 5.4. Upper jaw with evidence of leprosy from Roman Gambier Parry Lodge, Gloucester, Gloucestershire, England. Photo by Mary Lewis.

is from a seventh-century Irish missionary who used the word "lepra," but that word could cover a "multiplicity of terms" (Lee 2006, 72). Of particular interest in this period was a man from Beckford who was buried with two small dogs and a large number of grave goods but was placed at the margin of the associated cemetery (Wells 1996) (Figure 5.5). A greater number of individuals (and cemetery sites) with individuals with leprosy have been excavated in the east of England. St John, Timberhill, Norwich, Norfolk (tenth to eleventh centuries AD) is a significant site in that region. Anderson's detailed study identified a large number of skeletons with possible and definite leprosy, including nine nonadults (1996, 1998, 2009a; see also Bayliss et al. 2004 and Popescu 2009 on the complexity of the dating of this site). The cemetery may have served a leprosy hospital in Norwich. Apart from the skeletal analysis that has produced evidence for leprosy in many individuals, Watson et al. (2009, 2010) did some analysis of *M. leprae* DNA in some individuals, some of which were positive.

Some individuals with leprosy in this period are of interest because their burials included grave goods (e.g., at Beckford) or unusual findings, but very few individuals in this early medieval era were buried in unusual places or locations in cemeteries. Most were associated with parish church graveyards. An unusual find was an individual excavated at Willoughby-on-the-Wolds with evidence of leprosy was in a multiple burial context that included another skull showing a trepanation and many grave goods. It may be that the maxilla that had bone changes of leprosy belonged to the person with the trepanned skull. One young adult woman with leprosy

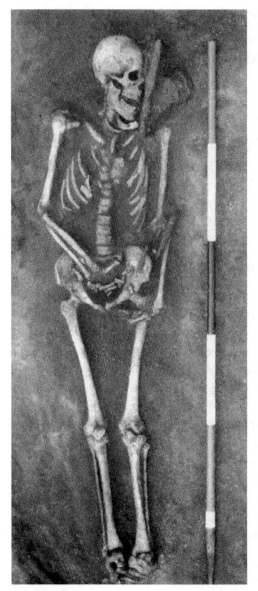

Figure 5.5. Skeleton with leprosy (A8) and grave goods at early medieval Beckford. Source: Wells (1996), reproduced with permission of Catrina Appleby of the Council for British Archaeology.

at Edix Hill, Cambridgeshire, was given a "bed burial" accompanied by a range of grave goods, suggesting she was in some way special (Duhig 1998) (Figure 5.6). Whether this was because she had leprosy, was of high status, or had some other identity is not known. Similarly, a man from Great Chesterford with evidence of leprosy was buried with a spearhead and a knife, among other grave goods. Strontium and oxygen isotope analysis suggested that he may have originated outside Britain, and DNA analysis

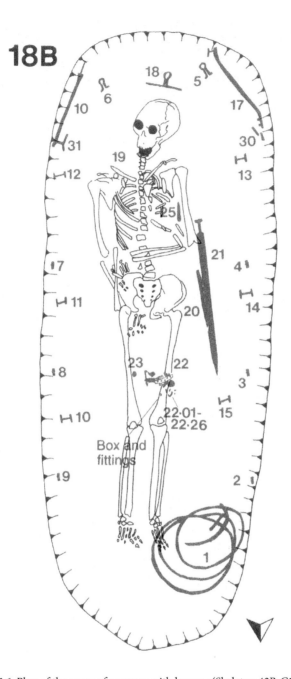

Figure 5.6. Plan of the grave of a person with leprosy (Skeleton 42B-G18) given a "bed burial" at Edix Hill, Cambridgeshire, England. Source: Malin and Hines (1998). Reproduced with permission of Catrina Appleby of the Council for British Archaeology Research and Paul Spoerry of Oxford Archaeology.

and lipid biomarker analysis both showed positive results for leprosy. DNA analysis identified the 3I strain of the bacteria, which is seen today in the Americas and Morocco (Inskip et al. 2015; and see the section in Chapter 1 relating to armadillos). This finding has been replicated more recently from a female skull from Hoxne, Suffolk, which also dates to the early medieval period (AD 885–1015; Inskip, Taylor, and Stewart 2017). In a recent study of aDNA from two skeletons (5029 and 5046) from a churchyard cemetery at Raunds, Northamptonshire, England, dating to the mid-tenth to mid-twelfth centuries AD, Type 3, subtype K of *M. leprae* was identified in 5046, even though 5029 also had bone changes of leprosy (Kerudin et al. 2019). Although it has been assumed that in this early medieval period, people with disease and/or impairment were treated as "lesser human beings . . . or even excluded" (Lee 2006, 59), Lee notes that people with leprosy were integrated with their community, as they were in the subsequent later medieval period, in contrast to the belief that all "leper burials were associated with leprosy hospitals" at the time (ibid., 69).

More individuals with leprosy have been identified in the late medieval period (mid-eleventh to mid-sixteenth centuries AD). This was the time when the majority of leprosaria were founded. While the increased population density associated with urbanism facilitates the transmission of the bacteria responsible for leprosy via exhaled droplets, in this period, the cemetery sites where people with leprosy were buried are in both rural and urban environments. As a comment to Monot et al. (2005), Pinhasi, Foley, and Donoghue (2005) suggested that leprosy was probably a recent disease that developed with urban communities and high-density living. While the bioarchaeological evidence of leprosy does increase in the period when people started to live in urbanized communities, this may reflect the fact that it is mostly urban cemeteries that have been discovered and excavated. The extant bioarchaeological evidence indicates leprosy in both rural and urban communities, in contrast to Pinhasi, Foley, and Donoghue's claim that there is no evidence of leprosy before urban life (2005).

In this late medieval period, the numbers of people affected by leprosy are distributed more evenly around Britain and into Ireland. Of particular importance in the south-central area of England is the large leprosarium of St James and St Mary Magdalene, Chichester, Sussex (Lee and Magilton 1989; Magilton and Lee 1989; Magilton et al. 2008). This is because the associated cemetery contained a large number of burials and the site has recently been the subject of a detailed study. It is the largest leprosarium cemetery excavated in England to date and is also one of the largest in the world. It is likely the hospital was founded "sometime before 1118" (Magilton 2008b, 57) for around eight people with leprosy. Evidence exists that people without leprosy started to be admitted at the end of the twelfth century, at least from the absence of evidence of leprosy in their skeletal remains. Many written sources exist for the hospital in the fourteenth century and we know that women were being admitted by the end of that century. Bequests continued through the fifteenth through seventeenth centuries, but by the end of the eighteenth century it was closed. The "Leper Cottage" is all that survives on the site today (Magilton 2008b). It was last referred to as a place for people with leprosy in a will of 1418 and had ceased to be a hospital by the end of the seventeenth century.

Most people at this site were buried in coffins, and interments occurred from the early twelfth century until the mid-seventeenth century. People with leprosy were buried there up to the fifteenth century. The cemetery is divided into Areas A and B and appears to have developed from southwest to northeast over time (Magilton 2008a, 91). Three hundred and seventy-four graves were excavated and 384 skeletons were studied (Lee and Magilton 2008b). In addition to a monograph describing the study of the site and the skeletons found, a number of other studies of the health and well-being of the population have been completed, illustrating the value of this cemetery for research (e.g., Ortner, Manchester, and Lee 1991; Knüsel and Göggel 1993; Lewis, Roberts, and Manchester 1995b). In a recent study of aDNA from two skeletons dating to the fourteenth to seventeenth centuries AD from Chichester, Type 3, subtype I of *M. leprae* was found (Kerudin et al. 2019). It is the first time this subtype has been identified in Britain; before this study, Denmark, Hungary, and Turkey were the only other countries with this subtype in archaeological skeletons.

St Mary Magdalen, in Winchester, an early leprosy hospital (late eleventh to mid-twelfth century AD), is a similarly important site that is located near Chichester. Thirty-seven skeletons at this site were diagnosed with leprosy (Roffey and Tucker 2012). Recent research targeting the aDNA of *M. leprae* from nine of the skeletons has revealed genotypic data (Taylor et al. 2013; Schuenemann et al. 2013; Mendum et al. 2014), and stable isotope analysis has indicated that the skeletons analyzed were mostly those of people who had been born and raised locally. One skeleton has also revealed a strain of *M. leprae* that is a close relative to the strain found in red squirrels today on Brownsea Island, a location close to the site (Avanzi et al. 2016). Roffey et al. (2017) have also identified the skeleton of a young adult male from this site as a pilgrim: biomolecular analysis has shown that he was not local

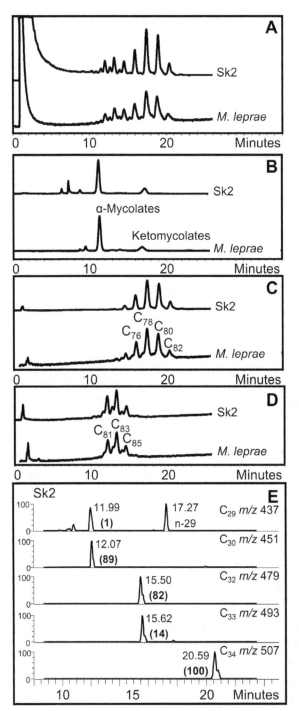

Figure 5.7A. Winchester, England; Skeleton 2. (*a*) High-performance liquid chromatography of fluorescent derivatives of total mycolic acids, separated according to size. (*b*) Total mycolic acid derivatives, collected from (*a*) and separated by high-performance liquid chromatography, according to α-mycolate and ketomycolate type. (*c*, *d*) Components of α-mycolates and ketomycolates, respectively, separated by size on HPLC. (*e*) Profile of mycocerosate derivatives examined by gas chromatography-mass spectrometry, showing a C34 component only found in *M. leprae* (see Figure 5.7B [c] for the *M. leprae* standard). Reproduced with permission of David Minnikin, University of Birmingham.

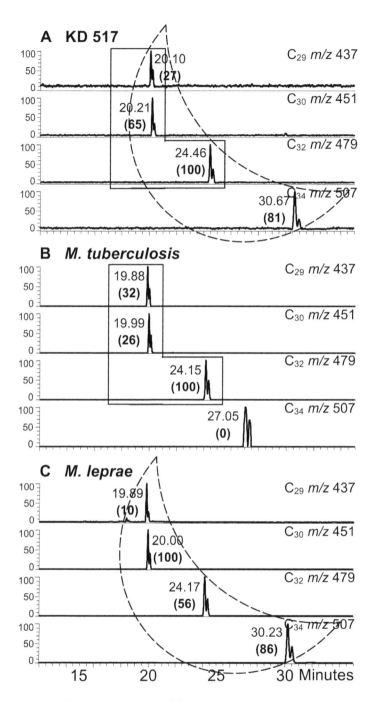

Figure 5.7B. Kiskundorozsma, Hungary; Skeleton KD 517 mycocerosates. (*a*) Mycocerosate pattern from KD 517, confirming leprosy based on the presence of high proportions of a specific C34 component. (*b*, *c*) Standard mycocerosate profiles for *M. tuberculosis* and *M. leprae*, respectively. Reproduced with permission of David Minnikin, University of Birmingham.

and that he carried a strain of leprosy consistent with one strain found in south-central and western Asia today. Furthermore, mycolic acid analysis of skeleton 2 from Winchester revealed an excellent profile correlation between the bone sample analyzed and a standard from *M. leprae* (Figure 5.7a; Taylor et al. 2013; Donoghue et al. 2017). Mycolic and mycocerosic acid profiles can also distinguish leprosy from tuberculosis, as seen for a seventh-century AD skeleton from Kiskundorozsma-Daruhalom dűlő II, Hungary (Figure 5.7b; Lee et al. 2012; Donoghue et al. 2017). Another cemetery site that has recently been excavated in Cambridgeshire (Midland Road, Peterborough) was also likely associated with a leprosy hospital in its early phase (eleventh to thirteenth centuries AD). Eleven skeletons with definite evidence of leprosy have been identified in the first phase (three females, six males, and two unsexed adults) and three (two females and one male) have been identified in the second phase (thirteenth to sixteenth centuries AD). More individuals (ten from the first phase and eight from the second phase) may also have experienced leprosy, but this possibility is based on bone changes that are less clear (Boyle 2017; Caffell 2015).

Five sites in the east of England (one churchyard, two leprosaria, and two general hospitals) have revealed skeletons with leprosy-related bone alterations. Three individuals were diagnosed with leprosy at the non-leprosy hospital site of St Leonard's, Grantham, Lincolnshire, and two non-leprosarium sites in Ipswich, Suffolk, have revealed evidence of leprosy (Mays 1989 [Blackfriars]; Brown et al. forthcoming [Stoke Quay]). At the Blackfriars site, all four individuals who were diagnosed with leprosy were buried in the nave or the south aisle of the friary church, which suggests higher status (Figure 5.8). In the north of England and Scotland, there is greater evidence for leprosy than in previous periods, although again this cannot be translated into real frequency data. Of the nine sites, one was associated with a general hospital that was located near a river crossing and next to a medieval road (St Giles by Brompton Bridge in North Yorkshire). One individual at that site had clear signs of leprosy (late twelfth to late fourteenth centuries AD; Cardwell 1995, 214–218; Figure 5.9). Of particular interest in northern England are the three contemporary sites in York, all of which are "normal" cemeteries in the Fishergate area but contained people with leprosy. While leprosy hospitals are documented for York at this time, it appears that these people were buried with the rest of their community and not in leprosaria.

The late medieval period has produced the greatest number of individual skeletons excavated with leprosy. The sample is biased somewhat by the

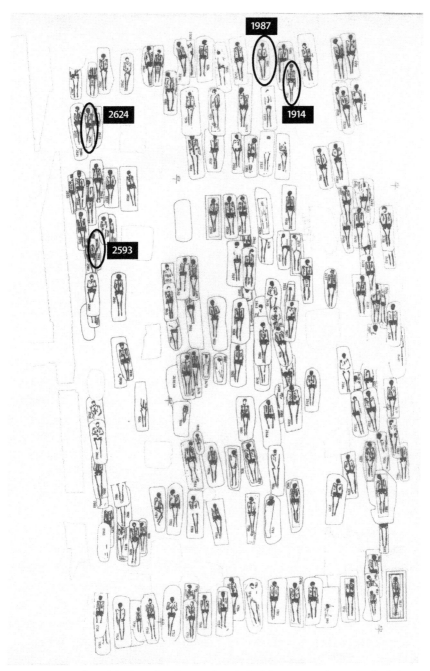

Figure 5.8. Site plan of the burials at late medieval Blackfriars, Ipswich, England. Skeletons 1914, 1987, 2593, 2642 are circled. Credit: Abby Antrobus, Suffolk Archaeological Unit.

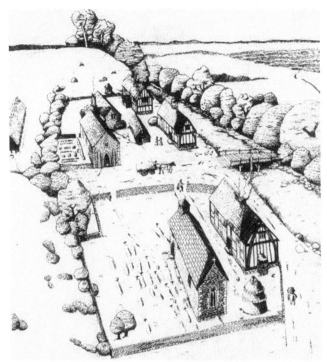

Figure 5.9. Reconstruction of the site of the medieval hospital of Brough St Giles, North Yorkshire, England. Credit: Reprinted by permission of Taylor & Francis Ltd, http://www.tandfonline.com, on behalf of the Royal Archaeological Institute.

sites at Chichester, Norwich, and Winchester. Twelve cemeteries have associated hospitals, seven of which were leprosaria, although most individuals were buried in non-leprosy hospital cemeteries and in a way that did not differ from the norm for the period and location. Most of the individuals whose skeletons have evidence of leprosy in this period lived before the fifteenth century. Manchester (1984, 1991) has suggested that leprosy declined beginning in the fourteenth century AD in Britain, and the evidence from skeletal remains shows that leprosy was not common in the post-medieval period (AD 1550–1850).

Evidence for leprosy in skeletal remains in England, Scotland, and Ireland, as described here and previously (Roberts 2002), suggests that the disease appeared in these locations as early as the fifth or sixth century (or even earlier if the prehistoric evidence from England and Scotland is accepted) and persisted to the nineteenth century. In Scotland, leprosy became more frequent during the time that it was declining in England (Browne 1975b). It persisted in Scotland to some extent, especially on the Shetland Islands in the parish of Walls and on the island of Papa Stour, west of the main island (Browne 1975b). Papa Stour was the place that all people with leprosy in the western districts of the Shetland Islands were sent in the early eighteenth century. Documentary evidence describes monetary collections around the islands for the people segregated there (Tait 1939). Fieldwork has identified possible huts used by people with leprosy on the Hill of Fielie, Brei Holm, on Papa Stour (Brady 2002; Smith et al. 2004). The last person with recognized endemic leprosy in Britain was probably 28-year-old John Berns of these Islands, who was admitted to the Royal Infirmary in Edinburgh in 1798, long after leprosy had disappeared from the Scottish mainland (Richards 1977, 105; Browne 1975b).

Robert the Bruce, king of Scotland (AD 1274–1329), was historically described as having leprosy, although his skeleton has never been examined for the evidence. However, when his skeleton was exhumed in 1918, a plaster cast was made of his skull; the Anatomical Museum at the University of Edinburgh now holds it (Kaufman and MacLennan 2000). Møller-Christensen and Inkster (1965, 20), who examined the skull, concluded that it showed "clear signs of *facies leprosa*, but to be 100 percent sure of the diagnosis of leprosy, we would have to unearth his skeleton once more and make a proper examination." However, Brothwell (1958, 289) was unsure: "It is difficult to say whether [the facial changes are] due to bad casting or [are] pathological." Kaufman and MacLennan (2000), who evaluated both the plaster cast and the documentary evidence associated with Robert the Bruce, concluded that he did not have leprosy. The features in the cast are indeed ambiguous (Figure 5.10).

In summary, the majority of skeletons with bone changes of leprosy in Britain and Ireland have been excavated from graveyards consistent with normal burial for the time and place. Twelve late medieval hospital cemeteries have been excavated, but only seven are known to have been serving leprosy hospitals. Relatively few of the small number of leprosy hospital cemeteries that have been excavated show widespread evidence of leprosy in skeletal remains (Satchell 1998). There are of course exceptions, such as at Chichester (Magilton, Lee, and Boylston 2008), Norwich (Anderson 2009a) and Winchester (Roffey and Tucker 2012). This evidence has implications for anyone who is analyzing how people with leprosy were treated in Britain and Ireland in the past. However, it is clear that focusing on leprosy hospital cemeteries will not detect all those who experienced leprosy.

While leprosy is still recorded in Britain today, it is usually seen in people arriving from countries where the infection is present (Figure 5.11). Browne (1975b) suggests that after the late medieval period, leprosy was not seen again until the nineteenth century. This is borne out in the skeletal evidence. Bhoyroo (1997) emphasizes that contact with British colonies and

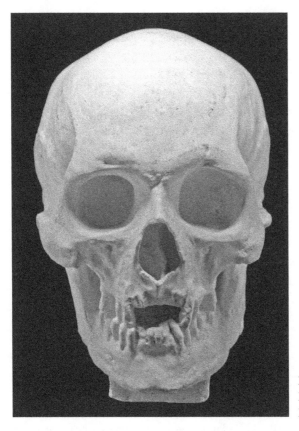

Figure 5.10. Plaster cast of Robert the Bruce's skull. Photo by Jeff Veitch.

the more widespread movement of people in the nineteenth century may have meant that leprosy in Britain became more common until treatment became available in the mid-twentieth century. By the 1860s, leprosy had become an increasing problem in those British colonies. At the start of the twentieth century, no leprosy hospitals, no beds for people with leprosy, and no leprosy register existed in Britain. However, while leprosy in Britain was reported in medical journals every year in the late nineteenth and early twentieth centuries, no numbers for people with leprosy were recorded formally until 1937. The government was probably not very aware of its frequency. It was not until 1951 that legislation included leprosy in the list of notifiable diseases (Bhoyroo 1997). Today there are very few references to people with leprosy in Britain apart from the study of Gill and colleagues (2005). From 1946 to 2003, fifty patients with leprosy were documented at the Liverpool School of Hygiene and Tropical Medicine. Nearly two-thirds had an origin in the Indian subcontinent. While leprosy is not common in Britain and Ireland today, the diversity of destinations of people traveling

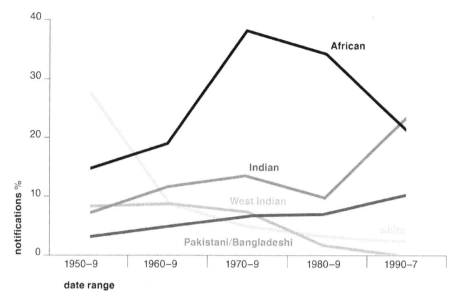

Figure 5.11. Notifications of leprosy by ethnicity in England and Wales in the period 1950–1997. Credit: Redrawn by Chris Unwin after Van Buynder et al. (1999).

for holidays and business around the world and rapidly increasing migration naturally provide leprosy with a possible route of transmission into these islands.

Croatia

In Croatia, evidence of leprosy is alluded to in two publications by Šlaus (2006a, 2006b). In addition, Watson et al. (2009, 2010) analyzed four bone samples for ancient *M. leprae* DNA from an individual dated to the eighth–ninth centuries with bone changes of leprosy from Radašinovci near Zadar, in the hinterland of Dalmatia; two samples proved positive. Šlaus (2010) reports more widely on the excavations at this site, indicating that two male and two female adults had bone changes of leprosy. None were buried differently from the rest of the ninety-two people in the cemetery, although the two women were buried in the same grave. More recently, Bedić et al. (2019) described the earliest evidence of leprosy in skeletal remains in Croatia to be the tenth and eleventh centuries AD (but see Watson et al. [2009, 2010]). They document two individuals with leprosy-related alterations buried at the site of Bijelo Brdo on the mainland of Croatia. Nevertheless, to date, we can conclude that there is little evidence of leprosy in Croatia in the published literature, perhaps reflecting the more recent development of palaeopathology there (Šlaus, Novak, and Vodanović 2011). However, Šlaus (2010) notes that "other sites" have reported evidence of leprosy and more are currently being excavated in Croatia that have already revealed skeletons with the infection. Leprosy is not common in Croatia today; only twenty people there were diagnosed with leprosy in the twentieth century. The last leprosarium closed in 1925 and the last indigenous "case" of leprosy in Croatia was in 1956 (Wokaunn, Jurćc, and Vrbica 2006; Bakija-Konsuo and Mulić 2011).

The Czech Republic

Other Northern European countries do not have nearly as much skeletal evidence for leprosy as England (or Denmark or Hungary). In 1993, the federal state of Czechoslovakia was dissolved and split into the Czech Republic and Slovakia. While there is skeletal evidence of leprosy in the Czech Republic, it was apparently less affected by leprosy than other European countries, although no reason for this has been suggested (Štěpán 1963). The most important medieval leprosy hospitals in the Czech Republic were in the northwest part of Prague and the southeast part of Brno (Dokládal 2002). In Prague, people with leprosy lived in the area of the original Romanesque church of St. Linhartus, which was established in the twelfth and thirteenth centuries. This was the first "asylum" for people with leprosy in Czechoslovakia. In the second half of the thirteenth century, the church was redeveloped and some affected individuals were moved to a new site outside the city walls, a leprosy hospital next to the chapel of St. Lazarus (ibid.). This was called the *domus leprosorum* (leprosy hospital). Both the chapel and the hospital were located on a busy road, giving those with leprosy the opportunity to beg from passersby. Following Emperor Charles IV's founding of Prague New Town in AD 1348, the two structures were incorporated into an area of newly erected houses (Strouhal et al. 2002). After that, it became more difficult to segregate people with leprosy. Two leprosy hospitals were sited in Brno. One, called Capella Leprosorum, was founded in the late thirteenth century on Krenova Street. The second, part of the hospital of St. Stephen's Church, was established in the first half of the fourteenth century outside the city walls (Dokládal 2002). In the late fourteenth century, the two Brno leprosaria were combined into the first town hospital (Zapletal 1952). Other leprosy hospitals were founded in the Czech Republic in the thirteenth and fourteenth centuries (ibid.), most by the French Order of St Lazarus, but by the sixteenth century leprosy is not described in documentary sources.

The first evidence for leprosy in skeletal remains comes from the fortified site of Prusanky, Prague, which dates to the ninth–tenth centuries. An individual 12–14 years old (grave 188) had bone changes to the facial area accompanied by cribra orbitalia. Ancient DNA analysis of four bone samples from different sites on the skeleton provided additional support for a diagnosis of leprosy (Strouhal et al. 2002; see also Taylor and Donoghue 2011). The greatest concentration of *M. leprae* DNA was in the fibula, followed by the nasal area, the radial epiphysis, and a rib. However, it is notable that this skeletal evidence is much earlier than the reference to leprosy's association with the Church of St. Leonard, Prague (1280) (Likovsky et al. 2006). A young adult from an ossuary at Křtiny, near Brno in Moravia, is also recorded (thirteenth–eighteenth centuries AD). Ancient DNA analysis of a bone sample from the nasal cavity of this skeleton was positive for *M. leprae*. In a similar study, Likovsky et al. (2006) found aDNA of *M. leprae* in two individuals from the twelfth-century Romanesque church cemetery of Chelčického nám at Žatec, in northwest Bohemia. One was 14–18 years of age (skeleton 973) and the other was a probable female adult (skeleton 9611; Likovsky et al. 2006). Evidence of leprosy has recently been reported from modern Slovakia, where a medieval female skeleton is recorded to have changes to the bones of the face and the feet (Hukelova 2018).

Denmark

Leprosy was present in the Nordic countries until the middle of the twentieth century. Ten thousand people with leprosy were recorded from Finland, Iceland, Norway, and Sweden in the period 1856 to 1956 (Richards 1977). This region was the "last part of Europe to lose endemic leprosy" (Ell 1988, 504). Even today, Denmark experiences leprosy through migrants from other parts of the world (Jensen et al. 1992). Leprosy could have come to Scandinavia from the west, the east, or the south (Richards 1960), maybe with the Vikings, first to Denmark and then to Sweden and Finland. The Vikings also visited the British Isles in the tenth through twelfth centuries (see Taylor et al. 2018 in relation to Ireland). In addition, recent aDNA evidence from skeletal remains excavated in Sweden indicates a strain of leprosy that may have derived from Asia (Economou et al. 2013a, 2013b). In Denmark, the first historical reference to leprosy is in the eleventh and twelfth centuries; the source refers to a woman from Norway (Richards 1977). There is much evidence for leprosy in skeletal remains in Denmark, where the earliest data dates to the late medieval period.

Møller-Christensen published widely on skeletal remains recovered from the late medieval leprosarium of St. Jørgen at Næstved located southwest of Copenhagen on the island of Zealand (1953a–c, 1967, 1969, 1978; see also Møller-Christensen and Faber 1952). Møller-Christensen also examined skeletons for leprosy from Æbelholt in the north of Zealand, a general hospital site; Åkirkeby on the island of Bornholm in the Baltic Sea, east of the rest of Denmark; Spejlsby, on the island of Møn in southeast Denmark; and Svendborg, on the island of Funen, west of Zealand. Åkirkeby and Spejlsby had fragmentary skeletal remains and Svendborg had poorly preserved remains (Møller-Christensen 1963). Thus, Næstved provided the key data for his analysis (Segal 2001). Leprologist Johs Andersen also used Møller-Christensen's work to make clinical correlations with bone changes in living people with lepromatous leprosy (cured and uncured) at the Purulia leprosy home and hospital in West Bengal, India (Andersen 1969).

Like the site of St James and St Mary Magdalene at Chichester, England, St. Jørgen (AD 1350–1550) is an extremely important site in the history of leprosy. Six hundred fifty skeletons were excavated there in the period 1948–1968. Eighty percent of the burials represented young and middle-aged adults. People up to 14 years old represented 7.7 percent of the population sample, and juveniles (14–20 years of age) made up 10.4 percent. Of the sexed individuals, 223 (50.2 percent) were men and 221 (49.8 percent) were women. In general, women with leprosy died younger than men (31.5 vs. 36.7 years) and men with no bone changes died slightly older (38.2 vs. 31.0 years; see also Blondiaux et al. 2016, who found a similar picture in medieval France).

Figures 5.12, 5.13, and 5.14 show line drawings of variations in expression of some of the bone changes identified at Næstved that Møller-Christensen identified. He also published individual "case" studies of skeletons from Næstved, notably of a woman with leprosy who had associated calcified cysts in her abdominal cavity. Weiss and Møller-Christensen (1971a) interpreted these cysts as signs of a parasitic infection caused by the tapeworm *Echinococcus granulosus*. Møller-Christensen also noted skeletal changes of both leprosy and tuberculosis in a skeleton from Næstved (Weiss and Møller-Christensen 1971b), evidence that is relevant to discussions elsewhere in this book about the decline of leprosy. Møller-Christensen's research has provided a legacy for scholars working on the bioarchaeology of leprosy, and the Næstved skeletons, which were originally stored at the Leprosy Museum Møller-Christensen created, are now housed at the University of Copenhagen for future research (Niels Lynnerup, pers. comm., August 2012).

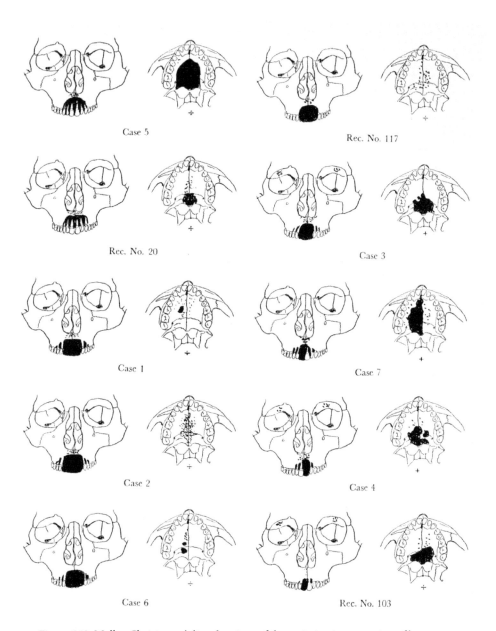

Figure 5.12. Møller-Christensen's line drawings of the variation in expression of bone changes of leprosy affecting the alveolar process of the maxilla and the oral surface of the palate based on skeletons from St. Jørgen's hospital, Naestved, Denmark. Source: Møller-Christensen (1953c, figure 94).

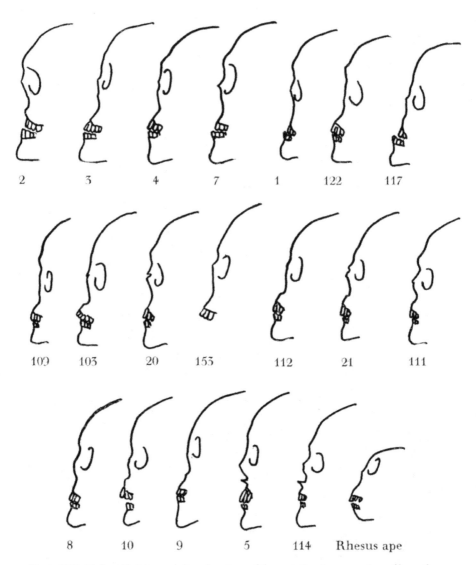

Figure 5.13. Møller-Christensen's line drawings of the variation in expression of loss of the anterior nasal spine based on skeletons from St. Jørgen's hospital, Naestved, Denmark. Source: Møller-Christensen (1953c, figure 98).

In more recent years, other authors have identified evidence for leprosy in Denmark. Segal (2001) studied skeletons from St. Jørgensgård, a medieval leprosy hospital in Odense on the island of Funen that dates to AD 1270–1542. The site was located outside the old eastern city gate along St. Jørgens Road. A church, a cemetery, and the remains of dwellings, wells, and workshops were excavated. People were buried supine (on their backs)

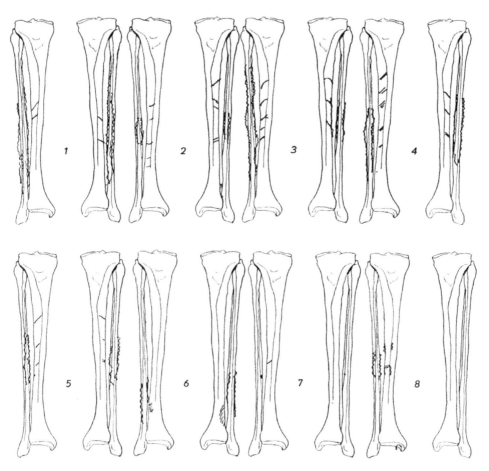

Figure 5.14. Møller-Christensen's line drawings of the variation in expression of bone formation on the lower leg bones based on skeletons from St. Jørgen's hospital, Naestved, Denmark. Source: Møller-Christensen (1953c, figure 106).

with their heads to the west. There was some evidence for coffins (108 burials) and a few grave goods (coins, potsherds, and pieces of metal) in grave fills of half of the burials. The skeletal remains of 485 people were studied (of the total of 924 that were excavated); 44 percent were men and 39 percent were women. Most were in the age class of 21–35 years old. Fifty-eight percent of people had some signs of leprosy; women were more affected. One hundred and seven (39 percent) of 277 individuals with preserved skulls had leprous bone changes, and 29 (9 percent) of 320 with preserved hand bones had affected metacarpals or phalanges. Seventy-six percent (22) of the 29 had foot bone changes too, and 48 percent of that 22 had associated facial bone lesions. Fifty-three percent of 341 people with preserved metatarsals or phalanges had bone changes. One hundred and twenty-five (69 percent) of those with foot bones affected had alterations in their lower leg bones. Fifty-five percent of 371 people (202) with at least one tibia or fibula preserved had periosteal lesions on the lower leg bones. In this study, cribra orbitalia was found to be associated with people with leprosy, as was also seen at Næstved.

Segal (2001) compared the findings from the St. Jørgensgård leprosy hospital cemetery to data collected from skeletons buried at the late medieval nonhospital sites of Västerhus in Sweden (Gejvall 1960), Odense (Boldsen and Mollerup 2006) and Æbelholt and Øm (Aarhus County, Jutland) in Denmark (Møller-Christensen 1952, 1958; Isager 1936), and the leprosy hospital sites of Næstved and Svendborg (Møller-Christensen, 1953a–c, 1967, 1978), and St James and St Mary Magdalene in Chichester, England (data from Lee and Magilton 1989). Compared to Næstved, St. Jørgensgård had fewer people with facial bone changes but more with hand lesions. Only one individual each from Æbelholt and Øm showed any bone change of leprosy (Møller-Christensen 1952). Segal (2001) suggests that the lower rates seen in her study may reflect the more stringent diagnostic criteria used to segregate people with leprosy in Odense, Funen, the location of her study. An individual (G483) from a late medieval leprosarium in Odense (assumed to be from St. Jørgensgård but the authors do not specify) was also the subject of an aDNA study that gave a positive result for *M. leprae* (Watson et al. 2009, 2010; Monot et al. 2009; Schuenemann et al. 2013). Brander and Lynnerup (2002) conducted the only other study in Denmark of skeletons associated with a possible leprosarium. The site, which was located at Stubbekøbing on the island of Falster, South Zealand, dates to AD 1481–1505. Four individuals (of a total of sixty) at this site had some leprosy-related bone changes, but they were too nonspecific to be definitely related to leprosy.

Boldsen (2005; see also Boldsen 2009) has considered the relationship between leprosy and the risk of dying in people who were buried in the church cemetery of Tirup (near Horshens, eastern Jutland) in the midtwelfth to mid-fourteenth centuries. Six hundred and twenty skeletons were excavated, representing the whole of the cemetery (which was demarcated by a boundary ditch; Kieffer-Olsen 1993). To diagnose leprosy, Boldsen used seven "leprosy-related skeletal lesions" of specific bones or parts of bones from 345 individuals (Boldsen and Mollerup 2006, 345;

Boldsen 2001). Scores were transformed into a statistic λ, which was used to show the likelihood that a person had experienced leprosy. On the basis of the findings, Boldsen diagnosed leprosy in 26 percent of the adults. He also found an association between dental enamel hypoplasia and leprosy, suggesting that stress predisposed people to leprosy later in life (Barker's [1994] developmental origins hypothesis), but there was no correlation between cribra orbitalia, another early life stress marker, and leprosy. The lack of an association between cribra orbitalia and leprosy at Tirup and the presence of a high association at Næstved suggest that the cribra orbitalia at Næstved may have been caused by something other than leprosy. An unpublished study by DeWitte (2015) of skeletons excavated from two Danish cemeteries from the twelfth and thirteenth centuries found no significant associations between any indicators of childhood stress (e.g., dental enamel hypoplasia) and having two or more signs of leprosy. However, it did find a significant association between periosteal lesions of the tibia and leprosy. The study concluded that "childhood stress is not predictive of developing lepromatous leprosy . . . but . . . stress that can occur later in life might be. However, several of the leprosy-lesions have low specificity and thus the co-occurrence of periosteal lesions and signs of leprosy might reflect other conditions rather than indicating that previous physiological stress increases the risk of lepromatous leprosy lesions" (DeWitte 2015). A related study by Kelmelis, Price, and Wood (2017) focused on the effect of leprosy on risk of death in 311 medieval adult skeletons from rural Øm Kloster in northern Jutland (AD 1172–1536). Kelmelis and colleagues concluded that the "mortality hazard" of people with lesions related to leprosy was greater than it was for those who did not have such lesions, suggesting that having leprosy increased the risk of dying.

Leprosy has also been recorded in people buried in non-leprosarium cemeteries in Odense that date from AD 1000 to the early nineteenth century (Boldsen and Mollerup 2006; and Segal 2001). The cemetery samples came from the Greyfriars monastery (Odense Gråbrødre, where 243 of 543 skeletons were studied), the Blackfriars monastery (Odense Sorbrødre, where 372 of 624 skeletons were studied), the St. Albani parish churchyard (where 152 of 398 skeletons were studied), and the St. Knud Cathedral (where 66 of 193 skeletons were studied). Using the same diagnostic criteria and a Bayesian approach (see Boldsen 2005), Boldsen and Mollerup (2006) found an average of 3.7 leprosy-related lesions in the Greyfriars monastery, 4.3 at Blackfriars, 3.6 at St. Albani, and 3.6 at St. Knud. The two cemetery samples of skeletons from the early part of the later medieval period (St. Albani and St. Knud) had a higher prevalence of leprosy (13–17 percent) than the later sites (after AD 1400), which were mostly post-medieval.

Boldsen and Mollerup also found a much lower than expected prevalence of individuals with the facial lesions of leprosy. They suggest that people with "predominantly facial signs of the disease were selectively removed from the population" (2006, 349), as Segal (2001) has described. Andersen (1969) noted that in the medieval period, diagnosis of leprosy focused on facial changes, which he suggested must be the reason for the absence of people with facial bone damage. However, exploring this further is challenging for a bioarchaeologist because of the difficulty of interpreting what a person looked like to their communities when all that remains is their skeleton. While concentrating on the facial appearance in leprosy was practiced in late medieval Europe, this does not mean that the practice always occurred or that all people with facial bone damage had a visually recognizable "leprosy face." It is true that the skeletal remains from the St. Jørgensgård leprosarium site in Odense had four times more people with facial bone changes than individuals buried at the village churchyard cemetery of Tirup, and therefore Boldsen and Mollerup's (2006) hypothesis may be correct but cannot be proved. They hypothesize that "skeletal lesions of leprosy in the Middle Ages were probably different from manifestations in modern-day people" (although see Andersen 1969, 101) and that "the evolutionary dynamics of infectious diseases makes individual diagnosis at the very best very difficult, and in reality unachievable" (349). However, it can be argued that when a complete skeleton is analyzed and the suite of accepted bone changes of leprosy are present, a diagnosis of leprosy can be made. Boldsen (2009), who considered evidence for leprosy in over 3,000 skeletons from eleven medieval Danish cemeteries and 99 skeletons from the Swedish site of Västerhus, again observed his seven skeletal changes of leprosy. He found that leprosy declined earlier in urban communities than in rural communities and suggested that this was because of segregation in leprosaria in urban situations. Again, if this is correct, then we can assume that leprosaria contributed to the decline of leprosy, but much data relating to a lack of segregation suggests that this is probably unlikely.

As in England, Denmark opened many leprosy hospitals in the medieval period and named many after St. George (who represents triumph over the dragon or overcoming odds; Segal 2001, 6; see also Richards 1960 for Norway; and Figure 5.15). Andersen (1969, 89) provides a useful list of thirty-eight leprosy hospitals in Denmark that were founded from the

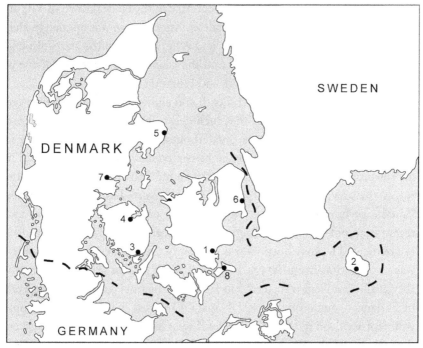

Figure 5.15. Locations of the Danish sites that have revealed evidence for leprosy. Credit: Figure by Yvonne Beadnell using data from Boldsen (2009), Matos (2009), Møller-Christensen (1978), and Segal (2001).

twelfth to the sixteenth centuries AD. Overall, the evidence of leprosy in the late medieval period of Denmark is plentiful and convincing and Denmark has extensive evidence of the founding of leprosy hospitals, as is the case in England and Ireland. The reported skeletal evidence is undoubtedly attributable to Møller-Christensen and Andersen's original scholarly work (see Bennike 2012).

Finland

There is no published evidence of skeletal leprosy in Finland. Environmental factors that lead to the poor preservation of remains probably account for the lack of evidence (Vuorinen, pers. comm., 2013), but the more recent development of paleopathological study in Finland is also likely a contributory factor (Núñez et al. 2011). However, much historical documentation indicates that leprosy was present (see Richards 1977). The first historical sources recount a leprosy hospital dedicated to St. George in Turku on the southwest coast in AD 1355 (Vuorinen 2002). In 1455, another hospital was founded in the southeast part of the country in the coastal town of Viipuri that was dedicated to St Mary Magdalene. Another one was opened in Helsinki in the mid-sixteenth century, and others dated to the sixteenth and seventeenth centuries, for example on the island of Gloskär (1651), one of the Åland Islands between Sweden and Finland (Richards 1977).

Leprosy hospitals in the seventeenth and eighteenth centuries consisted of several wooden buildings that included a dwelling house, a sauna, a church, and a barn. They were situated outside towns, as was the case in many other European countries. Leprosy frequency peaked in Finland in the seventeenth century, and during the eighteenth and nineteenth centuries all the surviving leprosy hospitals were converted to what were termed "lunatic asylums" (Vuorinen 2002). Reliable and accurate diagnosis and registration of leprosy began in the late nineteenth century in Finland. The last dedicated leprosy hospital was located in Orivesi in central Finland; it was in operation from 1904 to 1953 (ibid.). The last three people with leprosy at Orivesi were transferred to Helsinki; the final diagnoses of leprosy in Finland occurred in the late 1950s. Of interest is the fact that people with leprosy lived in the west and southwest of Finland almost exclusively through the centuries; virtually no evidence was reported for the eastern areas (ibid., Figure 1). Perhaps this reflects the spread of leprosy from the west, including from Norway and Sweden. Most leprosy hospitals were also located on the coast or on islands; only four were located inland, and the majority of them date to the nineteenth and twentieth centuries.

Germany

Surprisingly few skeletons have been diagnosed with leprosy in Germany, even though Galen reported leprosy there in his writings in the second century AD (Richards 1960, 105). This is even more surprising considering the long history in Germany of interest in palaeopathology (Teschler-Nicola and Grupe 2012). The first evidence, which was published in the 1970s, related to the excavation of part of a thirteenth- to sixteenth-century leprosy hospital cemetery in Melaten, Aachen, in western Germany in the 1960s and 1970s. Bone changes of leprosy affected a large majority of the forty-one skeletons excavated at this site (Schmitz-Cliever 1971, 1972, 1973a, 1973b). The skeletal remains from Melaten are reported to be now curated in the Institute of Anatomy at the University of Aachen. The Melaten site was located in an isolated area outside the city walls. Schmitz-Cliever (1971) noted that it very likely represented a large leprosarium that probably contained more than 900 skeletons, assuming a use of 320 years and three burials a year. In 1971, he reported that a one-meter-wide test trench had been excavated east of the chapel's apse to explore whether skeletons were buried at the site and whether any skeletons excavated had bone changes of leprosy. He found a total of ten skeletons, eight of which were buried in the medieval tradition in an east-west orientation with their heads to the west.

A more extensive excavation in 1972 revealed a total of forty-one skeletons (Schmitz-Cliever 1973b). Each was buried in an individual pit in a coffin and was oriented west to east with hands crossed at the pelvis. Schmitz-Cliever used Møller-Christensen's diagnostic criteria for the bones of the face, hands, feet, and lower legs to diagnose leprosy. Although the burial conditions did not preserve the facial bones of the skeletons well, 11 of 25 skulls with preserved facial bones (44 percent) had bone changes, 13 of 21 individuals with preserved tibiae and fibulae (62 percent) were affected, 16 of 26 with preserved foot bones (62 percent) were affected, and 19 of 23 with preserved hand bones (83 percent) were affected. Only three of the forty-one individuals did not have bone changes of leprosy, which suggested to Schmitz-Cliever (1973b) that diagnosis was accurate in the medieval period and that it led to segregation. Eighty-two percent of the forty-one individuals had lesions on the bones of the lower legs and feet, and the hand bones of two-thirds of that group were affected. Less than half of the 82 percent had associated changes to the facial bones. It is surprising that so many hand bones were affected, considering the relative lack of evidence in skeletons from other burial contexts described in Chapter 4.

One of the burials (113) is described more fully than the others (Schmitz-Cliever 1973a). A man with apparent leprosy was buried with two St. James perforated scallop shells (see also Roffey and Tucker 2012; Roffey et al. 2017). These are believed to identify the person as a Santiago de Compostela pilgrim who traveled to the capital of Galicia in northwest Spain.[4] This is a pilgrimage equal in importance to the hajj to Mecca in Saudi Arabia. The purpose of the pilgrimage was, among other things, to pray for good health, which may be the reason this man went to Spain. The pilgrim probably acquired the shells at the well in Santiago de Compostela and they were likely blessed before he brought them back; they were seen as a symbol of love and an amulet against illness and the evil eye. They were attached to the clothes and hats of pilgrims, as seen in illustrations of St James in which his attire is adorned with scallop shells. The earliest such illustration is from the twelfth century AD (Schmitz-Cliever 1973a; see also Hohler 1957). However, the evidence described for leprosy is not convincing and no skull was preserved for this skeleton. The changes to the bone that were accepted as relating to leprosy were to the distal left fibula and a fifth metatarsal (both of which exhibited periosteal new bone) and destructive lesions on the proximal joint surface of a first proximal foot phalanx (probably osteochondritis dissecans; see Chapter 4).

Leprosy has also been described in skeletons from the early medieval site of Lauchheim in southern Germany, which dates from the second half of the fifth century AD to circa 680 (Boldsen 2008). Although 1,308 burials were excavated, only 110 were analyzed for evidence of leprosy (see details of sample selection in Boldsen and Mollerup 2006). Sixteen percent of adults in this sample had bone changes of leprosy. Using the method he had implemented for other studies he published, Boldsen explored evidence for leprosy in skeletons from five sites located in Germany and Austria that dated from the twelfth through thirteenth centuries. He found a leprosy frequency of between 9 and 44 percent and that women were at a higher risk of death if they had leprosy. In four of the five sites, men with leprosy died at older ages than those without leprosy (compare this result to that of Kelmelis, Price, and Wood 2017, which showed the opposite result). Haas et al. (2000, 2002) has also provided more recently dated evidence for leprosy in Germany. They analyzed samples of the palate bones from two female skulls with bone changes of leprosy from an ossuary at Rain on the River Lech in Bavaria, West Germany (AD 1400–1800). The ossuary holds remains that represent over 2,500 people who had been originally buried in an urban cemetery. The town also had a leprosy hospital that functioned from AD 1481 to 1632. Haas and colleagues found positive results for aDNA of *M. leprae*. Nerlich, Marlow, and Zink (2006), who also analyzed bone samples from skeletons buried in this ossuary, found 59 bones with changes related to inflammation, 45 percent (24) of which had preserved pathogenic aDNA. Ten of these samples had preserved aDNA of the *M. tuberculosis* complex and five had the aDNA of *M. leprae*. One skeleton had coinfection of leprosy and tuberculosis.

Hungary

Medical historical data show that people in Hungary were aware of leprosy by the eleventh century AD (Fekete 1874), although there is evidence of the infection in skeletons from earlier periods (Pálfi et al. 2002; Köhler et al. 2010; and see below). In the eleventh century, people with leprosy began to be segregated outside towns and cities and leprosaria were established. The infection became more common from the thirteenth to the fifteenth centuries and began to decline after its peak in the fourteenth century (Zubriczky 1924). Leprosy hospitals were founded at later dates in the north and east of Hungary. However, only one leprosarium cemetery has been excavated, which Pálfi et al. (2002) suggest may explain the lack of skeletal evidence for leprosy in Hungary (although see discussion of the British skeletal evidence, where the majority of the data derive from non-leprosy hospital cemeteries). Nevertheless, there is much more evidence of leprosy in Hungary than in the majority of modern-day countries in the Old World. Hungary has also produced the most data from ancient biomolecular diagnoses of leprosy of anywhere in the world. Table 5.1 lists the sites where leprosy has been found.

Table 5.1. Bioarchaeological evidence for leprosy in Hungary

Name of Site(s)	Time Period(s)	Source(s)
Abony-Turjányos-dűlő	3700–3600 BC	Köhler et al. (2010, 2017)
Kiskundorozsma-Daruhalom dűlő II, Szarvas-Grexa, Orosháza-Béke Tsz, and Kiskundorozsma-Kettőshatár	Seventh–ninth centuries AD	Lee et al. (2012); Marcsik et al. (2010); Molnár et al. (2006); Pálfi et al. (2012); Taylor and Donoghue (2011)
Bélmegyer-Csömöki	Eighth century AD	Molnár et al. (2015)
Sárrétudvari-Hízóföld	Tenth century AD	Haas et al. (2000); Pálfi et al. (2002); Pálfi (1991)
Püspökladány-Eperjes and Ibrány-Esbohalom	Tenth–eleventh centuries AD	Donoghue, Marcsik, et al. (2005); Donoghue et al. (2002); Donoghue, Holton, and Spigelman (2001); Marcsik (2003); Marcsik, Molnár, and Ősz (2007); Pálfi et al. (2002); Taylor and Donoghue (2011)
Csengele-Bogárhát and Lászlófalva-Szentkirály	Eleventh century AD	Fóthi et al. (2001); Marcsik (2001); Marcsik (2001); Pálfi, Pap, and Fóthi (2001); Pálfi et al. (2010)
Orosháza	Eleventh–thirteenth century centuries	Balázs et al. (2019)
Székesfehérvár basilica	Eleventh–twelfth centuries AD and eleventh–sixteenth centuries AD	Donoghue and Spigelman (2008); Donoghue, Marcsik, et al. (2005); Éry et al. (2008)
Gerla monastery	Fourteenth–fifteenth centuries AD	Farkas, Marcsik, and Szalai (1991)
Szombathely Ferences templom	Fifteenth century AD	Marcsik, Molnár, and Ősz (2007)

Palaeopathological studies have been a focus in Hungary since the nineteenth century (Pálfi, Marcsik, and Pap 2012) and have produced much evidence for health and disease for many years. Research suggests that the first evidence of leprosy in Hungary dates to the Copper Age (3780–3650 BC; Köhler et al. 2010; and see below). Five individuals with leprosy were identified in pits at the site of Abony-Turjányos dűlő in central Hungary. Ancient DNA analysis did not provide any positive results, but it was possible to detect mycolic acids in the cell walls of *M. leprae* and proteins specific to that bacterium, the latter for the first time in bioarchaeology (Hajdu et al. 2010). Köhler et al. (2017) report more fully the findings from this important site for the early history of leprosy. One of the burials (Skeleton 20, a young adult male) was diagnosed with definite leprosy morphologically. He also had a healed depressed fracture on the frontal bone of his skull (see similar evidence in skeletons from Africa and Italy below). Ancient DNA analysis was unsuccessful.

Later evidence comes from sites dated to the seventh–ninth centuries AD (Marcsik, Molnár, and Ősz 2007; Pálfi and Molnár 2009):

> Kiskundorozsma-Daruhalom dűlő II (south, near Szeged): seven people (four older adult men, two middle-aged adult men, and one older adult woman) with bone changes of leprosy (of 94 examined).
> Szarvas-Grexa (southeast): one young to middle-aged woman.
> Orosháza-Béke Tsz (in the south, 100 km east of Szeged): one middle-aged man.
> Kiskundorozsma-Kettőshatár (in the south near Szeged): three older men.

Ancient DNA analysis for *M. leprae* in nasal cavity bone samples from six individuals from the Kiskundorozsma-Daruhalom dűlő II site revealed *M. leprae* DNA for three; co-infection of leprosy and tuberculosis was also found in one person (Molnár et al. 2006; Pálfi et al. 2012). Later analysis of aDNA and lipids of leprosy showed evidence for *M. leprae* infection in people buried at Kiskundorozsma-Kettőshatár (Marcsik et al. 2010) and Kiskundorozsma-Daruhalom (Taylor and Donoghue 2011). Lee et al. (2012) also conducted lipid analysis on skeleton 517 from Kiskundorozsma-Daruhalom dűlő II (which dates to the late seventh century) that confirmed the presence of both leprosy and tuberculosis. All of these early medieval individuals were buried normally for the culture and the period (e.g., see Figure 5.16). However, skeleton 21 from Kiskundorozsma-Daruhalom dűlő II had leprosy and was buried with a Byzantine coin on the margin of the

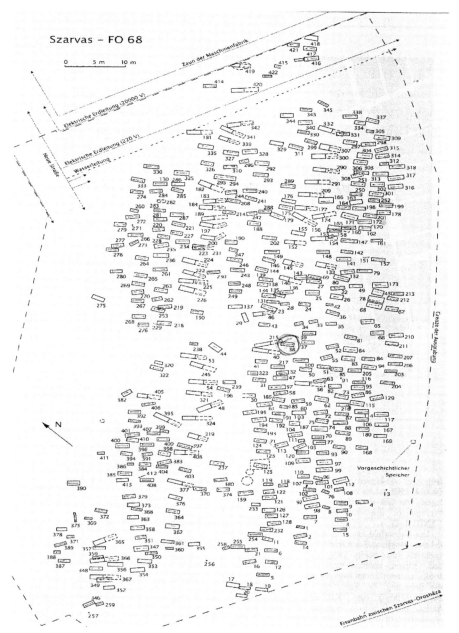

Figure 5.16. Plan of the Szarvas cemetery in Hungary. The burial of a person with leprosy is circled. Credit: Figure by Antonia Marcsik.

cemetery in one of the richest graves (Antonia Marcsik, pers. comm., July 2008). Molnár et al. (2015) further reports early data for leprosy from an eighth-century AD skeleton from Bélmegyer-Csömöki, where both *M. leprae* and *M. tuberculosis* aDNA were detected in a bone sample.

Pálfi et al. (2002) describe the extant skeletal evidence for leprosy in Hungary during the period of the Hungarian conquest in the tenth century AD. An elderly woman (S202) from the cemetery of Sárrétudvari-Hízóföld, near Debrecen in eastern Hungary, was affected with leprosy (Pálfi 1991). Ancient DNA analysis of a metatarsal showed *M. leprae*, confirming that the person had harbored the bacteria in life. A skeleton (S237) of a middle-aged adult male from the same cemetery showed less pathognomonic bone changes of leprosy in the feet (fusion of tarsal bones, bone formation around the joints). Ancient DNA analysis confirmed the presence of leprosy in his hard palate but not in a metatarsal sample. Haas et al. (2000), who analyzed metacarpals from these two skeletons and a bone sample from the hard palate of skeleton 237, found *M. leprae* DNA in skeleton 237 but not in 202.

There is leprosy evidence from four skeletons from the tenth–eleventh centuries from Püspökladány-Eperjes, near Debrecen (Marcsik, Molnár, and Ősz 2007; Pálfi et al. 2002). The nasal cavities of skeletons 11, 222, 429, and 503 were sampled for aDNA analysis; *M. leprae* DNA was found in 11 and 429 and both *M. leprae* and *M. tuberculosis* in 222 and 503 (Donoghue, Marcsik, et al. 2005; see also Donoghue et al. 2002; Taylor and Donoghue 2011). Donoghue, Holton, and Spigelman (2001), who focused on two samples from the nasal cavity of a skeleton with leprosy from the same site, found that both were positive for aDNA of *M. leprae*. Marcsik (2001) also identified bone changes of leprosy in the skull of an individual from Csengele-Bogárhát, an ossuary site near Szeged in southern Hungary (eleventh–thirteenth centuries AD; most likely the eleventh). Fóthi et al. (2001) and Pálfi, Pap, and Fóthi (2001) further described bone changes in the skull of a mature adult man from Lászlófalva-Szentkirály, 100 kilometers north of Szeged near Kecskemét (eleventh century, skeleton 79), and Farkas, Marcsik, and Szalai (1991) identified such bone changes in a skull from a mature adult man from the Gerla monastery in southeastern Hungary (twelfth century, skeleton 7). Analysis of the skull from Lászlófalva-Szentkirály has more recently revealed a co-infection of leprosy and tuberculosis (Pálfi et al. 2010). All the skeletons from all the sites dated to the tenth–eleventh centuries were buried normally for the period and the culture (Antonia Marcsik, pers. comm., July 2008). Marcsik (1998) also sampled a skeleton from the site of the Ópusztaszer monastery in southern Hungary near Szeged for aDNA analysis (fourteenth and fifteenth centuries; Marcsik 1998: grave 923) and skeletons 222 and 503 from Püspökladány-Eperjes. *M. leprae* DNA was present for 222 and 503 but not for 923 (Donoghue et al. 2002). Donoghue, Holton, and Spigelman's (2001) study of fibula and metatarsal samples from an individual with leprosy from the same site did not reveal any *M. leprae* DNA. Finally, Marcsik (2003) reports one individual with leprosy (258; male 35–39 years old) from the cemetery site of Ibrány-Esbohalom (tenth–eleventh centuries).

Later evidence comes from a variety of sites. For example, an eleventh–thirteenth century Muslim burial ground at Orosháza in southeast Hungary revealed a young woman with leprosy affecting the facial bones (Balázs et al. 2019). She had also sustained an injury to her cranium, but like so many burials of people in the past, she was buried among the rest of her community (see also sections in this chapter on Africa and Italy). In addition, a churchyard associated with a fifteenth-century leprosarium revealed skeletons with evidence of leprosy (Szombathely Ferences Templom, western Hungary). This is the only site in Hungary where skeletons have been found in association with a leprosarium (Antonia Marcsik, pers. comm., July 2008). Marcsik, Molnár, and Ősz (2007) identified five skeletons with leprosy (skeleton 3, 8–10 years old; skeleton 6, a young adult woman; skeleton 10, a middle-aged woman; skeleton 19, a possibly male adult; and skeleton 20, a young adult female). Individual 10 was subject to aDNA analysis, as were some individuals with no bone changes of leprosy. Skeleton 10 was positive for *M. leprae* (Donoghue, Marcsik, et al. 2005). Székesfehérvár-Basilica, a site in western Hungary that dates to the eleventh–fourteenth or sixteenth centuries, revealed three adult male individuals with leprosy (II/36, fifteenth–sixteenth century; VI/27, eleventh–twelfth century; IX/1, eleventh–sixteenth centuries). Ancient DNA analysis of two skeletons (VI/27 and IX/1) showed the presence of leprosy (Éry et al. 2008; Donoghue and Spigelman 2008). Donoghue, Marcsik, and colleagues (2005) also revealed leprosy aDNA in Skeleton 89 from this site following the analysis of individuals 79a and 89. Finally, Marcsik, Fóthi, and Donoghue (2006) conducted aDNA analysis of three eighteenth- to nineteenth-century skulls with bone changes suggestive of either leprosy or treponemal disease but did not detect any evidence of leprosy DNA. These were skulls of people originally from the Russian Khanty population in the region of the Ob River in western Siberia. This suggests either poor preservation of aDNA or that the individuals had treponematosis, not leprosy.

Clearly, Hungary has provided a wealth of macroscopic and biomolecular evidence for leprosy dating from prehistory through to the nineteenth century AD. These findings illustrate the long history of leprosy in this region. Pálfi et al. (2002, 208) suggest that leprosy arrived in Hungary in the Carpathian Basin with "ancient Hungarians from the East, around the ninth century," but there is clearly earlier evidence now (Köhler et al. 2010).

Iceland

Considering the abundance of documentary evidence for leprosy in some Scandinavian countries, the movement of people in the North Atlantic over time, and the survival of leprosy into the twentieth century in Norway, it is surprising that there is virtually no skeletal evidence of leprosy in Iceland. Although palaeopathology as a discipline has been a relatively recent development, all the extant skeletal remains excavated in Iceland have already been studied (Zoëga and Gestsdóttir 2011). In any case, there are relatively few remains that have been excavated anyway from this country. To date, only one individual has been identified with leprosy, a middle-aged woman of pre-Christian date (before AD 1000; probably Viking Age) from a single burial at Bessastaðir in Skagafjörður, northern Iceland (National Museum of Iceland, code BSS-A-1). This was the only burial at the site, which is consistent for the Viking Age when single or small groups of burials were the norm (Hildur Gestsdóttir, pers. comm., January 2013; Magnússon 1970).

Ehlers (1895, 155–156) perhaps explains the current lack of evidence. He suggested that leprosy was not present in the Icelandic population until the twelfth century, that it was brought there by people from Norway, but that the proportion of the population that was affected increased in the sixteenth century. There is plentiful documentary evidence for leprosy in Iceland in more recent centuries. In 1651, a royal edict pronounced that leprosy hospitals would be built in the four corners of the island to accommodate an increasing number of people with the infection. The hospitals were founded in the west at Hallbjarnareyra (Snefjell district), in the north at Madrefell (Örford district), in the east at Horgsland, and in the south at Klösterhole in Grimsnes. Although there is evidence to show that living conditions at the hospitals were very poor (Ehlers 1895), the island's fishermen were required to give up part of their catch for the hospitals periodically and the bishops and superintendents of each county had to provide patients with what they needed to survive. In 1707, many people with leprosy died during a smallpox epidemic on Iceland, as was the case during the measles epidemic of 1846. The southwest coast of the country had the highest frequencies of leprosy, along with other epidemic diseases. This was the area where the greatest density of people lived, where there was much poverty, and where many people from other countries gained access to Iceland. While immigration, other diseases that compromise the immune system, and high population density are all potential predisposing factors for leprosy occurrence and transmission, Ehlers (1895) also discusses diet as a risk factor, which included dried fish, sour milk, rancid butter or oil, mutton, and potatoes. Even though Ehlers was writing after the leprosy bacillus had been discovered in 1873, people's perceptions of leprosy and how it was contracted remained very varied, especially in more remote areas of the globe.

Although Richards (1977, 93) notes that people were not afraid of people with leprosy by the early nineteenth century, Iceland was the only country in northern Europe to pass a law (in 1776) that made it illegal for people with leprosy to marry (Ehlers 1895). By the early 1800s, a register of deaths in Iceland was kept, although Ehlers feels that there was confusion between leprosy and scurvy diagnoses. From 1800 to 1837, around 700 people are reported to have died of (or more likely *with*) leprosy, and by 1837, 125 people with leprosy had been recorded (Ehlers 1895; Richards 1977). By 1896, this number had risen to 236. In 1848, extant leprosy hospitals were shut down, having never provided shelter to more than about 5 percent of people with leprosy at any one time. This added to debates about whether people with leprosy should be segregated, and a new leprosy hospital was opened near Reykjavik in 1897 (Richards 1977). The number of people with leprosy declined from 96 in 1910 to eight in 1957; the last indigenous person was diagnosed in 1956 (Benediktsson and Bjarnason 1959).

Norway

Norway is extremely important in the history of leprosy, particularly in terms of Armauer Hansen's identification of the leprosy bacillus in the late nineteenth century and how this changed ideas about how leprosy was transmitted. The history of leprosy in Norway is well documented in the impressive St. Jørgen's Leprosy Hospital in Bergen, which is still standing (Figure 5.17).[5] Considering leprosy's long history in this country, it is surprising that few skeletons with leprosy have been excavated and identified in Norway, but this may be explained by the relative lack of paleopathologists working there (Holck 2012).

It has been suggested that leprosy spread to Norway from Britain and/or Ireland during the Viking period through seafaring contact, but the

236 · Leprosy

Figure 5.17. Reconstruction of one of the bedrooms in the St. Jørgen's Leprosy Hospital, Bergen, Norway. Credit: Bergen Museum.

infection could have derived from other regions (Vogelsang 1965, 1978; Richards 1960); e.g., leprosy was present in Scotland well into the eighteenth century. Faye (1882) reported that leprosy was "fairly common" in Norway by AD 1000. Most leprosy hospitals were dedicated to St. George and were founded in association with a church or a monastery. The oldest date to the twelfth century (in Nidaros, Trondheim, and Bergen). In the thirteenth and fourteenth centuries, two more hospitals (one for men and one for women) were founded in Nidaros and hospitals were established in Stavanger on the southwest coast, Tonsberg on the east coast, in Oslo, and at the inland eastern site of Hamar on the shores of Lake Mjøsa. In Bergen, the well-documented St. Jørgen's Hospital is mentioned in a will dated 1411, the year it was founded in conjunction with a monastery (Vogelsang 1965; see also Irgens 1973; and Meima et al. 2002). Leprosy declined in Norway from the fifteenth century (Vogelsang 1965), and by the sixteenth century was rare.

Most knowledge of leprosy in the medieval period is from Bergen and the St. Jørgen's Hospital (Stang 1895). Until the early sixteenth century, the hospital was solely for people with leprosy. It later became a general hospital, as was the case for many leprosaria elsewhere in Europe. In the seventeenth century, patients with leprosy became a small proportion of the hospital's admissions. Around 1702, the hospital burned down, as did many other buildings in Bergen. The hospital experienced accidental fires on many occasions but was always rebuilt (most recently in 2013). The last major rebuild was done in 1754, when the facility was replaced with a larger hospital to accommodate the increased numbers of people with leprosy. From the early nineteenth century, the hospital was reserved for people with leprosy as the infection increased again. The hospital eventually became a museum to educate the public about leprosy and the great contribution Norway has made to its decline.

While leprosy had an earlier history in Norway, it was in the nineteenth century that developments in knowledge of leprosy and its management became prominent throughout the world. Armauer Hansen's work transformed knowledge of leprosy and he eventually deemed it probably contagious. Hansen showed that leprosy declined in places where segregation was strictly enforced. His research changed ideas about how leprosy is transmitted between humans and contributed to establishing laws for its control. In the late twentieth and early twenty-first centuries, Lorentz Irgens was a key figure in developing scholarly work on leprosy in Norway, continuing Hansen's tradition (Irgens and Bjerkedal 2006). His work, which included an overview of the data contained in the nineteenth-century National Leprosy Registry of Norway (Irgens 1980; Irgens and Bjerkedal 2006), revealed that there are records for 8,231 patients and that leprosy was most frequent in rural northern and western Norway, especially in the coastal districts (67.5 percent in the rural north and 32.5 percent in the rural west) but was not common in eastern districts close to the sea. The epidemic peaked in the middle of the nineteenth century with a frequency rate of around

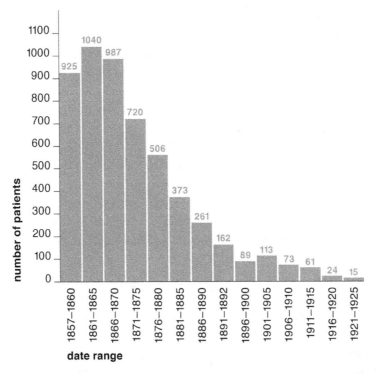

Figure 5.18. Frequency of leprosy in Norway, 1857–1925. Credit: Figure by Chris Unwin from data in Irgens and Bjerkedal (2006).

16.7 in 10,000 (Figure 5.18). Irgens found that the highest frequencies of leprosy correlated with the production of oats and milk on farms and with favorable conditions for the growth of mycobacteria in sphagnum vegetation. A sphagnum index was created to test whether sphagnum vegetation was a source of mycobacteria relevant to leprosy in humans. The variables Irgens recorded were sphagnum in the neighborhood, the height and orientation of vegetation above sea level, the distance between the vegetation and nearby farms, heights of farms above sea level, the orientation of farms, and the water supply for them (Irgens 1980, 1981; Irgens et al. 1981). A southern orientation at a low altitude favored sphagnum growth and a high relative humidity, a condition that is also needed for mycobacterial growth. The area in Norway with a mean relative humidity in July higher than 75 percent coincided with regions with a high frequency of leprosy, as did a high temperature (see Figures 5.19, 5.20, and 5.21). Irgens found the highest rates of leprosy in low tuberculosis areas but no evidence of leprosy in places where there was crowding. He suggested that a decline in leprosy

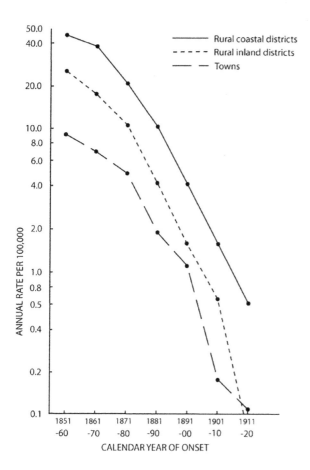

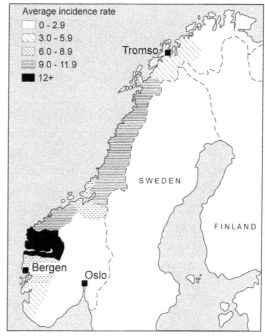

Above: Figure 5.19. Relative incidence rate of leprosy in towns, the rural coastal region, and the rural inland region of Norway, 1851–1920. Credit: Redrawn by Yvonne Beadnell after Irgens (1981, figure 3).

Right: Figure 5.20. Average incidence rates of leprosy in Norway, 1851–1920. Credit: Redrawn by Yvonne Beadnell after Irgens (1981, figure 2).

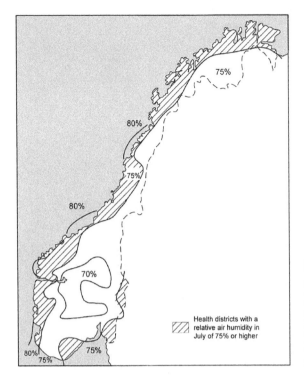

Figure 5.21. Average relative air humidity in July in Norway for the period 1874–1913. Credit: Redrawn by Yvonne Beadnell after Irgens (1981, figure 4).

was associated with an improvement in nutrition and a reduction in the influence of environmental mycobacteria because new ground was cleared, producing less favorable conditions for mycobacterial survival. Irgens and Bjerkedal (2006) note that the National Leprosy Registry enabled health care practitioners to evaluate trends and make plans for medical care. Their work shows how central patient registries can be useful in solving health problems.

The Leprosy Acts of 1877 and 1885 in Norway are believed to have contributed to the decline of leprosy by isolating infected people in hospitals, although there are other possible reasons, such as emigration to the United States of young and "presumably infective" patients in the nineteenth century, improved nutrition (which strengthens the immune system), and the protective role of tuberculosis cross-immunity (immunity from a disease that follows immunization against another disease) (Wood 1991, 428). Wood (1991; see also Manchester 1984, 1991) argues that tuberculosis is probably the most plausible explanation for a decline in leprosy, and we know that tuberculosis increased in Britain in the fourteenth century AD (see Chapter 6). This also occurred in Norway in the period 1850 to 1900, according to data from the National Leprosy Registry and a series of simulation models of leprosy transmission and control. Although Wood notes that the data for leprosy and tuberculosis are not strictly comparable, they show similar trends. Leprosy was common on the west coast and farther north in Norway, while tuberculosis was more common in the south and southwest. By 1920, new evidence of leprosy was rarely detected in Norway, thus showing that leprosy had declined before the advent of any effective chemotherapy (Meima et al. 2002). However, the decline cannot be completely explained by Norway's isolation policy because other factors were important, such as an improvement in living conditions.

Despite Norway's importance in the history of leprosy, there is surprisingly little to say about the bioarchaeology of leprosy in Norway. The first leprosy hospital was founded in Trondheim in the twelfth century (Vogelsang 1965, 30), indicating the longer history of this infection in the city. One female skeleton has been diagnosed with facial bone alterations from St. Laurentius Church in Oslo, an area today in the eastern part of the present capital known as the Old City. This skeleton was associated with a leprosy hospital that dated to AD 1200 to 1500–1600. Two more skeletons have bone changes related to leprosy; one, from St. Mary's Church in Oslo, has facial bone damage and the other, from St. Peter's Church in Tønsberg, has changes to the lower leg bones and the bones of the feet (Per Holck and Elin Brødholt, pers. comm., December 2012). The skeleton from Tønsberg (Number 6296, Schreiner Database) represents a woman approximately 25–30 years old, and the skeleton from Oslo (Number 4245, Schreiner Database) is of a woman 35–50 years old. The skeletons from St. Mary's generally date to the medieval period, or AD 1066–1537 (Brødholt 2007). Eighteenth- or nineteenth-century evidence in Norway is documented in a young adult female skeleton (383) from the West Front cemetery at Nidaros Cathedral in Trondheim (Kate Brayne, pers. comm., July 2008). The Norwegian State Archaeology Service excavated another three skeletons from that site in 1996 that may have evidence for leprosy but the disease could not be identified with certainty. This was a free burial ground in the eighteenth and nineteenth centuries that served the poorest people of the city.

Most recently, Pfrengle et al. (2018) reported aDNA analysis of one skeleton from Bergen with leprosy (14th century AD) and found that that person harbored the aDNA of *M. leprae* Type 3 (European), which is closest to DNA findings for medieval Britain.

Poland

There is as yet very little evidence of leprosy in Poland, even though palaeopathology has been studied there since the nineteenth century (Kozlowski 2012a). According to Wachholz (1921), the first skeletal evidence dates to the thirteenth century.

Most information comes from sources published in the 1970s and 1980s and in more recent years. For example, Gładykowska-Rzeczycka (1976, 1982) described a "medieval" middle-aged man with leprosy from the site of Suraż, Łapy County, Poland (Skeleton 20, 40–50 years old at death). Donoghue and colleagues (2002) also established a diagnosis of leprosy for this skeleton by identifying *M. leprae* aDNA in a bone sample from the nasal cavity. Donoghue, Holton, and Spigelman's (2001) study of bone samples from a metatarsal and the nasal cavity of the (assumed) same individual from this site revealed *M. leprae* aDNA only from the nasal cavity sample. Kozłowski (2012b) also discovered two skeletons with leprosy dated to the twelfth–thirteenth centuries from Kałdus and six skeletons dated to the twelfth–fourteenth centuries from Gruczno, both located in the Pomeranian region of north-central Poland, close to the River Vistula (Kozłowski and Drozd 2008). At Kaldus, the affected individuals were a mature man (#80) and an adult woman (#101) (Kozłowski 2012b). The latter had been buried in a different position from those in the rest of the cemetery. At Gruczno, the affected individuals were two mature adult men (#1156 and #259), three adult men (#236, #1515 and #1509), and one adult woman (#124). The skeletons from Kałdus and Gruczno are being analyzed biomolecularly at the Medical University in Łódź. The most recent information is that Kaldus skeleton 101 has preserved *M. leprae* DNA (Tomasz Koslowski, pers. comm., April 2013). Kwiatkowska (2005) also reports a skeleton with possible leprosy-related changes in the bones of the lower legs from medieval Wrocław-Ołbin (Silesia region), but as discussed in Chapter 4, lesions in the lower leg bones alone cannot be used as a diagnostic tool for leprosy because of their multiple etiologies. Justus and Agnew (2012) document unpublished but possible evidence for leprosy that affected foot bones of two adult skeletons from the eleventh–twelfth centuries in the Giecz Collection at the Muzeum Pierwszych Piastów in Giecz, which represents a site at an important trade center and military post in the Wiekopolska region. More work on the evidence is needed (Justus, pers. comm., November 2018). Unpublished data collected as part of the Global History of Health Project's European Module (Steckel et al. 2018) has recorded leprosy definitely in one skeleton (facial changes) and possibly in four others at the site of Poznan-Srodka (AD 950–1150; Tracy Betsinger, pers. comm., November 2018).

Sweden

Paleopathological study in Sweden dates back to the nineteenth century (Ahlstrom and Arcini 2012). In contrast to Finland, Iceland, and Norway, Sweden has revealed quite a lot of bioarchaeological evidence for leprosy. The oldest leprosarium in Scandinavia was founded in the mid-twelfth century in Lund (Arcini 2018). By the thirteenth century, Sweden had special hospitals called spitals that were run by the church for the poor. According to Mogren (1984), there are sixteen known leprosy hospitals in medieval Sweden and Finland, but they served the sick, the poor, and those with leprosy. The first historical descriptions of leprosy in Sweden date from the 1760s, but by 1940 only seven people in that country apparently had leprosy (Reenstierna 1941). The last person with leprosy in Sweden died in 1976 (Sundelin and Sörman 2004).

The first skeletal evidence, which comes from the Roman Iron Age in Halland in southern Sweden, is radiocarbon dated to AD 70–570 (Arcini and Artelius 1993). This adult woman was buried in a mound between two deposits of cremated bone. She was the only person at the site who had been inhumed rather than cremated, and Arcini (2018) suggests that this may have been because her external appearance was different from the rest of the community and warranted special treatment at death. After this early burial, there is no evidence for leprosy for the period 500–900 AD, perhaps because cremation was the normal funerary practice (Arcini 2018).

Arcini (1999) has described forty-three people with the infection who were buried in non-leprosaria contexts in Lund, Skåne, in the period AD 990–1536 (21 men, 15 women, and seven unsexed individuals; see also Arcini 2018). The total skeletal sample (3,305 individuals) was excavated from six cemeteries associated with four neighboring churches. Seven of the people with leprosy were buried in the period AD 990–1020/30 (all but one were located in the outer zone of the cemetery), twenty-seven were buried in the period 1020/30–1100, eight were buried in the period AD 1050–1100, and one was buried in the period AD 1300–1536. Most were associated with the wood-and-stone church of Trinitatis (ca. AD 990–1050) and the wooden church of Kattesund (ca. 1050–1100). The person buried in the period circa 1300–1536 was interred in the cemetery of Trinitatis church. There is also evidence in Lund from the cemetery sites of St. Andrew's, St. Michael's, and St. Martin's; at Tygelsjö (in the south near modern-day Malmö); at Fjälkinge (in the northeast); at Helsingborg (in the north); at Åhus (in the east); and at Löddeköpinge (in the west), where people with leprosy were buried both near the church and in the periphery of the churchyard (Arcini 2018). In addition to the evidence from Lund, there are several examples of affected skeletons from the Viking period in Skåne on the island of Gotland (Kopparsvik), Uppland (Skämsta), Östergötland (Klosterstad), and Uppsala (Arcini 2018). Later-dated examples of leprosy in human remains come from Örberga and Linköping (Östergötland), Skämsta (Uppland), Sigtuna (see below), and Västerhus (Jämtland; Arcini 2018). However, the majority of the evidence is from medieval Lund.

The cemetery associated with the medieval leprosy hospital of Lund and the purported leprosarium cemetery at Kvarteret St. Jørgen, Malmö, both in Scania, southern Sweden, have also revealed evidence (Arentoft 1999). Bone changes of leprosy in skeletons from St. Jørgen affected only 10 percent of individuals; this was from a sample of 200 skeletons (ages at death of 15 years or over) that Boldsen (2001) analyzed from a total of 1,600 graves excavated. This can be compared to the much higher percentage documented for the skeletons of people buried in the leprosaria cemeteries of Winchester in England and Næstved in Denmark. Of interest here is the small number of skeletons showing signs of leprosy in what is recorded to have been a leprosarium cemetery (Boldsen 2001, 382). Boldsen suggests that this could be because the site is dated to AD 1320–1520 and may represent a time when leprosy was declining. However, if the site at Malmö represents a lay cemetery rather than a leprosarium cemetery, it may be the case that people in Malmö with leprosy were selected out to be segregated and buried in a leprosarium cemetery that has not yet been discovered or in a leprosarium cemetery in another part of Sweden. Other evidence of leprosy comes from northern Sweden from the late tenth–late thirteenth centuries AD (Viking to medieval) site of Björned, Västerbotten. Analysis of a sample of bone from an individual (A4) revealed *M. leprae* DNA, but the skeleton had no skeletal lesions of leprosy (Donoghue, Marcsik, et al. 2005; Nuorala et al. 2004).

The site of Humlegården in Sigtuna, Uppland, has revealed six skeletons (of 227 excavated) dated to AD 1100–1300 with definite and possible skeletal evidence of leprosy (Kjellstrom and Wikstrom 2008; Kjellstrom and Linderholm 2008; Kjellstrom 2012). One person with leprosy also had possible tuberculosis-related spinal damage (Kjellstrom 2012, skeleton 3320), and a child 11–12 years old had the rarely seen damage to the central incisors (leprogenic odontodysplasia) and rhinomaxillary syndrome (skeleton 3092). The skeletons were buried in the outer area of this churchyard cemetery, which is where the authors of the publications from this site suggested the poor were usually placed. The site has also revealed a larger proportion of people with health problems than at other sites excavated in Sigtuna. Dietary studies of people with (six) and without (nineteen) leprosy through carbon and nitrogen isotope analyses have shown that the diets of both groups were the same (mainly terrestrial protein with a strong signal of freshwater fish). These people ate fewer vegetables but the same amount of freshwater fish as individuals in other studies in Sweden (Kjellstrom and Linderholm 2008). Linderholm and Kjellstrom (2011) also did a mobility history study of the skeletons from Humlegården and found no differences between the mobility history of those with and without leprosy. They concluded that both groups had been born and raised locally. Ancient DNA analysis has identified ancient *M. leprae* strains in skeletons from two cemeteries in Sigtuna, eight of which had bone changes of leprosy. Three gave strain data: one had a Type 3, subtype I and two revealed Type 2, subtype F (Economou et al. 2013a, 2013b). The latter strain is found mainly in Iran and Turkey today, which suggests that this subtype may have been transmitted to Scandinavia through the movement of people (see also Schuenemann et al. 2013, who analyzed the same skeletons as Economou et al. 2013a). This was the first time this subtype was identified in archaeological skeletons in Europe. Arcini (2018), who has done work on mobility using strontium analysis, reports that none of the individuals with leprosy she studied from Sweden were born and raised in the place where they were buried and that they did not have the same geographical origin. More studies such as these will be useful for exploring the mobility of the leprosy bacteria from its origin and thus the evolutionary history of the bacteria responsible for leprosy.

Finally, it should be noted that a skull from Nykjöping in Södermanland County, Sweden, that is housed in a collection in Norway also has bone changes of leprosy (Number 1652, Schreiner Database, University of Oslo). The sex and age are undetermined, but the skull appears to be female. The dating and context are listed as unknown (Elin Brødholt, pers. comm., December 2012).

In recent years, like in the UK, researchers in Sweden have produced some particularly innovative studies based on the remains of those who experienced leprosy there in the past. In doing so, they have created a nuanced picture of the impact of this infection on their past populations. It shows that leprosy was a disease that appears to have been accepted within their communities, as shown in the funerary record where people were, on the whole, buried normally for the time periods when they died.

Ukraine

Evidence for leprosy in skeletons in the Ukraine is limited to three reports, two of which are unpublished. One reveals evidence from the tenth century AD from the Michailovsky Gold Cathedral in Old Kiev and the other provides evidence from the fourteenth–sixteenth centuries (Inna Potenkhina, pers. comm., 2008; Kozak and Schultz 2006a, 2006b). The lack of evidence is inconsistent with the origins of paleopathological study of archaeological human remains in Ukraine (Potekhina 2011), but a number of other sites in Kiev have evidence of people with leprosy that have not yet been published (Inna Potenkhina, pers. comm., 2013).

The Mediterranean, Including Africa

In recent years, comparative modern genomic research has been used to explore the origin, evolution, and history of various diseases. Monot et al. (2005, 2009) have illustrated that *M. leprae* has undergone extensive "reductive evolution" (downsizing of a genome by removing genes) and that leprosy probably originated in Eastern Africa or the Near East and then spread as people moved across the world. Based on historical data, it has long been suggested that leprosy was brought to the Mediterranean from the Indo-Gangetic basin by the armies of Alexander the Great (356–323 BC), which were returning from the Alexandrian campaign (see Mark 2002). The Indo-Gangetic basin is defined as the north Indian river plain where the Indus and the Ganges drain the area. This includes most of north and east India, Bangladesh, and parts of Pakistan and southern Nepal today. But is there any early skeletal evidence for leprosy beyond Northern Europe, particularly in the Mediterranean and beyond, to support this hypothesis? The next section considers the skeletal evidence in the Mediterranean area, moves to the evidence from Africa, and then considers evidence from Asia. Today new "cases" of leprosy are rarely seen in the Mediterranean region, apart from the eastern area, where 4,356 were detected in 2018 (World Health Organization 2019). Leprosy is also seen in parts of Africa, including Egypt, the Democratic Republic of Congo, and Ethiopia (e.g., see Hotez, Savioloi, and Fenwick 2012), and in migrants from endemic areas.

Cyprus

The island of Cyprus has revealed only one individual with leprosy, the skeleton of a young adult woman from an early fifteenth-century basilica at Polis Chrysochous (Baker and Bolhofner 2014). She was buried with a coin and a pot handle and was not treated any differently from the rest of the burials there. The lack of evidence in Cyprus may be explained by the late development of palaeopathology there (Lorentz 2011), but the eastern Mediterranean area is almost devoid of evidence (see below).

France

Although France has a long history of paleopathological study that dates to the nineteenth century (Blondiaux, Dutour, and Sansilbano-Collilieux 2012), very little evidence for leprosy has been documented there either (Bret et al. 2013). The earliest evidence comes from a fourth-century site at Lisieux-Michelet, Normandy (Joel Blondiaux, pers. comm., February 2013). Blondiaux, Duvette, and colleagues (1994; see also Blondiaux et al. 2002) describe later evidence in two skeletons from two sites, one in the north at Neuville-sur-Escaut and the other in the south in Vaison-la-Romaine. A woman (skeleton number 67) had bone changes to her lower leg and foot bones and a man (skeleton number 23) had changes to his facial, lower leg, and foot bones (both are dated to ca. AD 500). Mortuary patterns and grave goods suggested a potential southern German origin for the people buried at Neuville-sur-Escaut (Blondiaux et al. 2002). The microradiographic method of Coutelier (1971) revealed that the microscopic changes in the bones of these two individuals were consistent with those seen in microscopic changes in images of the bones of amputated limbs of people with leprosy. Vendeuil-Caply, Oise, in northern France, a site dated to the fifth to early sixth centuries AD, has also revealed five individuals with leprosy (Joel Blondiaux, pers. comm., February 2013).

Saint-Martin et al. (2006) have described later evidence in skeletons from the Chapelle Saint-Lazare leprosy hospital in Tours, Indre-et-Loire, southern France, that date to AD 1359 and 1507. Another leprosy hospital cemetery at St. Thomas d'Aizier, Calvados, in northwest France that dates to the thirteenth–sixteenth centuries AD further has revealed skeletons with bone damage consistent with leprosy (Blondiaux et al. 2016). This site is particularly interesting because cementochronological study (which uses dental cementum increments as an indicator of age) has shown that women with leprosy had a lower survival rate than men with leprosy. In addition,

Ardagna, Bouchez, and Vidal (2012) have described leprosy in an older woman buried in a stone cist grave in the eastern part of the cemetery associated with the medieval Jean de Todon Chapel in Laudun l'Ardoise, Gard, in southern France. Finally, the most recent purported evidence of leprosy in France comes from the Paris Catacombs, which contain around six million people's remains that may date back to the fifteenth century and may include the remains of people transferred there from two leprosarium cemeteries (Deps et al. 2020). No evidence was found in a previous study of remains there in the 1960s (Møller-Christensen and Jopling 1964). However, of the 367 skulls of 1,500 considered for study, two showed bone changes of "probable" leprosy and twenty-one of "possible" leprosy (Deps et al. 2020).

Greece

Leprosy in Greece today is declining and rapidly dying out, as revealed by a study of new "cases" in the period 1988–2007 (Kyriakos 2010), but while leprosy has been eliminated in Greece as a public health problem, it has not been totally eradicated (Kyriakos 2003). A retrospective study of new leprosy occurrences has found that reports tend to be from rural areas, whereas people who have relapsed disease are from urban contexts. The paucity of evidence for ancient leprosy (Chryssa Bourbou, pers. comm., July 2008; Sherry Fox, pers. comm., July 2008) has probably been affected by the late development of paleopathological studies in Greece (see Roberts et al. 2005; Schepartz, Fox, and Bourbou 2009; Eliopoulos et al. 2011), although J. Lawrence Angel was an early pioneer (e.g., 1964). Possible evidence has been reported on the island of Kos from an ossuary dated to the fifth–sixth centuries (early Byzantine) that contained disarticulated remains, including two metatarsals showing characteristic bone change (Lagia, Petroutsa, and Manolis 2002; Lagia et al. 2006). Slightly later but almost contemporary evidence, again seen in metatarsals, has been found in comingled remains in tombs around the basilica church at Priniatikos Pyrgos in northeastern Crete that has been dated using pottery to the sixth century to the second half of the seventh century AD (Bridgford et al. 2014; Moutafi, pers. comm., February 2013). Bridgford et al. claim that this is "the first unambiguous evidence of osteological leprosy in Greece" (2004, 60). Crete also has a more recent history of leprosy on the island of Spinalonga, a place designated for people diagnosed with leprosy from the early twentieth century until 1957 (Bourbou 2003).

Israel

Leprosy is rare in Israel today but is noted in immigrants and workers from places where it is endemic (Lejbkowicz et al. 2001). Archaeological skeletal evidence for leprosy is also rarely reported, which may be explained by challenges Israeli bioarchaeologists face regarding the excavation and analysis of human remains (see Hershkovitz 2012). Israel's apparent biblical associations are relevant to discussions of evidence for leprosy there. It was once thought that leprosy was described in biblical texts, but it is now thought that this derivation occurred because of an incorrect translation of the Hebrew word *tsar'ath*. The word is now believed to be associated with a range of fungal and bacterial skin diseases that were prevalent in the eastern Mediterranean in the first millennium BC. Zias (2002, 261) has suggested that "modern day leprosy can definitely be excluded from the list of possible diseases subsumed under the generic term *tsar'ath*." However, he notes that what is accepted as undisputed leprosy did exist in this period and that a number of Greco-Roman authors mention it. Browne (1975a) also feels that true leprosy existed in Palestine and Syria and was likely described as "lepra." One of the Dead Sea Scrolls from Qumran (Yadin 1983), compiled in the second or third century BC, said that people with leprosy should be located east of the city of Jerusalem, where people with skin disorders had to live (this may have been the village of Bethany). Despite much discussion and the general conclusion that leprosy is not described in the Bible, this debate will continue.

There is little evidence for leprosy in excavated human remains in the Near East or Israel in particular. This may be because people with *tsar'ath* were buried in separate cemeteries (Zias 2002), as described in the Old Testament (II Chronicles 26:23). The earliest skeletal evidence in Israel comes from the first century AD in bones from a family tomb in a rock-cut grave in the Hinnom Valley, Akeldama, directly opposite the Temple Mount, Jerusalem (Gibson et al. 2002; Donoghue, Marcsik, et al. 2005). Over twenty ossuaries with human bones were excavated at this site. The site was radiocarbon dated using textile fragments from the grave of interest. A phalangeal sample from individual C1, who had been wrapped in a shroud and placed in a small chamber within the tomb, was analyzed for aDNA of leprosy. This showed that the person harbored both leprosy and tuberculosis during life. Leprosy is believed to have been fairly common in the Roman Empire during the Byzantine period (fourth–seventh centuries AD), and measures such as building special leprosaria were initiated (Zias

2002). However, in the Middle East in the Islamic and medieval periods, attitudes toward leprosy were generally tolerant (Dols 1979). This may be because the Koran says that segregation should not be practiced.

There is skeletal evidence for leprosy from three monasteries in the Judean desert east of Jerusalem (Martillous, Mar Theodosius, and Dier Hagla). Hospitals were commonly established alongside these ecclesiastical institutions and were isolated from the main communities. Martillous (or Martyrius; Zias 1991) was founded in the fifth century AD and was destroyed in AD 614. Zias (1985) reported that a common grave that dates to AD 492 revealed four people with leprosy, but later analysis suggested the evidence more likely suggested psoriatic arthritis (Zias and Mitchell 1996). Zias notes that people with psoriatic arthritis also have psoriasis of the skin, and people who "sought refuge in the desert monastery" may have been mistaken as having leprosy (Zias 1991, 198). Disarticulated skeletal remains with leprosy are described from the AD 600 mass grave at the Monastery of St. John the Baptist, located south of Jerusalem near the Jordan River (Zias 2002). The Jordan River, the place where John the Baptist is said to have baptized Jesus, was the "traditional site for the washing of the leper in Christian sources" (Hershkovitz and Spigelman 2007, 237). Ancient DNA analysis has been applied to the remains from this site with positive results (Rafi et al. 1994a, 1994b; a metatarsal), but later aDNA analysis of another metatarsal of the same skeleton did not show the same outcome (Donoghue et al. 2002). Another skeleton from Beit Guvrin in central Israel that dates to approximately the same period as the St. John the Baptist monastery (AD 300–600) was diagnosed with "Madura foot," a chronic disease caused by fungi or bacteria called actinomycetes (Resnick and Niwayama 1995b; Hershkovitz et al. 1992). However, Manchester (1993) suggested that the bone changes were more likely to be due to leprosy. While Hershkovitz, Spiers, and Arensberg (1993) agreed, they still favored a diagnosis of Madura foot. Later, using aDNA analysis, it was confirmed that the person had leprosy during life, although the result was a "weak positive" (Spigelman and Donoghue 2001, 2002; see also Donoghue, Holton, and Spigelman 2001; Donoghue, Marcsik, et al. 2005). Hershkovitz and Spigelman (2007) concluded that the person could have experienced both Madura foot and leprosy.

After the Byzantine period, hospitals were established for people with leprosy in the Near East, including during the Crusader period (eleventh–thirteenth centuries; Mitchell 2002). The Crusader states were established in 1099, when Jerusalem was captured at the end of the first crusade, but clearly leprosy was present in Israel before then (Zias 1991). King Baldwin the IV (1161–1185), ruler of the Latin Kingdom of Jerusalem, who was known for his military campaigns, had leprosy. His presence is of great importance in the history of leprosy in the Middle East. The Order of the Knights of St. Lazarus was made up of people with leprosy. During its 400-year history, it "advertised" people with the disease in order to gain gifts of land (Marcombe 2003). The order, which developed in a twelfth-century leprosy hospital in Jerusalem, catered mainly to members of the knightly class, who were believed to have a religious vocation. By the thirteenth century, the order had become militarized. While those with leprosy could fight in the army, they could not associate with the rest of the population (Mitchell 2002). According to Mitchell (2000a, 245), Baldwin was "one of the more remarkable kings in the medieval Christian world. . . . [He was] renowned because he developed leprosy and still maintained his position on the throne, becoming a successful ruler" (see also Mitchell 2000b). This picture contradicts ideas about the treatment of people with leprosy because this was a high-status person who became a king in AD 1174 at the age of 13. None of Baldwin's family had leprosy and he contracted it at a young age. When he became king, leprosy had led only to loss of sensation in his right arm and hand, but by his early 20s he had lost his sight, and loss of skin sensation had led to damage to his extremities.

Beyond the date of Baldwin, there is not much evidence of leprosy in Israel, except for a report of a sixteenth–nineteenth century skeleton from the Ottoman period coastal site of Dor in northwestern Israel (Faerman 2008). The bone changes were mainly in the postcranial skeleton. Ancient DNA analysis of eight bone samples from the skeleton, including from the inferior nasal conchae, the maxilla, a fifth metatarsal and a fibula, provided a positive result in four bones for *M. leprae*. The greatest concentration of DNA was in a fibula, followed by the nasal area, a radius, and a rib.

Italy

Leprosy is not a particular health problem in Italy today, apart from in people who have migrated there from endemic areas (Massone et al. 2012). Although paleopathological studies began there in the nineteenth century (Fornaciari, Giuffra, and Minozzi 2012), it has yielded a paucity of evidence for leprosy in archaeological human remains until more recent years.

Rubini, Zaio, and Roberts (2014) describe the earliest evidence from the Iron Age site of Corvaro (Central Italy; fifth century BCE). The next earliest data (second–fourth centuries AD) come from the Roman cemetery at Martellona, near Rome (Rubini, Dell'Anno, et al. 2012). The skull of a child 4–5 years old displays rhinomaxillary syndrome that suggests leprosy; no postcranial bones were preserved. An adult male burial (Tomb 74) in northern Italy dated to the end of the first half of the fourth century AD is also described with bone alterations related to leprosy. The bones exhibit probable cranial and postcranial bone changes. The evidence was found in the Celtic necropolis of Casalecchio di Reno, Bologna, and the person was buried normally for the time period and culture (Mariotti et al. 2005). Finally, Rubini, Zaio, and Roberts (2014) report evidence from the Roman site of Palombara (Central Italy; fourth to fifth centuries CE).

Rubini and Zaio (2009) describe skeletal remains from the central Italian cemetery of Campochiaro, Molise that date to the early medieval period, (sixth–eighth centuries AD). One of the cemetery's two areas (Morrione) revealed two skeletons with rhinomaxillary syndrome (Tombs 68, female and 108, male). Because of the absence of evidence of a permanent settlement in the area and because many graves included horse burials and goods relating to horses such as harness pieces, Rubini and Zaio (2011) have described the site as semi-nomadic and military. They suggest that skeleton 108, which revealed a perimortem wound to the frontal bone of his skull, represents a "leper warrior." If they are correct, this is analogous to trauma evidence seen in prehistoric and medieval Hungary (Abony-Turjányos dűlő and Orosháza, respectively) and Africa, described above and below. Ancient DNA analysis has revealed *M. leprae* DNA in both skeletons from this site (Rubini et al. 2017). Belcastro et al. (2005) also documents early medieval evidence in a skeleton of a man. He was buried in Tomb 144 in the second half of the seventh century AD in the necropolis of Vincenne-Campochiaro, Molise, in central Italy. Belcastro and colleagues noted bone changes of leprosy on both the cranial and postcranial skeleton. This individual was buried normally for the time and region.

Rubini, Erdal, and colleagues (2012) report evidence from the thirteenth or fourteenth centuries in Montecorvino, Foggia, Puglia (or Apulia), in southeast Italy; this is the first documented evidence from this part of Italy. Ten skeletons were excavated from graves near the main church and a female adult with leprosy (MCV2) had been interred in a brick tomb along with an iron belt buckle. In addition, the body of King Henry VII of Germany (1211–1242), the eldest son of Frederick II, was found buried in southern Italy (Fornaciari, Mallegni, and De Leo 1999). He was contained in a Roman sarcophagus in Cosenza Cathedral in Calabria on the Italian peninsula. He had rhinomaxillary syndrome and leprosy-related bone changes (diaphyseal remodeling) in the metatarsals and phalanges of his feet and periosteal lesions on his lower leg bones. He apparently committed suicide by falling into a ravine. Before that time he had been confined as a prisoner in the castles of southern Italy; the infection may have been one of the reasons for his confinement (and, indeed, his suicide).

Thus, the earliest evidence for leprosy in Italy is currently reported to be in the north, supporting Monot et al.'s (2005, 2009) work that indicates an African/Near Eastern origin for leprosy. It then likely spread as people moved east and west (Rubini, Erdal, et al. 2012). However, as more skeletal evidence is found in Italy, this hypothesis will no doubt be revisited.

Portugal

There is very little evidence for leprosy in skeletal collections examined in Portugal, which is surprising considering that paleopathological study there dates back to the nineteenth century (Santos and Cunha 2012). However, leprosy has been described in Portugal as early as the first century AD (Carvalho 1932). In the medieval period, around seventy leprosy hospitals were established in the country. One medieval site and one post-medieval site connected to leprosaria in southern Portugal have revealed skeletons with bone changes. The first, Ermida de Santo André, derived from a churchyard of articulated skeletons and comingled bones (Antunes-Ferreira and Rodrigues 2003; Antunes-Ferreira, Matos, and Santos 2013). The site is close to the Beja leprosarium, which functioned from the fourteenth to the sixteenth centuries AD. Neves, Ferreira, and Wasterlain (2012) and Ferreira et al. (2013) have noted two individuals from the second site, a leprosarium cemetery at Valle da Gafaria[6] ("leprosarium valley") in Lagos, that is historically dated to the fifteenth through seventeenth centuries AD. The individuals in this cemetery were buried in a different way from the typical Christian manner. Pfrengle et al. (2018) has recorded *M. leprae* DNA preserved in one individual from this site of Type 3 (European), which is closely related to medieval Danish DNA evidence. Radiocarbon dating revealed a date of AD 1340±48. Three other sites in Portugal have evidence of skeletons with leprosy (Vitor Matos, pers. comm., November 2018); two date to the Roman/Visigoth period (fourth to fifth centuries AD; Duarte 2015) and one to the post-medieval period (Melo et al. 2018).

Additional evidence comes from Matos et al. (2017), who describe three individuals from the Coimbra Skeletal Collection at the Museum of Anthropology of the University of Coimbra (Cunha and Wasterlain 2007; see also chapter 4) and one skeleton from the Luis Lopes Identified Skeletal

Collection in the Bocage Museum within the Museum of Natural History, University of Lisbon, Portugal). This collection consists of 1,692 skeletons with documented births from 1805 to 1972 and deaths from 1880 to 1975. This information from more recent evidence in skeletons is complemented by data based on X-rays of patients showing bone changes of leprosy from the well-documented Hospital-Colónia Rovisco Pais at Tocha in central Portugal, which was founded in 1947 for people with leprosy (Matos 2009). Today leprosy is not particularly common in Portugal, although a study of around 100 patients who were treated with lepromatous leprosy found that one third had migrated there and that the number of patients with lepromatous leprosy has increased since 1960 (Medeiros, Catorze, and Vieira 2009).

Spain

Evidence of leprosy described in skeletal remains is rarely described from Spain even though this country has a long history of paleopathological study (Rodríguez-Martín 2012). Leprosy is relatively rare there today (Alfonso et al. 2005). The earliest reported evidence in skeletons is Roman (second to fourth centuries) from a multiperiod cemetery in Pamploma, Navarra, in northern Spain, where a male skeleton appears to have convincing lesions on a foot bone (de Miguel Ibáñez, Ballestras Herrái, et al. 2008; de Miguel Ibáñez, Unzu Urmaneta, et al. 2011), although diagnosis based solely on one bone is highly questionable in any study. The foot bones of another man from the same location from the seventh–ninth centuries AD, or early medieval/Visigothic phase, were also affected. Exteberria, Herrast, and Beguiristan (1997) further described a woman from Gomacin, Puente la Reina (also in Navarra), with leprosy-related lesions to the bones of her face and left hand, and the distal ends of both her tibiae. Her skeleton dates to the eighth–tenth centuries AD, or the early medieval/ Islamic phase. Later evidence (twelfth century AD) comes from Capilla de San George, a fortress in Seville in southern Spain (Guijo, LaCalle, and Lopez 1999). A leprosy diagnosis was made from the metacarpal of a female adult skeleton (120) using aDNA analysis; this was one of four individuals with bone changes of leprosy at this site (Montiel et al. 2003). López Flores and Barionuevo Contreras (2009) also described possible evidence from skeletons 30 (a young man) and 68 (an older woman) from the southwest Spanish port city of Cadiz that dated to the eleventh–thirteenth centuries. The skulls of both individuals were not well preserved and there were no lesions consistent with leprosy on them, but the bones of one of the feet of skeleton 68 had leprosy-related changes. Finally, De Miguel Ibáñez and colleagues (2008) present very convincing evidence in the facial and foot bones of a man buried at Pamploma during the Islamic period (eighth–fifteenth centuries AD).

Africa

There is currently relatively little evidence for leprosy in human remains from Africa in general, and beyond Egypt and Sudan there is only one report. Additionally, Møller-Christensen and Inkster (1965, 5) reported a skull with bone changes of leprosy that the Anatomical Museum at the University of Edinburgh acquired in 1906 from southern Nigeria; the dating is unknown. They describe the facial bones as being affected by *facies leprosa* and the skull as having evidence of an antemortem wound (see also sections on Hungary and Italy in this chapter). Møller-Christensen (1967) also examined 1,844 skulls from Egypt but found no evidence of leprosy. This is surprising, considering, first, the immense archaeological activity over many years in Egypt that has concentrated on excavating and analyzing skeletal and mummified remains (see Baker and Judd 2012; Morris and Steyn 2012), and second, the presence of leprosy in living populations today. Nevertheless, Monot et al. (2005) has suggested an African origin for leprosy based on modern DNA evidence, something that Grmek (1989, 174) wrote about decades ago, when he cited the Middle or Far East as possible origins. In 2012, Southern Sudan, the Democratic Republic of Congo, the Côte D' Ivoire, Nigeria, and the United Republic of Tanzania were some of the few places in the world where the rate of detection of new "cases" of leprosy exceeded 1 in 100,000 population (World Health Organization 2013). In 2018, 20,590 new "cases" were reported (World Health Organization 2019), and the World Health Organization (2017) considered a number of African countries to have a high burden of leprosy.

Egypt

The first scholar to document leprosy in skeletal remains in Egypt was Dzierzykra-Rogalski (1980) and this is currently the earliest evidence. He suggested, with no supporting data, that the individuals affected with leprosy at Dakhleh Oasis were "white." The period of these burials from the Dakhleh Oasis in the Western Desert coincides with the dating of the Dead Sea Scrolls from Qumran (see Zias 2002). Four men dated to 250 BC (the Ptolemaic period) were affected. Molto (2002) has published the results of excavations of a later date at the Roman Kellis 2 (K2) Dakhleh

Oasis cemetery that present other important evidence for leprosy in Africa. Two young adult male skeletons radiocarbon dated to the early to mid-fourth century AD were affected; two other men with possible leprosy are as yet undated. The affected individuals were buried with the general cemetery population, possibly reflecting tolerance of people with leprosy in the Roman period in that region (Molto 2002). Two people had isotopic values indicating Nubian origins in the Nile Valley, suggesting that they had moved to Dahkleh during their lives (Dupras and Schwarcz 2001). The most "isotopically distinctive individual" had lepromatous leprosy; it was suggested that this adult man may have been exiled to the oasis because he had leprosy (ibid., 1199). Molto (2002) has surmised that the two individuals with definite bone changes of leprosy who were buried close together in the cemetery probably originated from another less arid region (citing Schwarcz, Dupras, and Fairgrieve 1999). While we do not know whether these two people were actually accepted by the Dahkleh community, it is possible that they had been ostracized from another region. Research using aDNA analysis has shown the presence of leprosy in ten skeletons with bone changes of leprosy from this cemetery and confirmed other diseases, including tuberculosis (Donoghue, Marcsik, et al. 2005, 390). Samples from skeletons B6, B116, and B222 all contained *M. leprae* aDNA, and B6 and B222 were also positive for tuberculosis. Skeletons B251, B265, B280, B392, and B437 were positive for leprosy and all except B437 were positive for tuberculosis. The skeletons at the Dahkleh Oasis site exhibited both specific (rhinomaxillary syndrome) and nonspecific lesions consistent with leprosy, the latter including palmar grooves and dorsal tarsal bars.

Nubia (Modern Northern Sudan/Southern Egypt)

The first bioarchaeological evidence for leprosy in Nubia was reported in the early twentieth century. Grafton Elliot Smith (1908; see also Smith and Derry 1910; Smith and Dawson 1924) described leprosy affecting the hands and feet of a Coptic Christian mummy from El Bigha in northern Sudan (ca. AD 500). The mummy was excavated in 1907 from a location a short distance from the temple of Biga, near Aswan. Later, Møller-Christensen (1967) reexamined the left hand and foot of the mummy to confirm the diagnosis. To my knowledge, this is the only evidence of leprosy in a preserved body anywhere in the world. Møller-Christensen and Hughes (1966) also discussed facial changes characteristic of leprosy in the skull of a female skeleton of the same period, an individual Smith and Jones (1910) had already referred to. The more recent remains of a Christian man

(3265) from Mis Island, excavated in 2007 from the 3-J-11 cemetery, appear to have evidence of leprosy (Ginns 2007; Rebecca Redfern, pers. comm., 2007; Daniel Antoine, pers comm., 2013). The site, which dates to any time from AD 500 to 1550, revealed 283 individuals during the 2006 and 2007 field seasons. All of the skeletons from the site are currently being analyzed at the British Museum in London (Daniel Antoine, pers. comm., 2016). Skeletal evidence that Kozak et al. (2008) reported from an Old Kingdom mastaba in Elephantine, Egyptian Nubia, is much earlier in date (around 2300 BC). They describe "probable" leprosy-related changes to the bones of the feet. Histological analysis of a section of bone from the right tibia also suggested changes attributable to chronic infective disease. Kozak and colleagues note that other skeletons from this site have evidence of leprosy, including damage to their facial bones. As this skeleton is so early in date and would put back the earliest date for leprosy in Nubia if accepted, work to confirm these findings is needed. Ardagna, Bouchez, and Vidal (2012) have further reported early evidence that is suggestive of leprosy from the site of Mouweis in the Shendi area (north of Khartoum). The site is archaeologically dated to the third century BC through the fourth century AD, although radiocarbon dating has generated dates of the thirteenth–fourteenth centuries AD. There are various methods for dating archaeological human remains within their funerary contexts, and sometimes dates using different methods for the same archaeological context can vary. Some scholars may be content with using only a radiocarbon date, while others are much more holistic in their approach and may use a number of tools to achieve a date for a skeleton, for example artifacts buried with the skeleton, stratigraphic information, and a radiocarbon date. A radiocarbon date and other dating evidence may also corroborate each other (e.g., see Cole and Waldron 2010, 2011, 2015). For a recently published example of this and some discussion about dating methods, see Baker et al. (2020).

Sub-Saharan Africa

Jopling (1982) suggested that leprosy was a recent introduction to sub-Saharan Africa via Portuguese settlers in Angola and Mozambique. Although the study of palaeopathology began in this region in the early twentieth century (Morris and Steyn 2012), there is only one report of leprosy in skeletal remains from sub-Saharan Africa, from Limpopo in the northernmost province of South Africa (L'Abbé and Steyn 2007). There, the graveyard of a rural Venda village community was excavated in 1999 in advance of the creation of a reservoir dam. One hundred and fifty-seven burials were excavated that dated to the period 1910–1999. The skeleton of one middle-aged man from this group was diagnosed with leprosy-related lesions of the facial bones.

The presence of people with leprosy in South Africa generated a dramatic social response in the nineteenth and twentieth centuries. Segregation of this group increased because clinicians declared that the appearance of people with leprosy was offensive and that society should be protected if leprosy was contagious (Deacon 1994, 2003). At that time, people believed that the disease was transmitted through inheritance. Because of this, people in authority worked to prevent sexual contact between those who were affected with leprosy and those who were not. In the nineteenth century, health care providers and the general public believed that "black" people were most susceptible to leprosy and that leprosy and "blackness" were closely related. This stereotype was analogous to the false, racist beliefs that native southwest African populations were idle, unhygienic, promiscuous, and had an unpleasant odor.

Robben Island, just off the south coast of Africa, became the place for segregation, first for political prisoners and then for people with disease. It was used to segregate people with leprosy from 1846 to 1931 (Deacon 1994). In 1855, almost three-fifths of people with leprosy on Robben Island were Khoi (native people from southwest Africa), two-thirds were men, and three-quarters of these men were from outside Cape Town. There were consistently more men than women. However, because of the belief that leprosy was contracted within families, authorities removed women with leprosy from Robben Island to the Old Somerset Hospital in Cape Town from 1871 to 1887. When they eventually returned to Robben Island, they were located at Murray's Bay, which was quite some way from the men. Until the early 1890s laws (see below), the number of people with leprosy on Robben Island remained lower than 100, but after those laws were passed, the number increased to 413. By 1915, the number of people had increased to 600. Before 1901, the criteria for admission to Robben Island included visible ulceration, deformity, destitution, and superficial skin problems (Deacon 1994, 52), but after 1901 the presence of leprosy bacilli was used to confirm diagnoses. People on Robben Island rejected the term "leper," preferring to call themselves "people," "sick people," or "prisoners." However, some people with leprosy were not segregated there because doctors helped conceal them, while others managed to evade being detected.

By the 1880s in South Africa, beliefs about the cause of leprosy had shifted away from the theory of hereditary transmission. In the early 1890s, the Leprosy Repression/Suppression Act took away civil rights from people with leprosy. Those with the infection were compulsorily segregated; this accompanied the social separation that generally eroded the civil rights of blacks after the 1890s. Two-thirds of South Africa's leprosy colonies were established in the period 1890–1901. This was the decade when the Parliament passed other acts related to the control of leprosy (the Cape Leprosy Repression Act of 1891, the Amendment Act of 1894, the Natal Leprosy Act of 1890, the Orange Rover Colony Leprosy Act of 1909, and the Transvaal Acts of 1904, 1907, and 1908, as amended in the Leprosy Act no. 14). Deacon (1994, 52) has noted that "few countries, even among the colonies, came out as early and as strongly on the contagion side of the transmission debate as South Africa did in the legislative arena."

Asia

China

Evidence for leprosy in skeletons from China is rare, but that may be because the development of paleopathological studies in that country is fairly recent (Pechenkina 2012). However, Zhenbaio (1994) has described the skeleton of a young adult female with characteristic bone changes to the facial region of the skull. The skeleton, which dated to the Han Dynasty (200 BC–AD 200), was excavated in Pinglu County in Shanxi Province, southwest of Beijing. Historical sources indicate an earlier date for the first evidence of leprosy in China, although Leung (2009, 3) maintains that "evidence in the old medical classics does not fully support the hypothesis of the existence of true leprosy in ancient China." Thus, perhaps very early evidence in human remains should not be expected from this part of the world.

By the sixteenth century AD in China, there was a resemblance between clinical descriptions of *mafeng* or *dafeng* and leprosy infection as it was described in the west, along with well-documented institutional strategies for coping with people with this disease (Leung 2009). Ideas of how leprosy was contracted in the late Imperial period (fourteenth century AD to 1912, when the Republic of China was founded) included contagion (spread of the disease through close contact), inheritance, and sexual intercourse, thus indicating mixed opinions about the epidemiology of the disease. By the early sixteenth century, people with leprosy were segregated in nonreligious institutions to prevent spread of the infection; these were organized and financed by local governments. People in China had learned that people with leprosy had been segregated in medieval Europe and believed that this had led to the disappearance of the disease there. While recent research shows that this was not what happened to everyone with leprosy in Europe, in China leprosy hospitals were used to segregate people with leprosy well into the twentieth century.

The history of leprosy in China might be considered within the global context of nineteenth-century colonialism and racial politics, although China was never a true colonial state and was never totally colonized by Japan. However, in the late nineteenth century, the diaspora of the Chinese was seen by many as the reason for leprosy's spread around the globe. As Leung has noted, "China was considered to be a global reservoir of the disease" (2009, 6; see also 2). By the turn of the twentieth century, the Chinese elite felt that leprosy was incompatible with modern hygienic living and that segregation of people with leprosy was essential. The People's Republic of China (founded in 1949) increased efforts to control the infection and reinstated a segregation policy for four decades, creating "leper villages" as part of the agricultural cooperative movement in the 1950s and 1960s (Shumin et al. 2003). Leprosy control was seen as part of China's complex rural reform, and in the 1990s Chinese authorities declared that leprosy had been eradicated there. However, even in 2005, 662 "leper villages" remained (Leung 2009, 209). In 2019, the World Health Organization reported 521 new "cases" of leprosy in China for 2018.

Japan

Even though paleopathological work has been done in Japan since the late nineteenth century (Giannakopoulou and Suzuki 2012), there is very little paleopathological evidence for leprosy in this country. The earliest evidence comes from a female skeleton dated to AD 1200–1600 from a burial at Yuigahama-minami, Kamakura, in the Kanagawa Prefecture south of Tokyo. The bone changes were identified in the metacarpals, the first metatarsal, the foot phalanges, and the tibia and fibula. As the skull was not affected, Hirata, Oku, and Morimoto (2000) suggested that this woman had tuberculoid leprosy. Morimoto (1995) reports another female skeleton from Neio, in the northernmost part of the main island. It was dated to the Edo period (around 1600) and had damage to the hard palate bones. Suzuki (2001) describes skeletons with leprosy from the Gokuraku-ji temple in Kamakura (also south of Tokyo) that dates to the twelfth to fourteenth centuries AD (no age or sex given) and another skeleton from the Edo period (seventeenth to nineteenth centuries AD) in the Kanto region, in the middle part of the largest island of Japan. Suzuki, Takigawa, et al. (2010) further describe another Edo period male skeleton from a burial at Hatani, in the Amori Prefecture (north main island), that had facial lesions consistent with leprosy. Ancient DNA analysis of this skeleton revealed a positive result for *M. leprae* from facial bone and fibula samples, the first such aDNA data in the Far East. Type 1 leprosy DNA was also detected, which is characteristic for Southeast Asia and India today. Suzuki and colleagues suggested that leprosy detected in people at this site could have originated from either of these parts of the globe. In more recent research, Suzuki et al. (2014) document additional evidence from the Edo period: two men with leprosy from different sites in eastern Japan (Tawara-ga-yatsu and Usukubo, both in Kanta—main island). *M. leprae* DNA was preserved in the bone samples analyzed. These men were buried with iron pots covering their heads. In this period, iron pots were considered to indicate important burials but were also thought to stop transmission of diseases such as leprosy. Because of the nature of the burial context, Suzuki and colleagues suggested that the community genuinely cared for these people. They were also buried in the same way as others in the cemeteries, except for the valuable iron pots that may have been a tribute to these dead men (ibid.).

India and Pakistan

There is a paucity of paleopathological data from South Asia in general and for leprosy in particular. However, figures for 2018 revealed that 120,334 new "cases" of leprosy had been detected in India that year, contributing 63 percent of new "cases" globally (World Health Organization 2019). While there are early historical data for the presence of leprosy in India, until recently there was no evidence for leprosy in any human remains analyzed from India or Pakistan (Subhash Walimbe and John Lukacs, pers. comm., July 2008; Robbins et al. 2009). This may be because of poor preservation, the impact of the funerary ritual of cremation on recognizing leprosy in the archaeological record, or the lack of paleopathological analysis in this part of the world.

In the first report of leprosy in human remains from India, Robbins et al. (2009, 6) described the skeleton of a middle-aged man dated to around 2500–2000 BC (post-urban Indus Age) with bone changes to the face; they posit that this is "the oldest example of lepromatous leprosy in the world." This evidence is approximately contemporary with the Scottish and Turkish skeletal data and the possible Nubian data, but is not as old as the Iranian evidence (see below).

The skeleton comes from Balathal (the contemporary state of Rajasthan) in northwest India. The man was buried in a stone enclosure filled with vitrified ash from burned cow dung. The burial practice suggests the Vedic tradition. At that time, it was customary to bury people with leprosy alive, although it cannot be proved whether this actually happened at this site (Cust 1881). Robbins et al. (2009, 7) suggest that the spread of *M. leprae* between Asia and Africa "most likely [happened] in the third millennium BC," a time when extensive contact through trade networks existed. Schug and colleagues (2013) have also reported evidence in Harappa in northeast Pakistan dated to the third to the second millennium BC (the period of transition from urban to post-urban). Three cemetery areas (R37: urban, 2550–2030 BC; H: post-urban/extension of urban, 1900–1300 BC; and G: ossuary; 2000–1900 BC) all revealed individuals with leprosy. Two people were buried at the R37 site, two at the H site, and five in the ossuary. The evidence for leprosy in the ossuary, which was located outside the city southeast of the city wall, suggests that people with the infection were excluded from their community, at least in death. This early evidence from Pakistan is approximately contemporary with possible evidence from India, Nubia, Scotland, and Turkey.

Data from neighboring islands (e.g., the Seychelles in the East Indian Ocean) show that leprosy was present in populations in the early modern period (early nineteenth century). From their first settlement in the early sixteenth century until 1903, the Seychelles were administered from Mauritius (Grainger 1980), and leprosy may have been introduced to Mauritius from France, which had previously administered the island. Colonial authorities isolated those affected with leprosy on the islands of Diego Garcia and Providence (Pridham 1846). In 1829, the Seychelles government acquired Curieuse Island, which it used as the islands' main leprosarium until 1900 (Grainger 1981). However, there were no laws in the Seychelles that required people with leprosy to go to a leprosarium until 1896. Curieuse Island was used again for people with leprosy from 1939 to 1968.

Concerns about the increasing visibility of leprosy in India in the second half of the nineteenth century (Kakar 1996) coincided with a concern for public health following the Indian Mutiny of 1857 and the British takeover of administrative authority from the East India Company (Buckingham 2002). The British focused on sanitary, medical, and social conditions for people living in India, clearly motivated by a desire to protect the health of the British army and British civilians. Of the three presidencies formed after 1857 (in Bengal, Bombay, and Madras), the Madras presidency (southern

India) has the most complete record of the activity of the British colonial government regarding leprosy care, even though the lowest number of people with leprosy were located there. It developed more care for people with the infection than the other two British presidencies and there are good records that describe the lived experience of these people (Buckingham 2002). The first leprosy census in India was conducted in 1872. However a report by the Royal College of Physicians (1867) said that because leprosy was viewed as a hereditary disease, segregation was not necessary. Despite improvements in knowledge about how leprosy was transmitted and the discovery of the leprosy bacillus in 1873, the Indian government ignored advice on segregation until leprosy was proved contagious. In 1889, the Leprosy Commission in India was founded (Kakar 1996). By 1899, there were nineteen leprosaria in India and by 1921 there were ninety-four. While controlling leprosy remains a challenge in India, data suggest that the number of new "cases" and the overall prevalence of the disease have declined markedly over the last twenty years or so. Since 2008, there has been an overall decline in new "cases" detected (from 134,184 in 2008 to 120,334 in 2018; World Health Organization 2019).

Iran and Uzbekistan

The only skeletal evidence of leprosy in Central Asia has been reported in Iran and Uzbekistan. In Uzbekistan, Blau and Yagodin (2005) identified a middle-aged woman with leprosy whose skeleton dates to the first to the fourth centuries AD. She was buried alongside an infant aged 6 to 18 months. The skeleton was one of fifteen individuals found at Devkesken 6 in a burial mound (Kurgan 5b) on the Ustyurt Plateau, which originally was part of the former Soviet Union (Blau and Yagodin 2005). The discovery of leprosy here suggests that trade routes, perhaps from India in the south, Iran in the southwest, and China in the east, brought the disease there. Of particular interest is that the people in this region were assumed to have been nomads. This contrasts with our understanding that often people with leprosy were members of sedentary agricultural communities with permanent settlements. However, there is no reason to suggest that nomadic populations could not contract leprosy because they are actually quite sedentary in one place at specific times in the yearly cycle, although in general their population densities are lower than for a sedentary population. *M. leprae* DNA has been also detected in skull and tibia samples from the woman with rhinomaxillary syndrome from Devkesken 6 (Taylor et al. 2009). The genotype of *M. leprae* in this individual was identified as Type 3, which is characteristic for Europe and North Africa (and exists in Central Asia today). Taylor et al. (2009), who compared the data with that revealed in an individual with leprosy from England that dates to the thirteenth to the sixteenth centuries AD (Blackfriars, Ipswich 1914), found the presence of unique SNP 3 in both skeletons, suggesting that it was ancestral to Type 3, subtype I, which is seen in the New World and Europe. Mycolic acid analysis of both individuals also confirmed the preservation of *M. leprae*.

Tuberculoid leprosy was very likely affecting a person whose remains were excavated from a site in Iran. The Neolithic mound site of Tappeh Sang-e Chakhmaq in northeast Iran is an early farming site. It includes two burial sites: East Tappeh (twelve skeletons) and West Tappeh (five skeletons). At the East Tappeh site, which dates to 6200–5700 Cal BC, the skeleton of a man 20–40 years old had bone changes consistent with leprosy (Tagaya 2014). Four right metatarsals displayed leprosy-related bone alterations on the distal ends; the bones were less than half the length of the corresponding bones on the left side. No phalanges of the feet were preserved and there were no alterations to the hand bones or on the skull, although the shafts of both of the lower leg bones (tibiae and the fibulae) had new bone formation on them, mainly affecting the right bones. As the East Tappeh site is dated to 6200–5700 Cal BC, that would make this skeleton the earliest evidence in the world if the diagnosis is accepted in a peer-reviewed publication. Of particular interest is that this person may have originated from Mesopotamia, west of Iran, as the person's skull shape suggests (Akira Tagaya, pers. comm., August 2016). However, other diagnoses should be considered for the changes in the foot bones, such as amputation of the forefoot due to diabetes. The latter is a more plausible alternative diagnosis if the bone changes are not considered to be due to leprosy. Diabetes usually affects the forefoot and midfoot bones—the tarsometatarsal, metatarsophalangeal, and intertarsal joints (Resnick 1995)—and may affect one or both feet. However, Matos (2009) and Matos and Santos (2013c) suggested that the combination of bilateral or unilateral involvement of the bones of the hands and feet and no rhinomaxillary syndrome would be acceptable for a diagnosis of tuberculoid leprosy. This skeleton had no evidence of rhinomaxillary syndrome and only one foot was affected; therefore this could be evidence for tuberculoid leprosy. However, a person with diabetes may be affected in only one foot too. To date, only one skeleton with possible diabetes has been diagnosed in palaeopathology (Dupras et al. 2010; see also Bieler-Gomez et al. 2018). Does the presence of periosteal lesions on the lower leg bones that may be the result of many possible underlying conditions (including leprosy) strengthen a leprosy diagnosis here?

Historical data suggest that leprosy was present in Iran by the ninth century AD. The first written evidence comes from Ali ibn-Rabban Tabari (AD 838–870). These data indicate that the Persian word for leprosy was *khoreh* and that it was later replaced by the word *jozam,* both of which mean an "eating away of the tissues of the body." In the nineteenth century, writers and travelers noted that people with leprosy were segregated with poor living conditions outside towns. By 1932, although 400 people with leprosy were officially recorded in Iran, many more were affected. Large leprosaria were opened at that time, for example one founded in Bababaghi in 1933 (Azizi and Bahadori 2011). It is believed that during the early twentieth century 12,000 people with leprosy were reported in Iran. By 1965 that number had decreased to 4,852 and by 1970 to 450 (Kohout, Hushangi, and Azadeh 1973). Although the World Health Organization considers that leprosy has been eliminated in Iran (using its standard of fewer than 1 in 10,000 people), Iran was one of seventy-seven countries that reported at least one new "case" of leprosy. It reported nineteen new "cases" in 2017 and twenty-nine in 2018 (World Health Organization 2018, 2019). It also has discriminatory laws against patients.

Turkey

There is evidence of leprosy in a very early skeleton from Karataş-Semayük in southwestern Turkey that was excavated from a site on the Elmalı plain (in the western Taurus Mountains, northwest of the coastal port of Antalya). The cemetery revealed a total of 540 individuals dating to the Early Bronze Age II period (2700–2300 BC). A middle-aged man (416ka, burial 286) who shared a grave with a woman is described with typical bone changes to the metatarsals of both feet (Angel 1969). While the image of the foot bones in Angel (1969) seem consistent with leprosy, it is difficult to be certain that this diagnosis is correct. If this person did have leprosy, then the evidence is fairly contemporary with possible evidence from India, Pakistan, Scotland, and Sudan (see above). Four skeletons diagnosed with leprosy have also been reported from the eighth- to tenth-century site of Kovuklukaya, near Sinop in northern Anatolia: two men, one woman, and one infant 4–5 months old (Yilmaz Selim Erdal, pers. comm., September 2008; Erdal 2004). Ancient DNA analysis revealed *M. leprae* DNA in two of the adult skeletons and in the infant (GM162; also reported in Rubini,

Erdal, et al. 2012). The remains of the child had no external signs of leprosy (Donoghue, Erdal, et al. 2005), and it was suggested that the baby had congenital leprosy based on the results of the aDNA analysis. The endocranial surface of the child's skull also displayed new bone formation, which may indicate a range of possible diagnoses (e.g., tuberculosis), but not leprosy (see Lewis 2004). More recent aDNA analysis of a bone sample from the skeleton from a mature woman (KK'02 20/1) buried at the same site also found *M. leprae* (Taylor and Donoghue 2011). The overall lack of evidence for leprosy in Turkey (only two sites) could be explained by the comparatively late development of palaeopathology in this country (Üstündağ 2011).

Southeast Asia

The history of leprosy in southeast Asia and the Pacific Islands is much more recent than for many other parts of the world, and some of these regions have not eliminated leprosy yet. In 2019, the World Health Organization reported that this part of the world had 114,004 registered "cases" and had detected 148,495 new "cases" during 2018, the most for any of the World Health Organization's regions (71 percent of the total new "cases"). The first observations of disease in human remains were made in the Pacific in the nineteenth century (Pietrusewsky and Douglas 2012). However, the development of paleopathology in Southeast Asia has been relatively slow (Tayles, Halcrow, and Pureepatpong 2012).

The following sections discuss some of the past and present evidence of leprosy in these regions. The modern-day country of Indonesia is the world's largest archipelago, consisting of a total of over 17,000 islands. Many of these islands are remote. There is no routine leprosy control program on them and access to health care is limited. This means that the frequency of leprosy can be high (e.g., see Naik et al. 1999; and Bakker et al. 2002). While skeletal remains of considerable antiquity have been found in Indonesia (Brown et al. 2004), as yet no skeletal evidence for leprosy has been reported. There is also no bioarchaeological evidence for leprosy in Malaysia, but there are more recent records of the segregation of people with leprosy. In the early twentieth century, a leprosarium was founded on Pangkor Island off Malaysia's northwest peninsula (Baird n.d.).

There is also no bioarchaeological evidence for leprosy on the archipelago of the Philippine Islands (over 7,000 islands in the western Pacific), but by 1906 there was compulsory segregation for people with leprosy in a leprosarium on Culion Island (Callender 1925). While the yearly average number of people reported with leprosy from 1906 to 1922 was 900 in the Philippine Islands overall, the actual number of people with leprosy on Culion Island was 5,600 in 1925 (ibid.).

To date, only two people have been archaeologically documented in Thailand with "probable" leprosy (Tayles and Buckley 2004). The Iron Age site of Noen U-Loke in northeast Thailand revealed the skeleton of a young adult man 16–20 years old (B107) dated to 300–200 BC. He was buried in a prone position (face down) with grave goods consisting of pottery, a fish skeleton, and a deer ulna. Another skeleton (B42), an older man dated to the first two centuries AD, was buried with four other people; the group was also interred with fetal age pigs. The bone changes of these skeletons strongly suggested leprosy or psoriatic arthritis. It could be hypothesized that skeleton B107 was buried face down because the individual had leprosy, but there are two other prone burials at the site, one in the same area as the older male. If the disease present in these individuals is leprosy, then it is possible, although cannot be proved, that the infection came from China or India, where there are earlier dated historical sources for leprosy.

There is also very little skeletal evidence for leprosy for the Pacific Islands. However, the majority of data attesting to leprosy's existence on some of the islands comes from the nineteenth and twentieth centuries, as described by Miles (1997). In the south-central Pacific, the best potential evidence for what appears to be leprosy is described in New Zealand in the eighteenth century. Montgomerie (1988) notes that leprosy was present in the Maori population there before European contact but that the evidence was "slim." There was also a delay of fifty to sixty years after European contact before definite leprosy was clinically recorded. However, visiting physicians who studied Maori people before contact (1840) had not recorded leprosy in that population. The first clear descriptions of leprosy in New Zealand date to the 1850s, which suggests that leprosy was introduced by travelers from other countries. It is believed that the disease was called *ngerengere* in New Zealand and that it could have been present in pre-European times. This was the name used in the southern part of the North Island, but it was sometimes called *tuwhenua* (*te waka tuwhenua* = canoe; i.e., brought by canoe; Gluckmann 1962). On South Island, the island which in some of the Maori legends existed before the North, it was called *tuhawaiki*. Indeed, there is convincing evidence from the nineteenth century that leprosy was introduced to New Zealand with Europeans as early as the end of the eighteenth century and that it was named *ngerengere* (Thomson 1854).

However, the term was used for other chronic diseases that caused skin problems, thus making specific diagnosis challenging from the evidence in these sources.

By the early twentieth century, three main areas of the Pacific were affected: Penrhyn, Manihiki, and Rakahanga, all in the Cook Islands. In 1926, the New Zealand government ordered a complete survey of leprosy for the Cook Islands and Fiji and agreed to send all people with leprosy to Makogai Island in Fiji for treatment. This was the practice until 1953, but after that it ceased because of the increasing success in treatment using new drugs and a lack of space at Makogai. After the Makogai option for segregation was no longer possible, all treatment was done in outpatient clinics. The presence of more people with leprosy in Aitutaki, a Cook Island, eventually led to the establishment of a Leprosy Treatment Centre there. The Bacillus Calmette-Guérin (BCG) vaccination, slum clearance, and an increase in health care staff led to a decline in leprosy from 84 new "cases" in 1957 to 1 new "case" in 1961 (McCarthy and Numa 1962). In the South Pacific, leprosy was also present on Easter Island (Rapanui) (García-Moro et al. 2000). It was said to have been introduced from French Polynesia at the end of the nineteenth century. It became endemic, and by 1917 a leprosy colony had been created there. By the 1980s, the infection had been eradicated on Easter Island and the last "death from leprosy" occurred there in 1988 (Cruz-Coke 1988). No skeletal remains from an archaeological site have been identified with leprosy on this island.

The only bioarchaeological evidence in the Pacific comes from the island of Guam, the largest of the Mariana Islands in Micronesia. Trembly (1995, 2002) identified at least six adult individuals with leprosy from pre-Spanish contexts. Some of these skeletons were radiocarbon dated to the seventh–eleventh centuries, the eleventh–thirteenth centuries, and the thirteenth–fifteenth centuries AD. Suzuki (1986), who also examined skeletal remains from Guam (but gave no information about provenance or dating), found three skulls with rhinomaxillary syndrome. Overall, leprosy in archaeological human remains on the Pacific Islands and in southeast Asia is rarely seen and the infection seems to have been a recent introduction, perhaps becoming more likely as explorers discovered these islands. The absence of evidence of leprosy also perhaps reflects the relative lack of archaeological exploration of the thousands of islands in this part of the world; poor preservation or nonsurvival of skeletal remains due to hot, humid conditions; and the absence of intensive development of palaeopathology in this region of the world.

There is no documented evidence for leprosy in skeletal remains from Australia or New Zealand (e.g., see Webb 1995). However, Møller-Christensen and Inkster (1965, 4–5) describe a "skull of a leprosy patient . . . sent to England in 1908" (a female adult) from Port Darwin, Australia, that is now in the Anatomical Museum at the University of Edinburgh in Scotland. Webb (1995) also reports no evidence of leprosy in any of the 4,500 "Aboriginal" Australian skeletons dated to before 1788 that he examined for disease. There appears to be no other author who refers to leprosy in Australia.

BIOARCHAEOLOGICAL EVIDENCE FOR LEPROSY IN THE NEW WORLD

Possible Skeletal Evidence in North America

Despite the long history of paleopathological studies in the New World (see Buikstra and Roberts 2012), leprosy is virtually absent in New World human remains, suggesting that the disease was a much later arrival in this part of the world. Most of the data in the New World that relate to leprosy are historical and from the post-Columbian period, most notably in the nineteenth and twentieth centuries. This finding corroborates Monot et al.'s (2005, 2009) research papers, which suggest that leprosy arrived in the Americas with the slave trade: "From West Africa, leprosy was then introduced by the slave trade in the 18th century to the Caribbean islands, Brazil, and probably other parts of South America, because isolates of *M. leprae* with the same SNP Type 4 are found there as in West Africa" (Monot et al. 2005, 1042; see also Ashmead 1895, 1896). In 1982, Hudson and Genesse claimed that there was no evidence of leprosy in Native American skeletal remains, and Hutchinson and Weaver (1998, 450) also noted that "at present there are no accepted pre-Columbian cases of leprosy in North America." This picture has not changed much since then.

However, Møller-Christensen and Inkster (1965) describe the skull of a mature "Comox Indian" found on Vancouver Island in Canada in the osteological collection of the Anatomical Museum at the University of Edinburgh (see also Ortner 2003, 266). The Comox (or Catlo'ltx) are a native group on the east coast of the island (Jerry Cybulski, pers. comm., March 2013). Møller-Christensen and Inkster (1965) claimed that the skull was "the first described case of *facies leprosa* in an American Indian skull." Dr. R. N. Turnbull gave the skull to the museum in 1866 (Accession Number 32.b.9). The images in Møller-Christensen and Inkster's 1965 article look

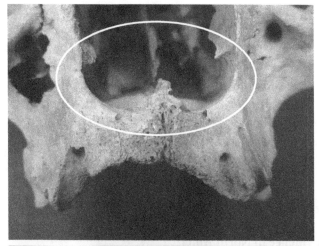

Figure 5.22. Alveolar process of the maxilla of the skull of a "Comox Indian" from Vancouver Island, Canada. Photo by Charlotte Roberts.

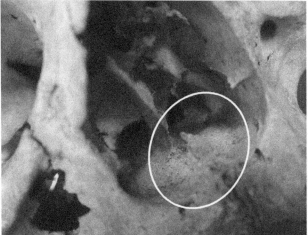

Figure 5.23. Nasal surface of the palate of the skull of a "Comox Indian" from Vancouver Island, Canada. Photo by Charlotte Roberts.

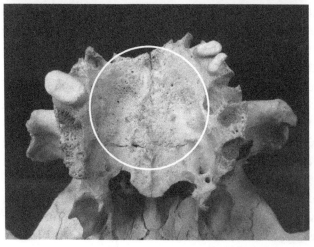

Figure 5.24. Oral surface of the palate of the skull of a "Comox Indian" from Vancouver Island, Canada. Photo by Charlotte Roberts.

convincing, and re-assessment in March 2013 confirmed that the skull of an adult woman displays bone changes that appear to represent rhinomaxillary syndrome (Figures 5.22, 5.23, and 5.24). This was corroborated by Keith Manchester (University of Bradford) and Vitor Matos (University of Coimbra) (pers. comm., August 2016). Inflammatory pitting of the oral surface of the hard palate, pitting and new bone formation on the nasal palatal surface, widening and remodeling of the nasal aperture, loss of the alveolar bone and anterior teeth, and loss and remodeling of the anterior nasal spine were noted. It is of particular interest that D'Arcy Island in the Haro Strait between Vancouver Island and the mainland of Canada was opened for Chinese immigrants with leprosy in the twentieth century (see below). Consultation (in October 2013) with Chris Dudar at the Smithsonian Institution in Washington DC and Jerry Cybulski at the Canadian Museum of History in Québec has explored whether the person was native to Canada, whether the skeleton is historic or prehistoric in date, and if the loss of the anterior teeth and recession of alveolar bone might be related to deliberate tooth ablation rather than leprosy, but there are "no references to tooth ablation in ethnographic or historical literature or in prehistoric skulls" for this region (Jerry Cybulski, pers. comm., March 2013). Origin and dating are not resolved.

It is noteworthy that in an article on smallpox on the Northwest Coast in the early nineteenth century, Gibson (1982–1983) mentioned leprosy as a possible problem for the native population. Suttles (1990) cites leprosy as a possible "aboriginal disease" in Northwest Coast groups in the *Handbook of North American Indians,* and R. Boyd's (2000) book on introduced infectious diseases among Northwest Coast native populations in the late eighteenth–late nineteenth centuries AD has an appendix on "unidentified Columbian leprosy." Ortner (pers. comm., August 2008) also discussed the skull of a woman with purported bone changes of leprosy that Aleš Hrdlička acquired from the Old Karluk (KAR-31) site on Kodiak Island off the south coast of Alaska (catalog number P366639, Smithsonian Institution). Photographs were sent to Keith Manchester, who agreed with this diagnosis, as did Vitor Matos (pers. comm., August 2016), but a differential diagnosis of treponemal disease should also be considered, at the very least. The dating of this skull is complex but the site was the first Russian settlement on the Karluk River (Cynthia Wilczak, pers. comm., 2018; Wilczak and Ortner 2013). Wilczak and Ortner also note that the skull is marked with a "red line," which was part of Hrdlička's method for differentiating Kodiak Island's cultural phases: red indicated one of two layers that Hrdlička's termed "pre-Koniag." While achieving precise dating is a complex task, Wilczak indicates that Hrdlička's "pre-Koniag" layers roughly correlate with the Early (4000–2700 Cal BP) and Late (2700–900 Cal BP) Kachemak phases. Ortner was of the opinion that "a pre-Columbian date is certainly plausible because the remains came from a very deep layer," although he notes that "Hrdlicka's stratigraphic and dating records were far from ideal" (pers. comm., August 2008). More recent research on this skull is ongoing and leprosy is a possible differential diagnosis, but "Hrdlicka's field notes, letters and the accession records" show that he likely did not excavate the skull himself and its condition suggests that it was exposed to the weather on Kodiak Island (Cynthia Wilczak, pers. comm., 2018).

Leprosy did not become a concern for the Americas until much later than the dates of these two skeletons. As early as 1639, immigrants from France settled in New Brunswick, Canada, and in neighboring Nova Scotia, and after 1755, many Scottish people arrived from the Outer Hebridean and Shetland Islands to the west and north-east of mainland Scotland, respectively (Jopling 1991). In the early nineteenth century, immigrants from Iceland, Russia, and Syria settled in Manitoba and Saskatchewan (Aycock 1940), but today most evidence of leprosy in North America comes from immigrant populations, especially in Texas and the Gulf Coast (Anderson et al. 2007). The average annual incidence for Texas was reported to be 1.9–2.4 "cases" per million. Forty-eight percent of 413 native-born Americans reported with leprosy in the period 1973–1997 were Hispanic (Taylor et al. 1999). California, Louisiana, New York, and Texas contributed 64.7 percent of the 133 individuals with leprosy reported in 2002 in the United States. New "cases" of leprosy in the Americas continue to be detected.

Canada

The first evidence of leprosy in Canada is described for New Brunswick (eastern Canada) in 1815, but there are various theories about how leprosy was transported to North America and particularly to Canada (Heagerty 1933). These include 1) after two sailors from the Levant landed there, they walked to Tracadie in New Brunswick and received hospitality from a French family, certain members of which contracted leprosy later; and 2) that a Scotsman came to Nova Scotia and brought leprosy with him.

By 1844, the people of New Brunswick were pressing the government for a leprosarium. The first one was created at Tracadie on the eastern coast.[7] Kalisch (1973, 489) notes that it was "dull, desolate, and monotonous"; it was effectively a barracks surrounded by a fifteen-foot-high fence, although it is also claimed that people were well looked after there compared to the people at D'Arcy Island (see below; Kula and Robinson 2013). Later, other provinces applied for people with leprosy to be admitted to Tracadie. After patients burned Tracadie to the ground in 1852, they spent the next winter in very poor conditions in the building described as the prison for criminals with leprosy. "They were left to rot for the winter months as the bodies of unattended dead patients were allowed to lie in bed for days at a time" (Kalisch 1973, 490). A new building was erected in 1853, and in 1880 New Brunswick appealed to the Canadian government to take over the running of Tracadie.

In 1906, the federal government passed the Leprosy Act, which mandated compulsory segregation for people with leprosy. D'Arcy Island off the east coast of the southern part of Vancouver Island in the Haro Strait, British Columbia, opened in 1891 for Chinese immigrants with leprosy. In 1906, it housed twenty-three patients with leprosy (Kula and Robinson 2013). It was administered variously by the Victoria municipal government, the British Columbia provincial government, and the federal government. In 1924, when it closed, the people with leprosy who were still on D'Arcy Island were moved to Bentinck Island, also off the coast of Vancouver Island. By late 1932, nine people with leprosy who had been born in Britain, France, and Russia were at Tracadie and five people were at Bentinck Island. Heagerty (1933, 466) wrote that leprosy was "virtually extinct" in Canada by 1933 and that any new evidence in people would be in immigrants.

United States

Kalisch (1973) documents the history of leprosy in the United States, particularly in Massachusetts, where the earliest record of leprosy dates to 1825. As the number of people with leprosy increased from the 1870s onward, a 68-acre farm near Brewster, Massachusetts, on Cape Cod was purchased for segregation. When local residents said it would do great damage to the tourist trade, the town of Brewster offered to re-purchase the farm and suggested that an island would be better for people with leprosy. Thus, the Massachusetts Board of Charities purchased Penikese Island in Buzzards Bay (Sabin 1981). The facility included a farmhouse for the medical superintendent and cottages for the patients and the patients were free to roam the island. Treatment included exposure to fresh air and sunlight, rest and a good diet; chaulmoogra oil was also used. In 1921, Penikese Island was abandoned and the remaining patients were transferred to Carville in Louisiana (Sabin 1981). Kalisch (1973, 501) argued that segregation at Tracadie

(Canada) and Penikese "had its origin and stimulus in Biblical precedent"; religious groups in New Brunswick and Massachusetts also believed that "lepers" in the Old Testament were afflicted because they had committed sin.

Some have suggested that migration of people with leprosy from Canada and elsewhere to the United States was a source of the infection in the United States in the past. For example, Aycock (1940) also reported that immigrants from Norway and Sweden introduced leprosy into Minnesota in the mid-nineteenth century. Feldman and Sturdivant (1975, 1976) recorded clear evidence of leprosy in Louisiana from the second half of the seventeenth century, a time when French families from Nova Scotia migrated to the colony (Meyer 1955; Jopling 1991). In the twentieth century, Hudson and Genesse (1982) describe seven centers of infection in the United States: the Upper Mississippi River Valley (Washburn 1950), Louisiana (from areas French Canadians migrated to), Texas (from people who had contact with French Louisiana), Florida (contact with the West Indies and the Caribbean), California (from Mexico, China, and the Philippines), Hawaii (mainly immigrants, particularly from the Philippines), and New York (immigrants, mainly from the Caribbean and Puerto Rico). From 1942 to 1982, the frequency of leprosy in the United States increased slightly, but 76 percent of the estimated 1,432 "cases" reported in the period 1967–1976 were born outside the United States; immigrants from Mexico and the Philippines contributed around half of the frequency rate. Of particular note is that in the late nineteenth century, when the number of people with leprosy increased in the US, the US Public Health Service opened a hospital in Carville, Louisiana (see excellent discussion in Martin 1963; Stein and Blochman 1973; Fairchild 2008; Ramirez 2008; Manes 2013; Atkin 2018). Leprosy in people living on the Hawaiian Islands of the United States also has a recent documented history, but there is no bioarchaeological evidence. From 1866, the settlement on the peninsula of Kalaupapa on the northern coast of the island of Molokaʻi was one of the largest "colonies" for people with leprosy in the world (see Law 2012; Tayman 2006).

Mexico and Central America

In Mexico and Central America, the only bioarchaeological evidence for leprosy is in Mexico and it has yet to be peer reviewed or published. Ruis González and Serrano Sánchez (2015) describe a young adult skeleton from Santiago Tlatelolco in Mexico City (colonial period: early fifteenth to early eighteenth centuries AD). This is interesting in light of the discovery in 2008 of *M. lepromatosis* in Mexican-born people (Han, Sizer, and Tan 2012). As is the case in the Caribbean (see below), it is believed that the Spanish introduced leprosy to Mexico in the early sixteenth century. The first leprosarium in Mexico was founded in 1528 in a country house in Txaplana (de las Aguas 2006). The second hospital was founded in the Mexican shipyards in 1572 by the Spanish doctor Pedro Lopez, suggesting a point of need for people with leprosy who were entering the country. The disease increased because of trade with the Philippines and because of the enslavement and transportation of people from Africa. In the Yucatán Peninsula, the first leprosarium was founded in 1793, but by the mid-nineteenth century, leprosaria were being closed there and the remaining sick were transferred to general hospitals. In Costa Rica, the first evidence of leprosy is from the early eighteenth century. A "national asylum for lepers of Mercedes" was founded in 1909 (de las Aguas 2006). In Nicaragua, there is evidence of leprosy in the nineteenth century; the sick were segregated on an island off the Pacific coast in 1893.

Caribbean

Cuba has no evidence of leprosy historically until the arrival of Spanish explorers in the sixteenth century (Gonzalez Prendes 1963) and much later when African slaves were transported there at the end of the eighteenth century. In the seventeenth century, segregation was practiced for people with leprosy, who often lived in huts in caves (de las Aguas 2006). Later, the first leprosy hospital (St Lazaro), a converted ranch, was founded in Havana. When it was destroyed by a cyclone in 1703, the sick had to be moved back to the caves. In 1708, King Philip V of Spain granted land for a new hospital and in 1714 he ordered that it be built. The British destroyed this hospital in 1862 when they invaded Cuba. It was rebuilt in 1898, but in 1899 the Americans decided to move patients out of the city to Loma la Vigía in Mariel, about 25 kilometers west of Havana. In 1917, a new leprosarium was built in El Rincon, located 35 kilometers from Havana, that served all of Cuba. In 1964, about 4,000 people had leprosy in Cuba, but the frequency of the disease decreased after drugs had been developed to treat it. The only skeletal evidence for leprosy in the Caribbean is of very recent date, from the site of a leprosarium in St. Eustatius in the Caribbean Antilles that was opened in 1866 (Gilmore 2008). Five skeletons were excavated at this site in 1989. Three of the five skeletons (a young adult female and two older adults) are described with leprosy-related bone changes.

South America

South America's experience with leprosy is similar to that of the United States. There is no bioarchaeological evidence for leprosy there, but leprosy is well documented historically, particularly from eighteenth- and nineteenth-century sources and briefly described below (de las Aguas 2006; see also Avelleira et al. 2014). The Portuguese brought leprosy to Brazil in 1496 and its frequency rate increased when they began enslaving and transporting Africans there. As the Portuguese set colonized indigenous people to work in various industries, the infection spread. Brazil, a pioneer in the fight against leprosy, built its first leprosaria in the sixteenth century. The name leprosy was also banished in that country in the 1980s in favor of the name Hansen's disease or Hanseniasis. Brazil has one of the highest prevalence rates of leprosy today; in 2018, 28,660 new "cases" were detected there (World Health Organization 2019), the second highest number in the world for that year. Brazilians have been very active in research on the infection and continue to contribute much to the clinical leprosy literature.

De las Aguas (2006) discusses the first evidence for leprosy in Argentina in 1792 due to contact with Spain. There is further historical evidence from the nineteenth century, when people with leprosy migrated there from Bolivia, Brazil, and Paraguay. In 1926, the Aberastury Law ordered that people with leprosy had to be segregated and prohibited them from marrying. In Venezuela, there is evidence that a leprosy hospital was founded in Maracaibo in 1841 and that two others were founded, one on the island of Providencia and the other in Cabo Blanco, near La Guaira (de las Aguas 2006). In Colombia, a leprosarium was founded in Cartagena in the late sixteenth century, but by the late eighteenth century it had closed and people with leprosy were moved to a leprosarium away from the coast at Caño del Oro. By 1929, there were three leprosaria in Colombia. The infection entered Ecuador via the slave trade from Colombia and Peru. In 1679, the sick were isolated in Guayaquil and Cuenca by royal warrant. In Peru, leprosaria were opened in the late nineteenth century in Cuenca in Azuay Province and in Miraflores in Lima Province.

Summary of Bioarchaeological Data for Leprosy

Most of the bioarchaeological evidence for leprosy is found in the Old World on three continents. There are only three possible instances of leprosy in skeletal remains in the New World, one in Mexico City and two on islands off the northwest coast of North America (Kodiak Island in Alaska in the United States and Vancouver Island in Canada). However, specific dates for these latter two individuals are not available. Leprosy appears to have been a disease of the northern and eastern hemisphere in the past. The data in the Old World is biased toward Northern European countries; less evidence comes from the Mediterranean and elsewhere.

Hunter and Thomas's (1984) study of leprosy's occurrence and distribution in Africa in area surveys dated from 1953 to 1970 is of interest for discussions of the global distribution of leprosy in skeletons. They reported that leprosy during this time period was most frequent in the higher latitudes of Africa but in tropical areas in the 1980s, as seen today. They suggested that the higher prevalence of leprosy in tropical areas might be because *M. leprae* might survive better outside the host in those regions. There also seemed to be an association between high rates of leprosy in Africa and heavier rainfall (1,016–1,778 millimeters annually) and a higher temperature (23.9–29.4° C, or 75.0–84.9° F); rainfall was the stronger variable of the two. However, Northern Europe, which has less rainfall compared to the African tropics and a lower temperature than Africa (but more rainy days), experienced a high frequency of leprosy in the past. However, they concluded that there is "some, possibly weak, degree of environmental influence at work affecting the geographic distribution of leprosy [at least] in Africa" (Hunter and Thomas 1984, 41). Nevertheless the environmental data analyzed do not corroborate the ancient skeletal evidence as it stands today.

While the evidence for leprosy in human remains in the Old World is quite extensive, the infection is absent in many parts of this region. However, the actual number of skeletons (and preserved bodies) that have been excavated and diagnosed with leprosy to date is relatively low. This probably reflects a number of factors, including the lack of development of paleopathological studies (and training) in those regions. Other factors could be that the infection is invisible (e.g., the practice of cremation and environmental factors that affect preservation could destroy affected bones or erode away leprosy-related damage on the bones of the face, hands, and feet), and the possibility that people with leprosy were not buried normally in the usual burial ground for the community. Or it may be the case that there really is no evidence.

278 · Leprosy

Table 5.2. Archaeological sites where skeletons have been buried "differently"

Country	Place Name	Notes	Source(s)
England	Beckford	Marginal but buried with many grave goods	Evison and Hill (1996)
	Bingham	Possibly at the top of a well; 10 meters away from the rest of the cemetery	Allen (2018)
	Raunds	Marginal	Boddington (1996)
	St Mary Spital, London	Marginal and in a mass burial	Connell et al. (2012)
	St Stephen's, York	Marginal	McComish (2015); Holst (2012)
Scotland	Kirk Hill	"Redeposited"	Lunt (1997, 2013)
Ireland	St Michael Le Pole, Dublin	Marginal	Buckley (2008a, 2008b)
Hungary	Kiskundorozma-Daruhalom	Marginal but one of richest graves of the cemetery	Pálfi and Molnár (2009)
Poland	Kaldus	Some buried differently	Kozłowski (2012b)
Portugal	Valle da Gafaria	Two individuals not buried in the normal Christian way	Ferreira, Neves, and Wasterlain (2013)
Sweden	Lund	Some buried in outer zone of St Trinitatis church	Arcini (1999)
	Sigtuna	Outer zone of churchyard	Kjellstrom 2012
Thailand	Noen U-Loke	Prone burial	Tayles and Buckley 2004

Conclusions

Our knowledge of the distribution of leprosy in skeletons from the past is influenced by archaeological excavation. In many countries, it is a legal requirement that sites be excavated archaeologically before land is developed. In most cases, land development takes place in urban settings. For example, in England, most development happens in the south and east and in towns and cities. This will bias data that is collected for leprosy from excavated skeletons. Where there is no evidence for leprosy, it may be because everyone in that part of the world was buried in leprosy hospital cemeteries because they were all segregated, or because remains of people who had leprosy have not yet been found. The actual distribution of leprosy in archaeological human remains will no doubt change as more evidence is found.

Although there is less bioarchaeological evidence for leprosy in southern latitudes of the world, there are many reasons for this situation, including poor preservation of human remains in tropical areas, a lack of training in bioarchaeology, and a low number of paleopathological studies. Although there is much bioarchaeological evidence for leprosy from Northern Europe, particularly in later time periods (the early medieval period onward), on the basis of the evidence to date, we know that leprosy had an early focus in the modern-day countries of Hungary, India, and Pakistan and possibly in Iran, Nubia, Scotland, and Turkey. These data now need to be related to what has been concluded from the genomic data for geographically related types and subtypes of *M. leprae* DNA that Monot and colleagues (2005, 2009) produced (see Chapter 6). The majority of the skeletons with

Table 5.3. Some scenarios for leprosarium and non-leprosarium burial contexts where skeletons with and without leprosy have been found

With bone changes	Without bone changes
SKELETONS IN LEPROSY HOSPITAL CEMETERIES	
Diagnosed correctly	Diagnosed incorrectly
Had low-resistant form of leprosy	High-resistant form
	Had only external signs and no bone change
	Had other diseases
	Diagnosed correctly but had no bone changes/died before bone changes occurred
	The person adopted a leprous appearance so they would be segregated into care
SKELETONS IN NON-LEPROSY HOSPITAL CEMETERIES	
Had high- or low-resistant form of leprosy	May have had leprosy at death but had no bone changes or external signs
Accepted in community and not stigmatized	May have had only external signs and no bone changes but were accepted in their community
Not diagnosed as leprous (no external signs)	
Evaded diagnosis	
Not accepted for admission to leprosy hospital (poor or too ill)	

bone changes of leprosy around the world have been found in non-leprosy hospital cemeteries and were buried normally for their communities in those regions and time periods. Although many locations of leprosaria are known from historical data, relatively few leprosy hospital sites have been excavated (see Appendix 4). There are also very few instances where individuals with leprosy were buried in a different way from the rest of the community (Table 5.2) or in isolated places such as islands.

Table 5.3 provides various interpretations of the bioarchaeological data. For example, in Britain just over 200 skeletons with leprosy have been identified from over fifty archaeological sites, seven of which were definitely leprosaria cemeteries. These findings contrast with those of Magilton (2008c, 23), who stated that there was an "absence of lepers from 'normal' cemeteries." The finding in the current study also contradicts the evidence in the vast majority of historical texts that infers that people with leprosy were stigmatized, ostracized, and marginalized or segregated in the past. Individual experiences of leprosy will have varied considerably over time, as would community reactions, which would have depended on many factors. If leprosy was as common as historical data show, it may have been considered a normal part of life in medieval Europe to have leprosy. Stigma and ostracism may not have been issues people with leprosy had to contend with. However, even if stigma and segregation was experienced in life, that did not mean segregation after death.

Leprosy could have a history of over 8,000 years, as seen in bioarchaeological evidence, but this picture will change as more evidence is uncovered. In the past, leprosy appears to have been a disease of the Old World and the northern and eastern hemispheres. The earliest evidence is possibly from Hungary, Iran, Pakistan, Scotland, Sudan, and Turkey, while Britain, Denmark, Hungary, and Sweden have the most skeletal evidence. Northern Europe has the most bioarchaeological evidence. None has been confirmed in the Americas, but there is evidence in more recent documentary data for leprosy in the Americas as a result of the slave trade and colonialism. Leprosy in nonadult skeletons is rare, and leprosy has been diagnosed only in one preserved body, an Egyptian mummy. The majority of skeletons with leprosy are buried normally for the time period, culture, and location. The bioarchaeological evidence does not corroborate the historical evidence for the frequency of leprosy, and some bioarchaeological evidence predates the historical data.

6

Reconstructing the Origin, Evolution, and History of Leprosy

> Ancient ships are commonly described as conveyors of merchandise, technology, and even ideas, but they are seldom mentioned as conveyors of disease in the literature. This is surprising because both passengers and crew typically had to endure poor hygiene, poor nutrition, and cramped quarters, especially on long voyages. All of these factors provide an excellent environment for the incubation and spread of disease.
>
> Mark (2002, 285)

In 1982, Jopling said that "nothing definite is known as to where and when Hansen's Disease (HD) has its origin, but it is generally believed to have originated in Asia. It was then introduced into Greece by returning soldiers and camp followers, after the wars in Asia in the fourth century BC, and from Greece it slowly spread throughout Europe" (10). He was of the opinion that leprosy came to the Caribbean and Latin America in very recent times and likely was introduced by European explorers and settlers from France, the Netherlands, Portugal, and Spain (as noted also by Møller-Christensen 1976) and later by immigrants from Australia, the Pacific Islands, and possibly China. The World Health Organization (1988b) also believed that leprosy reached Central and South America via European immigrants in the sixteenth century and that it entered North America in the eighteenth and nineteenth centuries as Norwegian immigrants entered Canada and French people moved into Louisiana. Dayal et al. (1990) and Lechat (1999) agreed with the European immigration hypothesis, but they also believed that slaves taken from Africa and laborers taken from China transported leprosy to the Americas. Møller-Christensen (1976) suggested that leprosy spread to West Africa from East Africa and Mark (2002) added the idea that Indian slave owners transported slaves to East Africa, possibly as early as the third millennium BC, thus suggesting that leprosy was present in Africa very early in time.

A number of scholars have provided more nuanced discussion about how leprosy may have been transported with humans around the world. Møller-Christensen (1976) discussed the hypothesis that leprosy originated in China (where the current earliest skeletal evidence dates to 200 BC) and India (where the current earliest skeletal evidence dates to 2000 BC) and was then taken to the Middle East at the time of Alexander the Great after he and his army campaigned (and contracted leprosy) in India in 327–326 BC (see also Hunter 1986; Dayal et al. 1990). According to this theory, Gnaeus Pompeius Magnus (Pompey the Great) and his army then took leprosy from the Middle East into Italy after his success in the Mithridatic Wars in the first century BC. However, contradicting this idea, the earliest skeletal evidence from this part of the world dates to the fifth century BC. Mark suggests that slaves with leprosy may well have been imported to the Mediterranean via this route; we know that Rome was one of the largest trading partners of India by the first century AD (Mark 2002). From Italy, invading Romans spread leprosy to Germany in the second century AD (but the current earliest skeletal evidence from Germany dates to AD 450) and Spain in the fifth–sixth centuries AD (but the first earliest possible skeletal evidence from Spain dates to around the turn of the first millennium). Another theory is that Saracens (Muslims) took leprosy to France in the eighth century AD, although the current skeletal evidence there dates to as early as AD 300 (see Hunter 1986). However, Alexander and his army were not the first to invade India, and any number of people could have been responsible for the transmission of leprosy along those routes to countries in the west. Grmek (1989) and Wilson (1966) hypothesized that Mesopotamia (modern-day Iraq) harbored leprosy in its population by the second millennium BC based on documentary sources, but they were not totally convinced. There is currently no skeletal evidence in Iraq, but older possible evidence has recently been reported for Iran (Tagaya 2014). Of note is early skeletal evidence for leprosy from India (2000 BC), and possibly Iran (6200–5700 BC), Sudan (2300 BC), and Turkey (2700–2300 BC). Clearly, some of the bioarchaeological evidence supports some of these theories, for example the earliest evidence of leprosy in India, but the evidence does not support a theory that leprosy originated in China. Of course, absence of evidence does not mean evidence of absence and the picture will change as more skeletons with leprosy are diagnosed. More comprehensive analyses of skeletal remains for identifying leprosy in China, India, Iraq, and Iran would contribute to better understanding of the early history of leprosy.

Leprosy and the Epidemiological Transitions

How does the evidence for leprosy relate to the three epidemiological transitions? The first took place 10,000 years ago, when humans began to have a more sedentary life, the size and density of the human population increased, animals and plants were domesticated, people began eating poorer and less varied diets, standards of hygiene decreased, animals transmitted diseases to humans (zoonoses), and the number of diseases that depend on population density increased (Barrett et al. 1998; Roberts 2015). Leprosy does not appear in the bioarchaeological evidence until this transition. As living conditions deteriorated in the late medieval period in urban situations (Pinhasi, Foley, and Donoghue 2005), the disease became more frequent. It may have affected the poorer sections of society disproportionately and an increased population density facilitated the transmission of the infection. Mobility and trade further facilitated the transmission of leprosy more easily across the globe. However, there is evidence of leprosy in skeletons before urban life, for example in the early medieval period in Britain.

The general picture reveals that by the time of the second epidemiological transition in the nineteenth and twentieth centuries AD, the incidence of non-infectious (or noncommunicable) degenerative diseases (e.g., heart disease and cancers) increased as people started to live longer, and the incidence of infectious diseases decreased (Harper and Armelagos 2010). Leprosy was an increasing challenge around the world in the nineteenth century, but after the advent of antibiotic treatment in the 1940s and 1950s and as lifestyles improved, the frequency of the disease eventually decreased. We are now in the third epidemiological transition, which started at the end of the twentieth century. This is when infectious diseases started to reemerge as resistance to antibiotics occurred through misuse and overuse (Barrett and Armelagos 2013). Although there is evidence of drug resistance in people with leprosy (see Williams and Gillis 2012), World Health Organization data indicate that the frequency of leprosy is continuing to decrease, at least for now.

While archaeological evidence for leprosy is focused in a number of regions, the majority of the evidence comes from later medieval urban northern Europe (see Chapter 5). Like tuberculosis, it appears to have been associated with urban life and likely with poverty (as Jopling 1982 noted for the tropics). Some of the bioarchaeological evidence of leprosy is earlier than the dates of accepted historical references, for example in India. However, some bioarchaeological data are later than historical sources indicate, for example in Greece and China. In addition, there are parts of the world where written documentary evidence of leprosy is plentiful but there is little or no bioarchaeological evidence (e.g., Finland, Iceland, and Norway).

While we know that leprosy flourished in Europe until the later part of the late medieval period, there is a lack of skeletal evidence for the end of the late medieval period (mid-sixteenth century AD) to the early nineteenth century, when it appeared in the Americas. For example, in England, only two archaeological sites have revealed post-medieval skeletal evidence (AD 1550–1850; see Chapter 5). While there is some evidence in a number of modern-day countries, there is a considerable gap in the historical data (Green 2012), and it is not until the second epidemiological transition that more evidence is documented historically for leprosy globally, at least in the nineteenth and early twentieth centuries.

Ancient and Modern DNA Evidence and the Movement of People with Leprosy

This section focuses on Monot and colleagues' (2005, 2009) theory of the origin, evolution, and transmission of leprosy based mainly on modern genomic data for leprosy from samples from patients. In 2005, Monot et al. suggested that leprosy originated in Eastern Africa or the Near East and spread via migration and that Europeans and North Africans probably introduced leprosy into West Africa and the Americas over the last 500 years (see also Mark 2017). Based on their data, they also showed that all present-day leprosy could be attributable to a single clone (an exact genetic copy) and that dispersal of the bacterial types and subtypes could be tracked by analyzing the DNA of *M. leprae* in patients. They also reported that *M. leprae* had experienced reductive evolution, or downsizing of the genome, and gained pseudogenes (sections of a chromosome that represent imperfect copies of a functional gene). They aimed to establish whether all strains of *M. leprae* had undergone similar events and to determine the level of relatedness of *M. leprae* strains by studying their evolutionary relationship. The modern *M. leprae* strains studied derived from patients in Brazil, Ethiopia, India, Mexico, Thailand, and the United States. The genomes of each strain were practically identical, but a correlation existed between where the patient with leprosy was located geographically and different leprosy types (single nucleotide polymorphism, or SNP), as follows (Monot et al. 2005, 1041):

The single nucleotide polymorphism Type 1 predominates in Asia, the Pacific region, and East Africa.

The single nucleotide polymorphism Type 2, the rarest, is detected only in Ethiopia, Malawi, Nepal/North India, and New Caledonia.

The single nucleotide polymorphism Type 3 predominates in Europe, North Africa, and the Americas.

The single nucleotide polymorphism Type 4 predominates in West Africa and the Caribbean region.

These results suggested two scenarios (Monot et al. 2005, 1042):

1. Type 2 spread east and preceded Type 1, and Type 3 disseminated west and then gave rise to Type 4.
2. Type 1 was the progenitor (an ancestor in the direct line) of Type 2, and Types 3 and 4 followed in that order.

Monot and colleagues hypothesized that infected explorers, traders, or colonists of European or North African descent likely introduced leprosy into West Africa rather than migrants from East Africa because Type 4 was much closer to Type 3 than it was to Type 1. The movement of people through the slave trade introduced Type 4 to the Caribbean and South America in the eighteenth century. The type of *M. leprae* responsible for leprosy in North America is thus closest to the European and North African variety—Type 3 (Monot et al. 2005). Thus, travel from the Old World to the New World likely contributed to the introduction of leprosy into the New World at a late date. Corroborating this intriguing theory, Truman et al. (2011) found the same previously undescribed and unique *M. leprae* strain (3I-2-v1) in most of the thirty-three wild armadillos they studied using genomic analyses and in almost two-thirds of thirty-nine people affected by leprosy who lived in areas of the southern United States, where they could have been exposed to *M. leprae* in armadillos.

In addition to the four main SNPs of *M. leprae* Monot et al. (2005) highlighted, Monot and colleagues (2009) identified some additional SNPs that enabled them to subtype the four main types into sixteen subtypes: Type I: A through D; Type II: E through H; Type III: I through M; and Type IV: N through P. These sixteen subtypes had "limited geographic distribution that correlate with the patterns of human migration and trade routes" (1283). Thus, the data led to a more nuanced understanding of how leprosy spread around the world from its origin. Monot and colleagues (2009) also analyzed ancient DNA of *M. leprae* from bone samples from twelve archaeological skeletons from Croatia, Denmark, Egypt, England, Hungary, and Turkey, all of which revealed Type 3 ancient leprosy DNA (which is said to be found in Europe, North Africa, and the Americas today) but with different subtypes according to country (I and J for Denmark; K, L, and M for Egypt; I for England; M and K for Hungary; and K for Turkey). This research provided new data that gave rise to two new suggestions regarding how leprosy spread into Asia. One hypothesis is that a northern route for transmission began in the eastern Mediterranean and spread through Turkey along the Silk Road to China, Iran, Japan, Korea, and Turkey. This theory involves the 3K strains that are seen in ancient Europe, central Asia, and Africa. A second hypothesis is that leprosy moved into Asia along a southern route from India, Indonesia, and the Philippines. This theory involves Type 1 strains from India that are seen in ancient Japan. However, as the authors pointed out, leprosy could have originated in Europe or the Middle East and then spread east (Monot et al. 2009).

Monot and colleagues (2009) suggest that leprosy likely spread to the west as people migrated there from Europe from the late nineteenth century onward. They note that the *M. leprae* strains in the Americas were of the European type/subtype 3I that were present in ancient Denmark, England, and Sweden. They concluded that it was highly unlikely that leprosy was taken to the Americas across the Bering Strait and that "it appears more probable that it was brought by immigrants from Europe because 3I is present in the Americas" (7). There is no definite skeletal evidence of leprosy before the late eighteenth century in the Americas. However, the discovery of two skulls with potential leprosy excavated from Kodiak Island, Alaska, in the United States and Vancouver Island in Canada, are in locations that are close to the Bering Strait. While the one from Alaska has been reburied, more work is needed on the other skull to establish precise dating. Research using stable isotope analysis could help establish the origin and movement of this person.

In 2018, Schuenemann et al. contributed findings relating to the origin and evolution of leprosy from ancient DNA analysis of seventeen skeletons with leprosy-related bone changes, some of which had been previously analyzed. Avanzi et al. (2016) also confirmed that the *M. leprae* type/subtype that is found today in red squirrels on Brownsea Island off the south coast of England is the same as the one found in some medieval skeletons in England and Denmark. In their analysis of ten new medieval *M. leprae* genomes, Schuenemann et al. (2018) found that there were four distinct leprosy types/branches that corresponded with specific geographic regions in the early medieval period and three branches in the high medieval period. However, they note that the branches are based on a limited number of SNPs. Their research classified *M. leprae* based on whole genome analyses of ancient skeletal samples from the Czech Republic, Denmark, England, Hungary, and Italy, revealing a much more complex phylogeography. On the basis of their research they confirmed the following branches:

> Type 0, the most ancestral branch of *M. leprae*, partially corresponds to Type 3, subtype K, which is found today in China, Japan, and New Caledonia in the South Pacific and was present in ancient times in skeletons in Denmark, Egypt, Hungary, and Turkey.
> Type 1 is found today in India and southern Japan and was present in ancient times in Japan.
> Type 2 is found today in the Near East and Asia and was present in ancient times in Austria, the Czech Republic, Denmark, England, Italy, Sweden, and Uzbekistan.
> Type 3 is found in Latin America and is also present in armadillos, red squirrels and non-human primates. In skeletal remains, it was present in Croatia, Czech Republic, Denmark, Egypt, England, Hungary, Norway, Sweden, and Turkey.
> Type 4 is found today in West Africa and South America.
> Type 5 is found today in Japan and the Marshall Islands.

Figure 2 in Schuenemann et al. (2018) shows the phylogenetic analysis of ancient and modern *M. leprae* strains. This research team found a close relationship between late medieval English strains and contemporary strains present in humans and armadillos in the southwest of the United States (Truman et al. 2011) and in red squirrels in the British Isles (Avanzi et al. 2016). They concluded that Type 2 *M. leprae* strains were possibly dominant in northwest Europe in the early medieval period and that strains were present from four distinct *M. leprae* types in the late medieval period, although Type 3 dominated at that time. In support of their theory that the diversity of *M. leprae* increased in medieval Europe, they showed that there were three different *M. leprae* types at one site (Odense, Denmark). There is evidence that humans can contract leprosy from armadillos, and there is no reason why close contact with red squirrels and their products could not have facilitated the transmission of leprosy from squirrels to humans in the past. Today red Eurasian squirrels (*Sciurus vulgaris*) are distributed across conifer forests and mixed woodlands in Northern Europe and Siberia but are declining due to loss of habitat and the increasing presence of the gray squirrel from North America. Whether red squirrels harbored leprosy in the past is unknown and whether they were or are capable of transmitting leprosy to humans has not been determined yet. While squirrel bones are not found frequently in archaeological contexts (but see Price 2003), this small woodland mammal may not appear in many contexts that are archaeologically excavated. However, they are present in areas where their fur was exploited. They were also kept as pets in the European medieval period (Richard Thomas and Peter Rowley-Conwy, pers. comm., 2018)

Research on the types/subtypes/strains of *M. leprae* identified in archaeological skeletons using aDNA analysis is particularly important for the history of leprosy in addition to the contributions it makes to our knowledge about the evolution of the modern strains. Table 6.1 shows the extant data.

A number of conclusions can be inferred from the aDNA analysis that has revealed type/subtype information:

Type 0 has been found in Denmark and Hungary.
Type 1 has been found in Japan.
Type 2F has been found in Denmark, England, Ireland, Italy, and Sweden (Taylor et al. 2018).
Type 3I has been found in Denmark, England, Ireland, and Sweden.
Type 3J has been found in Denmark.
Type 3K has been found in Denmark, Egypt, England, Hungary, and Turkey.
Type 3L has been found in Egypt and Uzbekistan.
Type 3M has been found in the Czech Republic, Egypt, and Hungary.
Type 4 has been found in the Czech Republic.

While distinct Type 3 ancient subtypes/strains (K, L and M) have been identified in ancient skeletons from modern-day Czech Republic and Hungary in Central Europe and in Egypt and Uzbekistan, Northern European countries (Denmark, England, Ireland, and Sweden) have revealed different subtypes (3I, J, and 2F). Type 2F is seen in the Near East and Asia today. Of particular interest are the links between the Near East and Asia and Northern Europe (2F is found today in both the Near East and in Denmark, England, Ireland, and Sweden), and between the Americas today and Northern Europe (3I is found in both the Americas and in Denmark, England, Ireland, and Sweden), supporting the idea of transmission of the 3I strain via people making transatlantic sea voyages. Schuenemann et al.

Table 6.1. Ancient DNA analysis of archaeological human remains revealing strain data

WATSON ET AL. (2009):
Type 3 (eighth–ninth centuries AD, Croatia)
Type 3 (thirteenth–sixteenth centuries AD, Denmark and 900 to 1000 AD England)

MONOT ET AL. (2009):
Type 3 (eighth–ninth centuries AD, Croatia)
3I and J (eleventh–fifteenth centuries AD, Denmark)
3K, L, M (fourth–fifth centuries AD, Egypt)
3I (900–1000 AD and thirteenth–sixteenth centuries, AD England)
3M (eleventh century AD, Hungary)
3K (eighth–ninth centuries AD, Turkey)

TAYLOR AND DONOGHUE (2011):
3 (900–1000 AD, England)
3I (thirteenth–sixteenth centuries AD, England; see also Taylor et al. [2009])
3K (eighth–ninth centuries AD, Turkey)
3M (ninth century AD, Czech Republic)
3L (first–fourth centuries, AD Uzbekistan; see also Taylor et al. [2009])
3K and 3M (tenth–eleventh centuries AD, Hungary; at the same site)
3K (seventh century AD, Hungary)

SUZUKI, TAKIGAWA, ET AL. (2010):
Type 1 (mid eighteenth–early nineteenth centuries AD, Japan)

ECONOMOU ET AL. (2013A, B)
Types 2F and 3I (tenth–fourteenth centuries AD, Sweden)

TAYLOR ET AL. (2013)
2F and 3I (eleventh–twelfth centuries AD, England)

SCHUENEMANN ET AL. (2013):
2F (1100–1300 AD, Sweden)
3I (1293–1383 AD, Denmark)
2F (1046–1163 AD, Denmark)
2F and 3I (eleventh–twelfth centuries AD, England)

MENDUM ET AL. (2014)
2F (tenth–twelfth centuries AD, England)

DONOGHUE ET AL. (2015):
Type 3 (eighth–ninth centuries AD, Croatia)
3M (ninth–tenth centuries AD, Czech Republic)
3K, 3M (tenth century AD, Hungary)
3K (eighth–ninth centuries AD, Turkey)
3L (first–fourth centuries AD, Uzbekistan)

GAUSTERER, STEIN, AND TESCHLER-NICOLA (2015):
Type 3 (100 AD, Austria)

(continued)

Table 6.1—*Continued*

INSKIP ET AL. (2015):
3I (fifth–sixth centuries AD, England)

ROFFEY ET AL. (2017):
2F (eleventh–twelfth centuries AD, England)

INSKIP, TAYLOR, AND STEWART (2017):
3I (885–1015 AD, England)

SCHUENEMANN ET AL. (2018):
Types 0 and 3, and 2F (1044–1383 AD, Denmark)
Type 4 (ninth–tenth centuries AD, Czech Republic)
2F, 3 (415–545 cal. AD and 995–1283 AD, England)
0 (seventh–eighth centuries AD, Hungary)
2F (mid- to late seventh century AD, Italy)
2F (1032–1115 cal. AD, Sweden)

TAYLOR ET AL. (2018):
2F (eleventh–thirteenth centuries AD, Ireland)
3I (twelfth–thirteenth centuries AD, Ireland)

KERUDIN (2019):
3I (fourteenth–seventeenth centuries AD, England)
3K (mid-tenth–mid-twelfth centuries AD, England)

(2018) theorize that Type 3 has been present in England since the fourth century, although the infection could have been present much earlier. The facts that *M. lepromatosis* has recently been identified as a cause of leprosy, originally in Mexico (Han et al. 2008), and that we know that it is related to *M. leprae*, suggest that *M. lepromatosis* may have been taken to Mexico along with *M. leprae* (subtype 3I) if it originated in Africa. Further aDNA analysis of other skeletons with leprosy will corroborate or change our ideas about the transmission pathways of leprosy.

There is no doubt that leprosy spread with people who traveled or migrated for whatever reason (military campaigns, trade, pilgrimages, etc.), but the details of this scenario are far from clear at present (see Donoghue et al. 2015; Donoghue 2019; Witas et al. 2015). However, this knowledge is beginning to be borne out by the bioarchaeological evidence from ancient DNA analysis, which also appears to support Monot et al.'s (2005, 2009) theories of transmission routes. There are exceptions to their model, however. Skeletal evidence from Hungary and India is early in date, as is possible evidence from Iran, Nubia, Scotland, and Turkey. Later data come from China, Egypt, Thailand, and Uzbekistan. Evidence from Micronesia and Japan comes next chronologically, followed by evidence from Israel and Italy. There is also bioarchaeological evidence from Italy, most other parts of Europe, and Scandinavia from the late centuries BC/early centuries AD.

THE DECREASE IN LEPROSY FREQUENCY

The incidence of leprosy allegedly decreased from the fourteenth century onward in Europe. Many reasons for this have been suggested over the years. These include refraining from eating fish; the assumption was that a diet containing fish caused leprosy (e.g., Hutchinson 1906). Hutchinson argued that leprosy occurred in coastal areas or close to rivers or lakes and that "the poison which produces the disease gains access to the body in the form of food" (10). He suggested that the bacillus was stimulated by some ingredient in fish in an early stage of decomposition and that leprosy remained in places where fish was a significant part of the diet, for example in Norway and Iceland. Thus, a decrease in the frequency of leprosy may have been associated with a greater reliance on other foods. Bergel (1960, 302) revisited this hypothesis, suggesting that "decomposing fish contain ... rancid fats and saturated fatty acid, which would be expected to induce a prooxidant condition ... [which] favours the growth of the Hansen bacillus." However, there has been no further exploration of this possible explanation for a decrease in leprosy frequency. Likewise, are there no good scientific grounds for linking drinking goat's milk to leprosy, another dietary explanation for the infection (Richards 1960). It is more likely that multiple factors that varied according to geographic location and time period led to the decrease in frequency of leprosy (Hunter and Thomas 1984). Other, more sensible, explanations for the decrease relate to what is known about its epidemiology, but as Ell (1987, 346) reminds us, "diseases do not exist as microbiological and immunological entities alone, but are part and parcel of the social circumstances of a given time and place."

The reasons that have been suggested for a decrease in leprosy also include a ban on sexual intercourse between those who were infected and those who were not (a policy that indicates little understanding of leprosy's transmission routes) and segregation (but it is well known that for a variety of reasons 100 percent segregation was not practiced). Some regarded these two factors to be key for reducing the incidence of leprosy, for example in Norway well before the "great decline" after 1868 (e.g., Lie 1929, 866). Other reasons include improved nutrition (e.g., in Norway in the late nineteenth century; see Irgens 1981). A better diet and a stronger immune system could have helped prevent infection. Carayan (1977) also suggests that expansion of agriculture to cover more land in Western Europe in the fourteenth century led to better nutrition and the disappearance of leprosy. Of note, the World Health Organization (1985) recorded a natural decrease in incidences of leprosy in Northern Europe, Hawaii, Japan, Venezuela, and the continental United States that began before drug therapy or Bacillus Calmette-Guérin (BCG) immunization were available that continued, even though new patients were migrating to these parts of the world. One theory is that improved living conditions, including better diets, increased people's resistance and lowered frequency rates regardless of whether leprosy control programs were present (World Health Organization 1985). This is supported by data from Japan, where the leprosy rate decline in the Okinawa prefecture lagged twenty years behind the rest of Japan because of slower economic development. However, living conditions in towns and cities in Europe in the fourteenth century AD were often very poor, so linking a decrease in leprosy frequency to an improvement in living conditions is not possible for that time period (Hunter 1986).

In fourteenth-century Europe, the plague led to a decrease in absolute population numbers and likely targeted weaker people with leprosy. However, it is clear that those who survived lived in better conditions that created a more challenging environment for leprosy to survive. They also had better wages because there were fewer people available to labor (see for example, Baten et al. 2018) and their health was probably better (e.g., see DeWitte 2014). Improved clothing, including greater access to woolen garments, and the increased availability of cheaper wood for fuel after the plague would also have made close physical contact for warmth less necessary and may have prevented the spread of leprosy from close contact (Hunter 1986). However, it is unlikely that this alone was responsible for what has been purported to be such a dramatic decrease in the frequency of the disease. The incidence of leprosy was also decreasing before the plague struck the European population (Ell 1987). In addition, leprosy can spread in less dense populations, as happened in Norway and Scotland in the nineteenth century (Molesworth 1933), and people with leprosy were probably no more vulnerable that any person with health problems. Some have also suggested that people with leprosy are immune to the plague (Richards 1990; Ell 1987). This has never been proved, but biomolecular analysis may help researchers explore this theory by looking at known susceptibility and resistance genes for plague in skeletons with leprosy around this time period.

Another possible explanation for the decrease in leprosy is the fact that harvest failures at the end of the fourteenth century led to higher morbidity and mortality, including people with leprosy (Tisseuil 1975). However, the argument could also be made that the plague reduced absolute numbers of people and that fewer people needed to be fed (Ackerknecht 1965). In any case, it is well known that harvest failures led to undernourishment and malnourishment, factors that would have compromised immune systems and predisposed people to disease. Indirect mortality associated with the plague and subsequent plague epidemics could also have contributed to the decrease in leprosy due to loss of charitable contributions and rents to support leprosy hospitals (Richards 1977), but we know that many people with leprosy did not rely on hospitals.

In listing suggested reasons for a decline, one possibility Fine (1982) mentions is that the number of people with leprosy who emigrated to Northern Europe after the last Crusade decreased in the later thirteenth century AD. However, skeletal evidence has revealed that leprosy was present in Europe well before the Crusades. Its presence was more likely attributable to urbanization and higher population numbers than it was to the Crusades (Mitchell 2002; and see Kelmelis and Pedersen 2019 for a study of skeletons from medieval Denmark that found the highest rate of tuberculosis in a rural site and the highest rate of leprosy in an urban site). Hunter (1986) has intimated that an antigenic shift (a genetic alteration in an infectious organism that causes a dramatic change to the protein that stimulates the immune system to produce antibodies) led to loss of pathogenicity and a decrease in the incidence of leprosy, but leprosy already has low pathogenicity (ability of an organism to cause disease). It has also been argued that mycobacteria in the environment (soil and water) could have modulated transmission and increased or decreased resistance to leprosy (or tuberculosis) and/or acted on immunological mechanisms (Rafi and Feval 1995; Lechat 1999; Fine et al. 2001). While it is the case that environmental mycobacteria may have affected immunity, this hypothesis has not been explored bioarchaeologically. Biomolecular analyses might be useful in that regard.

Changes in climate could have led to increases and decreases in the frequency of leprosy (Bayliss 1979; Duncan 1994). As an extension of the research Bayliss (1979) did on the link between leprosy and temperature in Europe, Duncan (1994) used mean monthly temperature records for Edinburgh, Scotland, from 800 to 1900 AD. Duncan plotted the frequency of British leprosy hospitals against temperature for each 100-year period from the twelfth through the seventeenth centuries AD, although she recognized that an increase or decrease in the founding of hospitals cannot be taken to mean that the numbers of people with leprosy correspondingly increased or decreased.

Data on temperature for the Medieval Warm Period (AD 1150–1300) show that the range of mean maximum summer temperatures was 24.4–25.2° C (75.9–77.4° F). In this temperature range, she indicated that *M. leprae* likely could have survived up to seven or eight days outside a human host, thus promoting the survival of the infection within the human population. In the period AD 1300–1500, called the Climatic Worsening, the range in average annual temperature decreased (to 22.8–23.8° C, or 73.0–74.8° F). At that range, *M. leprae* could probably have survived outside a host for 5.3 to 6.5 days. From AD 1500 to 1550, the average annual temperature rose a little, but in the period 1550–1700 AD (the Little Ice Age) it was cooler than it is today (21.8–22.6° C, or 71.2–72.7° F) and *M. leprae* was likely able to survive outside a host for only 4.2 to 5 days. This would not have favored the survival of leprosy. By coincidence, the greatest number of leprosaria were founded during the Medieval Warm Period and the smallest number were founded from 1550–1700 AD, during the Climatic Worsening and the Little Ice Age (Figure 6.1).

Leprosy has been found in skeletons in many parts of the Old World. Although most skeletons have been found in Europe, some have been found in the warmer Mediterranean region and in the cooler North Sea and Baltic Sea. Today leprosy is more frequent in the southern hemisphere (e.g., Brazil; World Health Organization 2019). This suggests that a decrease in temperature is not likely to have led to a decrease in leprosy frequency in the past. Natural selection likely played a role: where there was a high frequency of leprosy, people would eventually become resistant. People with no resistance would die, people with some resistance would contract leprosy in a more chronic form, and those with high resistance would most likely escape the infection. A greater proportion of the offspring of parents who were more resistant would survive than the offspring of susceptible parents. However, there would be differences in resistance for all offspring, and through the generations the number of people with leprosy would decrease and the manifestations of leprosy would become less chronic and pathogenetic (Molesworth 1933).

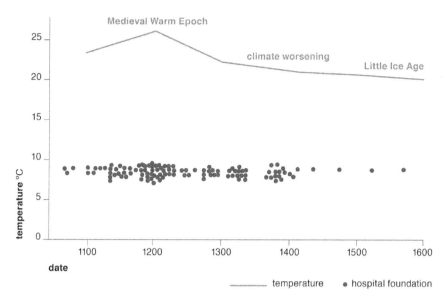

Figure 6.1. Plot graph of foundings of leprosy hospitals in Britain through time and according to maximum summer temperature. Credit: Redrawn by Chris Unwin after Duncan (1994).

Many have suggested that the frequency of leprosy decreased because of cross-immunity between leprosy and tuberculosis (e.g., Chaussinand 1953; Manchester 1984, 1991). Both are caused by mycobacteria, and the hypothesis is that a population's increased exposure to tuberculosis led to immunity to leprosy. This hypothesis has not been tested much in the archaeological record yet and bioarchaeological evidence has not yet been used in attempting to extrapolate absolute frequencies of leprosy and tuberculosis. The hypothesis is based on historical data for leprosy and tuberculosis, which indicates that tuberculosis was more common in later time periods than in earlier ones. It is important to remember that the diagnosis and interpretation of tuberculosis in skeletal remains has limitations (Roberts and Buikstra 2003, 110–114), as is the case for leprosy. However, the bioarchaeological evidence of tuberculosis actually follows the same general pattern of occurrence through time as that for leprosy (Roberts and Buikstra 2003). Although more data have been published on tuberculosis since Roberts and Buikstra (2003), including the earliest evidence in Dorset, England (Mays and Taylor 2003; Taylor, Young, and Mays 2005), the general trend has not changed. Many more skeletons have been recorded with leprosy in the early and late medieval periods than was the case when Manchester (1984) was published. The suggestion that the incidence of leprosy decreased as the incidence of tuberculosis increased thus cannot be supported with hard scientific evidence yet. That being said, if people became more and more exposed to tuberculosis, then tuberculin positivity could have increased and people could have developed protection against leprosy (Manchester 1991). The logic of this hypothesis is that the resistance to leprosy that developed meant that those who contracted the disease likely developed tuberculoid leprosy (the low-resistance form) rather than lepromatous leprosy (the high-resistance form) and the incidence of leprosy eventually decreased. Thus, Manchester's (1984) hypothesis that tuberculosis frequency increased as leprosy frequency decreased is based on temporal change in immune status modified by bacterial exposure, a theory that seems very plausible. He also proposed that the decrease in the frequency of leprosy in the fourteenth century AD may be related to increasing immunity and a shift to the tuberculoid leprosy end of the immune spectrum. If tuberculoid leprosy became more common, the skeletal evidence would be less common or more difficult to observe and interpret. Manchester suggested that this may explain why there is less bioarchaeological evidence for leprosy in later periods of time. It is interesting that macroscopic and biomolecular analyses have diagnosed some skeletons with both leprosy and tuberculosis (Table 6.2). However, co-infection was rare in the past (as it is today), contrary to Rawson et al. (2014). While the data for leprosy and tuberculosis affecting the same person might be rare, the data that do exist must not be forgotten when considering tuberculosis as a cause for the decrease in the frequency of leprosy. Donoghue et al. (2005) have suggested that people with lepromatous leprosy are more susceptible to tuberculosis, and of course that could lead to increased mortality from tuberculosis. Hohmann and Voss-Böhme (2013, 234) document that co-infection could also be a cause of the decrease in leprosy frequency and argue that it must be considered as another "significant hypothesis." The suggestion that leprosy and tuberculosis could co-occur in a person or a population means that the tuberculosis-leprosy cross-immunity and co-infection hypotheses could both be relevant. However, in research that revisited the cross-immunity hypothesis that used data from studies of human innate immunological responses, paleomicrobiology, bioarchaeology, and paleopathology, Crespo, White, and Roberts (2019) developed a multifactorial model. They concluded that past populations do not represent homogeneous immunological landscapes, and thus it is likely that the frequency of leprosy in medieval Europe did not uniformly decrease due to cross-immunity. This theory continues to be debated, but both the cross-immunity and co-infection hypotheses seem to be the most reasonable and logical explanations for the decrease in the frequency of leprosy (see also Leitman, Porco, and Blower 1997; Wilbur, Buikstra, and Stojanowski 2002). However, leprosy is present in some parts of the Old World today, such as Asia, where tuberculosis is also common.

It is appropriate here to document Fine's (1984) list of common characteristics of leprosy and tuberculosis: they are transmitted by the respiratory route, they have long and variable incubation periods and both have subclinical infection phases, their clinical pictures and transmission potentials vary, in the 1980s they were both disappearing from developing countries (this was more true for tuberculosis than leprosy), they both show drug resistance (this was less true for leprosy), both are also associated with poverty, both are stigmatized, and the responses of both to Bacillus Calmette-Guérin vaccination vary. The differences are that they may be contracted differently (e.g., from different animals), their rates in urban and rural contexts vary, and their relationship with latitude and BCG vaccination efficacy varies, which could indicate interactions with environmental mycobacteria.

Because the debate about the relationship between these two mycobacterial diseases has not yet been resolved, it is worth following it from the time of Chaussinand's studies in the 1940s (Chaussinand 1944, 1948; see also Lowe and McNulty 1953a, 1953b; Fernandez 1957). At that time, rapid spread of tuberculosis was believed to confer some resistance to leprosy and to lead to gradual elimination. Muir (1957) saw leprosy as a disease of villages and argued that exposure to tuberculosis increased resistance to leprosy and initiated a decrease in frequency. In field observations in Kenya, the Netherlands, and New Guinea, there was support for these ideas, and evidence showed that contact of rural people with urban centers led to a rise in the frequency of tuberculosis (Leiker 1977, 1980; Leiker, Otsyula, and Ziedses des Plantes 1968).

While this evidence supported a hypothesis of cross-immunity, many other factors were involved, and some field studies found negative or inconclusive results when they tested this hypothesis (see also Wilbur, Buikstra, and Stojanowski 2002). In addition, not all researchers have agreed with its premise (e.g., see Newell 1966). Chaussinand (1955) recommended that the use of the BCG vaccination should be studied in areas where leprosy was frequent. However, the idea of using it as a preventative measure has occurred before the 1950s. Fernandez (1939) first reported lepromin conversion, or the conversion of lepromin negativity to positivity in humans after

Table 6.2. Skeletons with both leprosy and tuberculosis

	Source(s)
AUSTRIA	
Zwölfaxing: Skeleton 88	Donoghue et al. (2015)
DENMARK	
Naestved: Skeleton 363	Weiss and Møller-Christensen (1971b)
EGYPT	
Kellis 2, Dakhleh Oasis: Skeletons B6, 222, 251, 265, 286, 392	Donoghue, Marcsik, et al. (2005)
ENGLAND	
St James and St Mary Magdalene, Chichester: Rib lesions and leprosy in Skeleton 291 and in seven other skeletons that had "suggestive" TB bone changes (53, 106, 150, 239, 307, 331). TB may be a cause of rib lesions and these people may also have had leprosy, as they were buried in a leprosarium.	Lee and Boylston (2008a)
GERMANY	
Rain	Nerlich, Marlow, and Zink (2006)
HUNGARY	
Bélmegyer-Csömöki: Skeleton 22	Molnár et al. (2015)
Hajdúdorog-Gyúlás: Skeleton HG56	Donoghue et al. (2015)
Kiskundorozsma-Daruhalom dűlő II: Skeleton 517	Lee et al. (2012); Molnár et al. (2006); Pálfi et al. (2012)
Lászlófalva-Szentkirály: Skeleton 79	Donoghue et al. (2015) Pálfi et al. (2010)
Püspökladány-Eperjesvölgy: Skeletons 222 and 503	Donoghue, Marcsik, et al. (2005)
ISRAEL	
Akeldama, Hinnom Valley: Skeleton C1	Donoghue, Marcsik, et al. (2005)
SCOTLAND	
Dryburn Bridge (DNA of TB present but leprosy diagnosed macroscopically; no *M. leprae* DNA preserved)	J. Roberts (2007)
SWEDEN	
Björned: Skeleton A4	Donoghue, Marcsik, et al. (2005); Nuorala et al. (2004)
Humlegården, Sigtuna: Skeleton 3320 (possible TB evidence)	Kjellstrom (2012)
TURKEY	
Kovuklukaya: Skeleton 24/1	Donoghue, Erdal, et al. (2005)

NB: At all sites except Chichester and Naestved, diagnosis was established using aDNA and/or lipid analysis for either leprosy or TB.

BCG vaccination. (Lepromin is an extract of leprosy bacilli from armadillo liver tissue or a lepromatous nodule.) Although several field studies of the BCG vaccination had the same outcome in the period 1939–1959, there were problems with some of them (Newell 1966). After Shepard and Saitz (1967) found that BCG almost completely suppressed *M. leprae* multiplication in mice, experiments began around the world to cultivate *M. leprae* to identify strain differences (Rees et al. 1967). By 1968, BCG vaccination of mice conferred a tenfold to hundredfold reduction in leprosy bacilli. Three major field trials of BCG were done in the 1960s. In a trial involving 19,000 children in the Teso District of Uganda, the vaccine provided a high degree of protection against contracting leprosy, but there was no evidence that it gave people who already had lepromatous leprosy additional protection. Another trial involving 5,000 people of all ages in the Karimui-Nomane District in the Eastern Highlands of Papua New Guinea yielded a leprosy protection rate of approximately 50 percent. In the third trial, with 23,000 Burmese children, the protection rate ranged from 12 to 54 percent. The variation in vaccine efficacy may be explained by the different numbers of people with either lepromatous leprosy or tuberculoid leprosy in these three studies, as could differences in the BCG strains used in the vaccines, regional differences in patterns of infection, and differences in strains of *M. leprae* and *M. tuberculosis* and environmental bacteria present (Fine 1984). In their review of these trials, Hunter and Thomas (1984, 33) note that "over the years, the results have been negative or puzzling at best." More recent data from India on BCG efficacy in preventing leprosy have shown a 54 percent rate of protection against lepromatous leprosy, a 68 percent rate of protection against tuberculoid leprosy, and a 57 percent rate of protection against single skin lesion leprosy (Zodpey, Ambadekar, and Thakur 2005). A study in Rio de Janeiro, Brazil, has further reported a 56 percent protection rate (Düppre et al. 2008).

Hunter and Thomas (1984, 47) found that today, in general, "an increase of 5 percent in the level of urbanization in a country is associated with a decrease of 500 cases in the rate for leprosy." They suggest that many variables in modern urban situations probably contribute to a decrease in leprosy frequency: less frequent concealment of infection, less tolerance to people with leprosy, earlier diagnosis, easier access to treatment, better treatment and compliance, reduced infection and fewer carriers, living conditions that are less favorable to transmission, and more BCG vaccination of children. They also reported that an increase of 10 percent in the level of urbanization led to additional tuberculosis in their study and concluded that leprosy was associated with climate, that tuberculosis was associated with urbanization, and that urbanization was correlated with a decrease in leprosy frequency. These various studies show inconsistent support for the theory that exposure to tuberculosis was responsible for decrease in leprosy frequency in the past and suggest that the factors relating to this decrease may vary in different regions of the world and in different time periods.

Applying this logic to the distant past is interesting because urban contexts do seem to have favored the existence of leprosy, but its decrease in the fourteenth century seems, on the basis of the TB hypothesis, to have co-occurred with an increase in urban living and an increase in the frequency of tuberculosis (although, as discussed in chapter 5, more human remains tend to be excavated from urban contexts and therefore we might expect more evidence of leprosy from those funerary contexts even during the fourteenth century). Lynnerup and Boldsen (2012) have suggested that rural people may have been more likely to contract tuberculosis (*M. bovis*) and to have become immune to leprosy because of their closer contact with animals that can harbor tuberculosis, such as cattle (see also Kelmelis and Pedersen 2019). However, while evidence in early medieval (rural) England shows the presence of co-infection in skeletons from cemeteries (Roberts and Buikstra 2003; and this study), it does not show both leprosy and tuberculosis in individual skeletons. Five sites have revealed skeletons with either leprosy or tuberculosis in England (Bedhampton, Cannington, Edix Hill, Great Chesterford, and School Street, Ipswich). However, the sites of Addingham, Chichester, Eccles, Raunds, and York Minster have individuals with both leprosy and rib lesions, and the latter could be related to tuberculosis but also to many other lung diseases (Roberts, Lucy, and Manchester 1994). Even if the ribs were affected only in tuberculosis, this still does not provide much evidence for co-infection in individual skeletons. Of relevance here is Wilbur, Buikstra, and Stojanowski's (2002) study of the frequency rates for leprosy and tuberculosis from the Gulf Coast of North America using two datasets from Texas (1938–1980 and 1991–1998). The data could not support the cross-immunity hypothesis because as the incidence of tuberculosis decreased, the incidence of leprosy increased, but if the data were analyzed in a different way there were indications that both were rising/declining together. While the data provided evidence against the cross-immunity hypothesis, inaccurate diagnosis and underreporting may have confounded these modern data. However, the co-morbidity hypothesis of Hohmann and Voss-Böhme (2013) remains relevant here (see above).

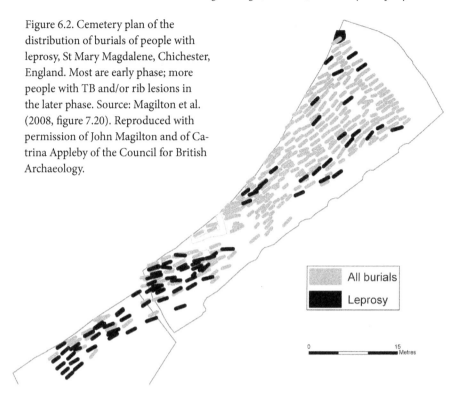

Figure 6.2. Cemetery plan of the distribution of burials of people with leprosy, St Mary Magdalene, Chichester, England. Most are early phase; more people with TB and/or rib lesions in the later phase. Source: Magilton et al. (2008, figure 7.20). Reproduced with permission of John Magilton and of Catrina Appleby of the Council for British Archaeology.

Of all the data considered that might explain the decrease in the frequency of leprosy, there are strong indications that tuberculosis probably played a strong part in Europe in the fourteenth century AD but that multiple factors also likely contributed, depending on location and time period. A biomolecular analysis of large samples of skeletons is needed to corroborate the hypothesis that tuberculosis played a major role in the decrease of leprosy frequency. This has already begun to a certain extent, but all skeletons need to preserve *both* leprosy and tuberculosis aDNA well at any one site. Finally, a large dating program using data from archaeological site stratigraphy, radiocarbon dating of skeletons, and well-dated archaeological materials (e.g., grave goods) plus the biomolecular evidence for tuberculosis and leprosy before, during, and after the fourteenth century AD would make a detailed examination of the frequency of these two diseases possible in these periods. The data could potentially document the absolute rise and fall of these related infections. While some cemetery data suggest more leprosy than tuberculosis in their earlier phases (Figure 6.2), these are not absolute frequency data, and to date such data have been rare.

Conclusions

Modern genomic data for leprosy are helping researchers evaluate the origin, evolution, and spread of leprosy. Ancient DNA evidence for the strains of the bacterium is also furthering knowledge of the spread of leprosy, but much more work in this vein is recommended, alongside stable isotope analysis, which can provide information about the mobility and dietary histories of people in the past, and mitochondrial DNA to document ancestry and contribute data on origins (see relevant chapters in Brown and Brown 2011). The historical evidence suggests that the frequency of leprosy decreased in the fourteenth century, but at the moment archaeological evidence is lacking to support such a hypothesis. Many reasons for this have been suggested, but cross-immunity created by exposure to tuberculosis remains the strongest possibility. The two diseases have many characteristics in common, and tuberculosis and leprosy have been found together in skeletons in a number of instances. Co-infection with tuberculosis may be another hypothesis to consider as an explanation for the decrease in the frequency of leprosy.

Conclusions

A Future for Leprosy: Clinical and Bioarchaeological Perspectives

Elimination of infection . . . is currently beyond us without more knowledge of the routes of transmission and understanding potential environmental reservoirs. However, in the meantime, we must keep looking for the cases, and the fewer we find, the harder we must look.

Medley et al. (2017, 3)

Overview of Findings, Some Limitations of the Data, and Addressing the Myths of Leprosy

As I wrote this book and surveyed the published literature on leprosy in clinical, historical, and bioarchaeological domains, my perspectives on leprosy changed considerably. Like others, I have viewed opinions about how people with leprosy were viewed and treated to be extreme at times. This book aimed to address ingrained myths about leprosy in order to change people's views of those affected with this infection.

Both the historical and bioarchaeological data suggest that leprosy should be interpreted with respect to time and place. The historical literature records a long history for this infection. Some of the earliest accepted references to it are from China and India. However, how people with leprosy were diagnosed is unclear and researchers today debate whether diagnosis was done accurately in the past. Many historical sources also present the view that people with leprosy were segregated from their communities. This does not fit with the bioarchaeological evidence for people with leprosy because the majority of skeletons with bone changes were buried in community cemeteries consistent with the time period and location; however, some leprosaria cemeteries have been excavated that contain skeletons with leprosy. In addition, few skeletons with leprosy that have been found in funerary contexts were buried differently from their neighbors. However, skeletons with no bone changes of leprosy have been also located in leprosaria cemeteries, suggesting a number of scenarios. It could be argued that "normal" burial practices for the time period and location support the hypothesis that people with leprosy were more accepted within their societies than has previously been assumed and asserted. Indeed, when a disease is common in a population, having it may become the norm (Metzl and Kirkland 2010). Additionally, the people who are affected with the disease might not even consider themselves to be ill (KM Boyd 2000).

However, interpreting evidence for disease in skeletal remains is challenging, particularly because only a small proportion of people with untreated leprosy will develop bone changes. We know that many hospitals were founded for people with leprosy in medieval Europe, although the rules for a number of them suggest that segregation was not strict and that people without leprosy were admitted. Diagnosing the specific and nonspecific bone changes associated with leprosy is also challenging, but advances in ancient DNA and mycolic acid analyses of human remains has increased our knowledge about the presence of this disease in human skeletal remains. These analytical procedures have their limitations and are not accessible to the majority of bioarchaeologists globally due to cost and/or access to facilities, but they enable some bioarchaeologists to detect leprosy in skeletons without recognizable bone changes. The bioarchaeological evidence for leprosy shows a long history and is mainly found in the Old World, especially Europe, and in the northern hemisphere. There is no confirmed skeletal evidence in the New World. Some of the bioarchaeological data are older than the accepted historical evidence. Comparing modern with ancient leprosy genomic data, particularly the subtypes of *M. leprae* that affect people, has increased our understanding of the origin, evolution, and spread of leprosy and is helping us answer questions such as how, why, and when leprosy was transmitted in the past. The decrease in the frequency of leprosy during the late medieval period may be attributable to an increase in tuberculosis, but proving this association is difficult from a bioarchaeological perspective.

Everybody should be aware of the persistent myths about leprosy. The evidence in this book shows that all of them are largely untrue. Here is a list of the myths again, and the facts.

It is easy to contract. Leprosy is not easy to contract and is not passed from one person to another easily.

Leprosy can be passed from one person to another rapidly. Prolonged and long-term exposure to the bacteria is needed for transmission from person to person and the majority of people exposed to leprosy only develop subclinical infection and not the disease itself.

Leprosy is incurable. Leprosy can be cured using antibiotic therapy, which has been free for all who need it since 1995.

The fingers and toes of people "drop off." This is incorrect. The finger and toe bones that are affected by *M. leprae* are damaged and shortened through destruction of bone. The skin adapts by contracting around them, but the toenails and fingernails are often retained.

Leprosy is described in the Bible. Leprosy as we know it today is not described in the Bible. This misconception is related to a mistranslation of a Hebrew word, *tsar'ath*—a range of skin diseases that may have included leprosy.

Leprosy can be transmitted via sexual intercourse or by touching. There is no secure, convincing, or plentiful evidence for these transmission routes. However, they cannot be entirely ruled out.

Leprosy can be inherited. There may be a genetic component to contracting leprosy, but there is still debate about this issue. Researchers do not know how and when leprosy is inherited and which of all the leprosy-associated genes are necessarily important. However, as more knowledge about the genetics of leprosy is gathered, it is increasingly being reported that genes exist that make people more or less susceptible to leprosy and that these genes may be inherited through families.

Leprosy is a tropical disease today. While the majority of populations experiencing leprosy today do live between the Tropic of Cancer (23.5° north) and the Tropic of Capricorn (23.5° south), the bioarchaeological evidence suggests that in the past, the majority of people with leprosy lived north of the Tropic of Cancer, and particularly in Europe. The association of the tropics with leprosy today is probably a result of risk factors such as poverty and poor access to medical facilities.

Leprosy is not a problem for people today. While figures from the World Health Organization indicate that new "cases" of leprosy have shown a steady decline since the late 1990s, the legacy of leprosy (impairment, stigma, isolation) remains. Helping people with leprosy remains a huge challenge in the world.

All people with leprosy were segregated from society in the past. The bioarchaeological data show that not everyone with leprosy was segregated in the past and that people with leprosy likely remained part of their community.

A Future for Leprosy in History and Bioarchaeology

As Scollard (2006, 339) notes, "Leprosy is not going to disappear anytime soon." Scholars say that if it exists in animals then it cannot be eradicated and the situation regarding the epidemiology of leprosy in animals is still being debated (Saunderson 2008). As recently as 2014, both *M. leprae* and *M. lepromatosis* were discovered in red squirrels in the British Isles, and we need to learn much more about this relationship. Whether leprosy could reemerge as a major infectious disease like tuberculosis remains to be seen, but improvements in socioeconomic conditions for people will help control leprosy (Stearns 2002). Indeed, ridding the world of poverty would help considerably (Wilkinson and Pickett 2019). Strong and continued political will is needed to eradicate leprosy and this might wane as governments see it as officially "eliminated" in their countries (Britton and Lockwood 2004). When some sources state that leprosy is eliminated in many parts of the world, the general public comes to believe that leprosy is a disease that does not need resources targeted at it. This problem is exacerbated as awareness of and knowledge about leprosy decreases in the clinical world. Leprosy remains and will continue to be a target of interest, especially for fiction writers and the media, but often that interest may not be particularly well focused and can do more harm than good in terms of how the public understands the history of leprosy and how we view it today. Because people in the Western world are not generally exposed to leprosy unless they travel to countries where the disease remains present, views tend to be "static and clichéd" (Edmond 2006, 246). Managing leprosy is not just about drug therapy treatment (the clinical perspective), it is also about looking after the whole person from a social perspective.

However, it is hoped that leprosy will be eradicated one day, and it is increasingly essential to ensure that there is a well-documented accurate resource about its history, from its origin to the day it is announced that it is totally eliminated. As Kofi Annan, the seventh secretary-general of the United Nations, said in 1998, it is not possible to have a future vision without a sense of history. It is important to look back to the past, as this book has done. This has also been seen with the International Leprosy Association's global project on the history of leprosy, which seeks to ensure that "the experience both of those affected by the disease and of those who have tried to respond to their needs and to the medical challenge should not be lost or easily forgotten" (Robertson 2002, 3).[1] It is especially important to retain archives from places where people with leprosy lived, especially the oral histories of people with leprosy (i.e., those who have been cured but remain in leprosaria today). However, while the focus of the International Leprosy Association is on the period from 1847 (when we have the first systematic description of leprosy; see Danielsson and Boeck 1847a) to the present, the project is also gathering information on resources relating to leprosy in earlier periods. It was originally funded by a three-year grant (1999–2002) from the Nippon Foundation in Japan that was administered through the World Health Organization. The three original aims were to identify existing historical resources and make links between them; provide advice on selecting, storing and cataloging materials (International Leprosy Association and the International Council on Archives 2000); and compile information on discriminatory legislation and customs. Archives naturally include academic publications that provide a record of leprosy reaching back hundreds and even thousands of years, especially in bioarchaeology and medical history. It is important, however, to reiterate that ideas about the history of leprosy are changing since scholars have begun to interrogate data in more detail instead of accepting the negative view of leprosy that has been carried down the centuries. It is especially important to consider evidence for leprosy before the mid-nineteenth century so that the biased view of the many characteristics of its long history does not prevail in the documentation of this disease, as Green (2012, 32) has noted.

Bioarchaeology will likely play a key role in providing much more nuanced views about the experience of people with leprosy in the past than has been possible before (notwithstanding the limitations of this field, as Wood et al. 1992 describe). This is partly because of the application of new analytical techniques to archaeological human remains (e.g., aDNA analysis) and partly because bioarchaeologists are performing more holistic and multidisciplinary led studies. It perhaps should be restated that evidence in human remains for leprosy is the primary evidence in the past, not forgetting that other disciplines such as history can provide narratives that can be used to contextualize the bioarchaeological evidence. Future bioarchaeological studies should include data not only from the writings of medical historians but also from the fields of medical anthropology and clinical and evolutionary medicine. It is also essential that more scholars (and hopefully as an extension, the media) move away from medicalizing and objectifying people with leprosy and begin approaching the subject from a more humanistic point of view that looks at the life experience of the people and the populations who are, and have been, affected with leprosy (see also Kumar et al. 2019 on the "New Face of Leprosy," and the need to stop using images of people with leprosy who are severely disabled to attract charity donations).

It is also important that bioarchaeologists take a cautious approach to inferring impairment from bone changes of leprosy seen in human remains, as is assuming that people with leprosy experienced it and any related disability similarly (see Roberts 2017). One of the many things I learned when I was a nurse is that patients with health challenges are individuals whose experience of even the same disease can differ (see Roberts 2017). This understanding must also be applied to past societies. There has also been a tendency to talk about individual diseases rather than the person or people who experienced or experiences them (Erueti et al. 2012), and this can still be the case in clinical medicine, bioarchaeology, and the media (Egnew 2009). Many times while nursing in the 1970s I heard "the appendicitis in Bed X" or "the colon cancer in Bed Y," which dehumanized those who were affected. Dehumanization and objectification of disease remains present in societies to this day. Physicians such as Hippocrates (fifth century BC) felt that it was more important to know what sort of person has a disease than to know what sort of disease a person has, and William Osler (nineteenth century) suggested that we should be caring for the individual patient rather than for the special features of a disease. This is because no two bodies are the same and no two individuals respond to a disease in the same way (Hong and Oh 2010). Bioarchaeologists should also avoid the word "leper" and attempt to avoid using the words "case," "sample," "specimen," or "subject" to describe people in the past with leprosy (this language conjures up images of things in jars in a museum that become objects of curiosity) or indeed the words "assemblages," "materials," or "series" to describe cemetery populations. Changing the language you use will go a long way toward ensuring that the remains of our ancestors are not regarded as objects of curiosity. Using appropriate language ultimately lends itself to preventing the development of stigma and isolation today and to avoiding assumptions that all people with leprosy in the past were stigmatized.

Evolutionary medicine is one of the most important fields to which the bioarchaeology of leprosy can contribute (Nesse and Williams 1994; Elton and O'Higgins 2008; Gluckman et al. 2009, 2011; Stearns 2012; Taylor 2015; Stearns and Medzhitov 2016). As Nunn et al. (2015, 748) have observed, "Evolution is central to understanding and improving health." Thus, bioarchaeologists may help scholars from other disciplines and clinicians understand leprosy better. Their research may even contribute to eliminating this infection. The Paleopathology Association's motto, "Let the dead teach the living," is very relevant here. Paleopathology can provide a contextualized and deep temporal and geographical perspective for the global history of disease that is not possible in clinical medicine. Evolutionary medicine explores the evolution of disease to provide explanatory models for human health today (see Trevathan et al. 2008 on the increase in publications in this field). Palaeopathology fits perfectly within evolutionary medicine as a key source of information about the origin and evolution of disease over thousands of years (e.g., see Rühli and Henneberg 2013; Zuckerman et al. 2012). It is a "set of concepts and approaches with which to analyze many different parts of medical science" (Stearns 2012, 1). Biological aspects of human beings alter relatively slowly compared to cultural changes. Because the epidemiological transitions that led to settled agriculture, to living in towns, and to experiencing industrialization were so rapid, subsequent human populations have experienced poorer health. In particular, especially today in the third epidemiological transition, people around the globe have been subject to infectious disease (Gluckman and Hanson 2006). I believe that palaeopathology's main remit should be to contribute to understanding health and well-being in the human species today. However, there are caveats to this opinion, and I will close this book with them.

    Bioarchaeologists could reflect more on the ethics surrounding the excavation, study, and interpretation of human remains and consider increasing the dissemination of their findings to a wider range of beneficiaries, including the public. This is particularly pertinent in work on skeletons with leprosy. It could be easily argued that excavating and studying skeletons with leprosy (and doing destructive analysis, such as ancient DNA analysis) could harm a proper understanding of leprosy in the context of the many world cultures experiencing it today. This could happen if findings are presented in a sensational way instead of with sensitivity. Thus, when working with skeletons with leprosy-related bone changes, we should be mindful of the potentially damaging outcomes of presenting our findings (see also Lambert 2012). Studies are not purely about catching the media's attention. They are more about showing ethically and sensitively what bioarchaeology can contribute to understanding leprosy in the past and present and about helping us all appreciate the impact of this infection on people in antiquity and today. One might argue that as a white female middle-class British academic, I have no understanding of the real costs of leprosy today. I acknowledge that I may not have the most appropriate background to have written this book, but I hope that I have presented a reasonably balanced view of this infectious disease and its effect on people around the world, past and present. I believe that approaching an understanding of any disease benefits hugely from a synthesis of information from different disciplines. Historians and bioarchaeologists (aka paleopathologists) have in their various ways provided the deep time perspective on this disease and how people with leprosy were understood, cared for, and treated in the past. Bioarchaeologists have particularly benefited from appreciating the knowledge we have about leprosy today as they have interpreted their data, and bioarchaeological work has illustrated the benefits of considering leprosy from an evolutionary perspective and how that can help correct past misunderstandings and myths. In the future, palaeopathology will benefit the present more if this sort of approach is taken more often. Bioarchaeology is a discipline that has an exciting future. While its practitioners use multidisciplinary approaches and big data to undertake ambitious projects, they also produce nuanced narratives for individual skeletons.

I sincerely hope that as research on leprosy develops, this infection will increasingly become a more accurately depicted disease in the eyes of everyone, from academics to the public. There has been and continues to be a tendency in Western society to focus on dramatic images of people with leprosy and to regard them as objects of curiosity and blame. It is time to change such attitudes and to take a more responsible approach that will lead to a better understanding of the real lived experiences of people with leprosy, both in the past and in the present. I hope that this book has provided some food for thought and that it has increased understanding of leprosy.

Appendix 1

Full List of Acknowledgments

Thank you to all the colleagues, acquaintances, friends, and organizations below who have helped in so many ways and who have answered the most inane questions, more often than not very promptly!

ARCHAEOLOGICAL SITES USED IN THIS BOOK

Caroline Ahlström Arcini, National Heritage Board, Lund, Sweden
Sue Anderson, Spoilheap (http://www.spoilheap.co.uk/)
Abby Antrobus, Suffolk County Council, UK
Brenda Baker, Arizona State University, USA
Iain Banks, University of Glasgow, UK
Naraa Bazarsad, Mongolia
Zsolt Bereczki, University of Szeged, Hungary
Tracy Betsinger, State University of New York, Oneonta, USA
Vera Tiesler, Universidad Autónoma de Yucatán, Mérida, México
Susi Ulrich Bochsler, University of Bern, Switzerland
Chryssa Bourbou, 28th Ephorate of Byzantine Antiquities, Hellenic Ministry of Culture, Greece
Anthea Boylston, formerly University of Bradford, UK
Kate Brayne, Rudyard Consultancy, UK
Elin Therese Brødholt, University of Oslo, Norway
Jo Buckberry, University of Bradford, UK
Alexandra Buzhilova, Institute of Archaeology, University of Moscow, Russia
Niamh Carty, University College, Cork
Mario Castro, University of Chile, Santiago
Jenny Cataroche, Public Health Directorate, Health and Social Services, States of Guernsey
Andrew Chamberlain, University of Manchester, UK

Tina Christensen
Margaret Cox, Knight Ayton Management, UK
Natasha Dodwell, Cambridge Archaeological Unit, UK
Kirsten Dinwiddy, Wessex Archaeology, UK
Andy Dunwell, Centre for Field Archaeology, Glasgow, Scotland
Yilmaz Selim Erdal, Hacettepe University, Turkey
Marina Faerman, Hebrew University of Jerusalem, Israel
Kori Filipek-Ogden, Durham University, UK
Gordon Findlater, University of Edinburgh, UK
Gino Fornaciari, University of Pisa, Italy
Sherry Fox, Arizona State University, USA
Hildur Gestsdóttir, University of Iceland
Nia Giannakopoulou, Imperial College, London, UK
Gisela Grupe, University of Munich, Germany
Dawn Hadley, University of York, UK
Sue Hirst, Museum of London Archaeology, UK
Per Holck, University of Oslo, Norway
Malin Holst, York Osteoarchaeology, UK
Zuzana Hukelova, Institute of Archaeology, Slovak Academy of Sciences, Nitra, Slovakia
Sarah Inskip, University of Cambridge, UK
Margaret Judd, University of Pittsburgh, USA
Hedy Justus, Joint POW/MIA Accounting Command, Hawaii, USA
Anna Kjellstrom, University of Stockholm, Sweden
Vasiliki Koutrafouri, University of Edinburgh, Scotland
Tomasz Kozlowski, Nicolaus Copernicus University, Toruń, Poland
Christiane Kramar, University of Geneva, Switzerland
Anna Lagia, Albert-Ludwigs-Universität, Freiburg, Germany
Oona Lee, University of Birmingham, UK
Mary Lewis, University of Reading, UK
Louise Loe, Oxford Archaeology, UK
John Lukacs, University of Oregon, Eugene, USA
George Maat, University of Leiden, The Netherlands
Roberto Macchiarelli, University of Poitiers, France
Jennifer Macey, Portsmouth Museum, UK
Antonia Marcsik, University of Szeged, Hungary
Vitor Matos, University of Coimbra, Portugal
Simon Mays, Historic England, UK
Lauren McIntyre, Oxford Archaeology, UK

Jackie McKinley, Wessex Archaeology, UK
Sheilla Mendonca de Souza, Fiocruz, Rio de Janeiro, Brazil
David Minnikin, University of Birmingham, UK
Theya Molleson, Natural History Museum, London, UK
Ioanna Moutafi, University of Cambridge, UK
Eileen Murphy, University of Belfast, Northern Ireland
Marc Oxenham, Australian National University
Anastasia Papathanasiou, Ephorate of Paleoanthropology and Speleology, Athens, Greece
Megan Perry, East Carolina University, USA
Michael Pietrusewsky, University of Hawaii, USA
Elizabeth Popescu, Oxford Archaeology, UK
Sarah Poppy, Cambridge Archaeological Unit, UK
Inna Potekhina, Academy of Sciences, Kiev, Ukraine
Natasha Powers, Allen Archaeology, UK
A. Prescher, RWTH Aachen University, Germany
Rebecca Redfern, Museum of London, UK
Gwen Robbins Schug, Appalachian State University, Boone, North Carolina, USA
Julie Roberts, Liverpool John Moores University, UK
Conrado Rodriguez Martin, Instituto de Bioantropología, Tenerife, Spain
Aida Romera, National Museum of Scotland, Edinburgh
Jerry Rose, University of Arkansas, USA
Kati Salo, University of Helsinki, Finland
Alison Sheridan, National Museum of Scotland, Edinburgh, Scotland
Lucy Sibun, University College, London, UK
Andy Simmonds, Oxford Archaeology, UK
Mario Slaus, Croatian Academy of Sciences and Arts, Zagreb, Croatia
Andrei Soficaru, Institutul de Antropologie "Francisc J. Rainer," Croatia
Maryna Steyn, University of Pretoria, South Africa
Rebecca Storm, University of Bradford, UK
Eulalia Subira, Universitat Autònoma de Barcelona, Spain
Takao Suzuki, Tokyo Metropolitan Institute of Gerontology, Japan
Akira Tagaya, Japan
Nancy Tayles, formerly University of Otago, New Zealand
Maria Teschler-Nicola, Natural History Museum, Vienna, Austria
Chris Thomas, Museum of London, UK
Steve Timms, Mike Griffiths Associates, UK

Carmelita Troy, Rubicon Archaeology Ltd, Midleton, County Cork, Ireland
Handan Ustundag, Anadolu University, Turkey
Subhash Walimbe, Deccan College, Pune, India
Don Walker, Museum of London, UK
Cynthia Wilczak, San Francisco State University, USA
Moushira Zaki

Questionnaire Groups

Greger Larson, Becky Gowland, Tom Moore, and Robin Skeates for handing out questionnaires at Durham University; Alyson Caine and Stewart Gardner for processing the data

Groups

Students in forensic vocational course, students in my course Bones and Human Societies, students in my course Archaeology in Action, students who participated in Romanian field course, students in MSc Archaeological Science and MSc Palaeopathology programs, attendees at Medieval and Renaissance Studies conference in Durham, Friends of Swaledale Museum, East Witton Women's Institute, archaeology staff at Fulling Mill Museum (Skeleton Science visitors) in Durham, visitors to Oriental Museum in Durham, East Witton campanologists, personal friends, members of the public who attended the Wellcome Trust Noah's Ark evening event, and fellow flower arrangers.

Images

American Association for the Advancement of Science/Rightslink
Johs Andersen (late), Copenhagen, Denmark
Abby Antrobus, Suffolk County Council, UK
Catrina Appleby, Council for British Archaeology
Yvonne Beadnell, formerly of Durham University, UK, who redrew Figures 2.1, 2.3, 2.6, 2.7, 4.9, 4.11, 5.19, 5.20, and 5.21 and drew Figure 5.15
Jean Brown, formerly of the University of Bradford, UK
Tjasse Bruintjes, The Netherlands
Saul Crawshaw
Hans Davanger, Bergen Museum, University of Bergen, Norway
Catherine Hellings, Leprosy Mission International

Catherine Hubert-Kazmierczyk, Bibliotèque Nationale de France
Stefanie Knab, National Library of Medicine, USA
Tomasz Kozlowski, Nicolaus Copernicus University, Toruń, Poland
Greg Lasley, Greg Lasley Nature Photography, USA
Mary Lewis, University of Reading, UK
Diana Lockwood, London School of Hygiene and Tropical Medicine, UK
John Magilton (late), archaeological consultant
Keith Manchester, University of Bradford, UK
Antonia Marcsik, University of Szeged, Hungary
Vitor Matos, University of Coimbra, Portugal
David Minnikin, University of Birmingham, UK (who also provided captions)
Don Ortner (late), Smithsonian Institution, USA
Scottish News Services
Svein Skare, Bergen Museum, Norway
Paul Spoerry, Oxford Archaeology East, UK
Taylor and Francis Limited
Chris Unwin, Durham, who redrew Figures 2.8, 2.9, and 4.1, and drew Figures 5.1, 5.2, 5.3, 5.11, 5.18, and 6.1
Jeff Veitch, Durham University, UK
Geoff Warne, Leprosy Mission International
Glenn Waterman, The Leprosy Mission Canada

Colleagues Who Have Helped with Various Queries

Zahra Afshar, formerly of Durham University, UK
Irene Allen, LEPRA
Richard Annis, Archaeological Services, Durham University, UK
Abby Antrobus, Suffolk County Council, UK
Iain Banks, University of Glasgow, UK
Paul Barnwell, Oxford University Department for Continuing Education, UK
Colleen Batey, University of Glasgow, UK
Michaela Binder, formerly of Durham University, UK
Sue Black, Lancaster University, UK
Jane Buckingham, University of Canterbury, New Zealand
Jane Buikstra, Arizona State University, USA
Jennie Coughlan
Jerry Cybulski, Canadian Museum of Civilization

Franklin Damann, National Museum of Health and Medicine, Armed Forces Institute of Pathology, Washington DC, USA
Patricia Deps, Universidade Federal do Espírito Santo, Vitória, Brazil
Marta Diaz-Zorita Bonilla, University of Tubingen, Germany
Stephen Driscoll, University of Glasgow, UK
Chris Dudar, Smithsonian Institution, USA
Bassey Ebenso, University of Leeds, UK
Grete Eilertsen, Leprosy Museum, Bergen, Norway
Jacqueline Eng, Western Michigan University, USA
Frank Garrod, Archaeological Society, Cyprus
Roberta Gilchrist, University of Reading, UK
Will Goodwin, University of Central Lancashire, UK
Becky Gowland, Durham University, UK
Anne Grauer, Loyola University of Chicago, UK
Sonia Guillen, Centro Mallqui, Lima, Peru
Ingrid Haugrønning, formerly of the Leprosy Museum, Bergen, Norway
Catherine Hellings, Leprosy Mission International
Cassandra Hodgkiss, Durham University, UK
Tina Jakob, Durham University, UK
Clark Larsen, Ohio State University, USA
Anwei Law, International Association for Integration Dignity and Economic Advancement, USA (http://www.idealeprosydignity.org)
Christine Lee, California State University, USA
Oona Lee, University of Birmingham, UK
Diana Lockwood, London School of Hygiene and Tropical Medicine, UK
Niels Lynnerup, University of Copenhagen, Denmark
Patrick Mahoney, University of Kent, UK
Keith Manchester, University of Bradford, UK
Antonia Marcsik, University of Szeged, Hungary
Christine McDonnell, York Archaeological Trust, UK
David Minnikin, University of Birmingham, UK
Scott Norton, Honolulu, Hawaii
Patrick Ottaway, PJO Archaeology, UK
Julie Peacock, Durham University, UK
Kate Pechenkina, Queen's College, City University of New York, USA
Paola Ponce, University of York, UK
Elizabeth Popescu, Oxford Archaeology, UK
Simon Roffey, University of Winchester, UK
Peter Rowley-Conwy, Durham University, UK

Mauro Rubini, Anthropological Service, SBAL, Rome, Italy
Ana Luisa Santos, University of Coimbra, Portugal
David Scollard, National Hansen's Disease Programs, Louisiana State University, Baton Rouge, Louisiana, USA
Sarah Semple, Durham University, UK
Pauline Stamp
Richard Thomas, University of Leicester, UK
Richard Truman, National Hansen's Disease Programs, Louisiana State University, Baton Rouge, USA
Val Turner, Shetland Amenity Trust, UK
Glen Waterman, Leprosy Mission International
Darlene Weston, University of British Columbia, Canada
Alicia Wilbur, University of Washington, USA
Richard Wright, University of Sydney, Australia

Appendix 2

Questionnaire about Leprosy

The Questionnaire

This anonymous questionnaire aims to capture information about your knowledge of the disease of leprosy, the results of which will be incorporated into a book I am writing on the past and present of leprosy. Many thanks for participating.

Please answer the following questions by ticking what you think is the answer:

1. What is leprosy?
   - ☐ a tumour
   - ☐ an infection
   - ☐ a disease affecting the joints
   - ☐ don't know

2. What pathological organism ("germ") causes leprosy?
   - ☐ a virus
   - ☐ a parasite
   - ☐ a bacterium
   - ☐ don't know

3. What part of the world is leprosy most commonly found today?
   - ☐ North America (USA and Canada)
   - ☐ Asia
   - ☐ Scandinavia
   - ☐ Eastern Europe
   - ☐ don't know

4. How does a person "catch" leprosy?
- ☐ by somebody with leprosy touching them
- ☐ by breathing in the pathological organism
- ☐ through sexual intercourse
- ☐ through the bite of an insect
- ☐ don't know

5. What mostly makes somebody more likely to "get" leprosy?
- ☐ if they have done something wrong
- ☐ if somebody in their family has leprosy
- ☐ if they are poor
- ☐ if they drink contaminated water
- ☐ don't know

6. Is leprosy described in the Bible?
- ☐ yes
- ☐ no
- ☐ don't know

7. What parts of the body are affected in leprosy?
- ☐ brain
- ☐ hips and knees
- ☐ nerves
- ☐ don't know

8. Do the fingers and toes "drop off"?
- ☐ yes
- ☐ no
- ☐ don't know

9. Can leprosy be cured today?
- ☐ yes
- ☐ no
- ☐ don't know

10. What happened to people who had leprosy in past times?
- ☐ they could live normally within their communities
- ☐ they were all put into isolation in leprosy hospitals

Appendix 2 · 321

☐ it depended when and where they lived
☐ don't know

FINALLY: What does the word "leper" mean to you? Give up to *FIVE* keywords:

Groups of People That Completed the Questionnaire

First-year undergraduate archaeology students, Durham University
MSc Palaeopathology students, Durham University
Students on field course in Romania
Staff in the Department of Archaeology, Durham University
Visitors to Durham Museums (Oriental Museum and Fulling Mill Museum of Archaeology)
Attendees at a public lecture, Friends of Swaledale Museum, Reeth, North Yorkshire
Attendees at the Wellcome Trust's Noah's Ark evening, London
A group consisting of East Witton, North Yorkshire, campanologists, East Witton Women's Institute, fellow flower arrangers, and the local community lower Wensleydale community, North Yorkshire

Questionnaire Results

The six most common words conjured up by the word "leper," in order of frequency used

Isolation (including separated, removed, banished, outsider, ostracized, social pariah, segregation, colony, secluded, expelled, hermit, alienated, stay away, marginalized, outcast, someone to be avoided, forbidden, alone, abandoned, shunned)
Diseased (including sickness, unwell, ill, illness, pathological, disease)
Disfigured (including deformed, ugly, malformed, crippling, grotesque, horrible, unsightly)
Stigma
Infectious (including contagious)
Poverty (including rags)

Other Words (Negative)

Unlucky, evil, depraved, deprived, mental illness, wild, contaminate, other countries, ignorance, damage, abnormal, demeaning, deadly, unwanted, ignored, distress, sight problems, underground, punishment from God, unapproachable, discrimination, misery, not welcome, worthless, destitute, other, tumour, incurable, tragic, disability, Third World, charity, bells, prejudice, debilitating, disadvantage, unfortunate, pity, misunderstood, Ben-Hur, skin, untouchable, lonely, bandages, suffering, feared, missing digits/limbs, dirty/unclean, beggar, pain, scabs, sores, sad, old times, ancient, death, early death, dying, terminal, skin falling off, rotting, nonexistent, stagnation

Other Words with No Negative Undertones

Antibiotic, medieval, Princess Di, compassion, help, curable, long history, biblical, holy, Jesus, rare, Dark Ages, person with leprosy

Internet Definitions of "Leper" (access date November 2018)

http://www.webster-dictionary.org/ — A person afflicted with leprosy; synonym = lazar. Related words include castaway, displaced person, evictee, exile, expatriate, expellee, outcast of society, outlaw, pariah, persona non grata, social outcast, unacceptable person, undesirable, untouchable.

http://www.yourdictionary.com/ — A person having leprosy; a person shunned or ostracized because of danger of moral contamination, and also a pariah, untouchable, outcast, and leprosy case.

http://www.thefreedictionary.com/ — A person affected by leprosy, and a person who is avoided by others, or a pariah

http://www.medicinenet.com/ — Someone with leprosy (Hansen's Disease). The term leper is now in disfavour; a pariah; a person who is avoided or shunned by society

http://www.merriam-webster.com/ — A person affected with leprosy; a person shunned for moral or social reasons; synonyms: castaway, castoff, outcast, offscouring, pariah, reject; related words: untouchable; outsider; deportee, exile

http://ldoceonline.com — Someone who has leprosy; someone that people avoid because they have done something that people disapprove of

http://dictionary.reference.com/ — A person who has leprosy; a person who has been rejected or ostracized for unacceptable behavior, opinions, character, or the like; anathema; outcast

Appendix 3

Skeletons from Archaeological Sites in the British Isles with Leprous Bone Changes

Site	N	Date	Features	Reference
PREHISTORIC				
Dryburn Bridge, East Lothian, Scotland	1	2300–2000 cal BC	Cist burial	J. Roberts (2007)
Kingsmead Quarry, Horton, Berkshire	Redeposited skull	No later than 1600–1100 BC	Ditch	McKinley forthcoming
ROMAN				
Chapel Lane, Bingham, Nottinghamshire	1	Postdates Roman period (AD 200–250)	Top of disused well in boundary ditch; slightly separate from 51 other burials	Allen (2018)
EARLY MEDIEVAL				
SCOTLAND				
Hallow Hill, St Andrews, Fife	1	8th century AD	Cist lined with stone slabs; 5 cover slabs	Lunt (1996, 2013)
Kirk Hill, St Andrews (St Mary's Church), Fife	1	5th–12th century AD	Disarticulated; re-deposited	Lunt (1997, 2013)
ENGLAND				
York Minster, Yorkshire	1	Early medieval (between AD 410–1050)	"Early medieval" cathedral	Manchester and Roberts (1986)
Wharram, Percy, Yorkshire	1	960–1000 AD	North churchyard	Mays, Harding, and Heighway (2007); Taylor and Donoghue (2011)

(continued)

Site	N	Date	Features	Reference
Norwich, St John the Baptist, Timberhill, Norfolk	Possibly 35	980–1050 AD	North churchyard Possibly served leprosy hospital	Anderson (1996, 1998, 2009a); Bayliss et al. (2004); Popescu (2009)
St Margaret, Ormesby, Norfolk	1	11th–14th centuries AD	Edge of north side of cemetery but delimited by modern road	Anderson (2009b)
Thorpe, St Catherine's, Norfolk	1	"Late Saxon"	—	Wells (1962)
Barrington, Edix Hill, Cambridgeshire	1	500 AD–1st half 7th century AD	Bed burial; grave goods	Duhig (1998)
Burwell, Cambridgeshire	1	7th century AD	Large grave	Møller-Christensen and Hughes (1962); Lethbridge (1931)
Ipswich, School Street, Suffolk	1	10th–11th century AD	Pre-friary cemetery; "pillow" stones	Mays 1989
Ipswich, St Augustine, Stoke Quay, Suffolk	1 and possibly 4 others	10th–11th century AD	—	Brown et al. (forthcoming)
Hoxne, Suffolk	1	885–1015 AD	From modern garden	Inskip, Taylor, and Stewart (2017)
Raunds, Furnells, Northamptonshire	2	8th–10th century AD	Marginal in cemetery	Powell (1996)
Willoughby-on-the-Wolds, Nottinghamshire	1	Late 5th–early 7th century AD	Multiple burial, many grave goods; associated trepanned skull	Roberts (1993); Harman (1993)
Eccles, Kent	1	7th century AD	Pit	Boocock, Manchester, and Roberts (1995)
Great Chesterford, Essex	1	450–600 AD	Grave goods	Waldron (1994b); Inskip (2008); Inskip et al. (2015)

Site	N	Date	Features	Reference
Bevis's Grave, Bedhampton, Hampshire	1	Late 6th to early 11th century AD	—	Manchester and Roberts (1986); Shennan (1978)
Collingbourne Ducis, Wiltshire	2	Early 6th century AD 5th–7th century AD	Jewelry	Dinwiddy (2016)
Beckford, Herefordshire	1	550 AD	2 dog skeletons; grave goods; marginal in cemetery	Wells (1962, 1996)
Cannington, Park Quarry, Somerset	3	Between 3rd–early 8th century AD	Metal tool with one? Imported Eastern Mediterranean pottery at the site	Brothwell, Powers, and Hirst (2000)
Tean, Isles of Scilly, Cornwall	1	7th century AD	—	Brothwell (1961)

LATE MEDIEVAL

SCOTLAND

Site	N	Date	Features	Reference
Deerness, Newark Bay, Orkney Islands	1	13–14th century AD	Chapel underground passage	Taylor et al. (2000); Brothwell (1977); Brothwell and Krzanowski (1974)
Aberdeen, Aberdeenshire	1	14th–17th century AD	Carmelite friary	Cardy (1996)
Glasgow, Strathclyde	1	14th–15th century AD	Cathedral	King (1994, 2002)

ENGLAND

Site	N	Date	Features	Reference
St Giles by Brompton Bridge, Yorkshire	1	Mid-13th–late 14th century AD	Next to road and river crossing general hospital	Cardwell (1995)
Scarborough, Yorkshire	1	After 1066 AD	Graveyard associated with chapel	Little (1943); Brothwell (1958)
Runcorn, Norton Priory, Cheshire	1	Early 14th century AD	Sandstone coffin in Chapterhouse of priory	Manchester and Roberts (1986); Brown and Howard-Davis (2008)

(continued)

Site	N	Date	Features	Reference
York, All Saints, Fishergate, Yorkshire	4	11th–16th century AD	—	McIntyre (2016); McIntyre and Chamberlain (2009)
York, Fishergate House, Yorkshire	1	12th–16th century AD	—	Holst (2005)
York, St Stephen's Church, Fishergate, Yorkshire	1	Late 11th–mid-14th century AD	Marginal; copper alloy pin	McComish (2015); Holst (2012)
Grantham, St Leonard's, Lincolnshire	3	Likely 13th century AD	General hospital	Boulter (1992); Hadley (2001); Trimble, Unsworth, and Hurley (1991)
Huntingdon, St Margaret's, Spittal's Link, Huntingdonshire	1	12–14th century AD	Leprosarium	Duhig (1993)
South Acre, St. Bartholomew, Norfolk	6	1100–1350 AD	Leprosarium	Wells (1967)
Ipswich, Blackfriars, Suffolk	4	1263–1538 AD	Nave margins; nave; and 2 in south aisle	Mays (1991); Taylor et al. (2006); Taylor and Donoghue (2011)
Midland Road, Peterborough, Cambridgeshire	26 definite/possible adults and 3 definite/possible non-adults	11th–16th century AD	2 phases; buried normally	McComish, Millward, and Boyle (2017); Caffell (2015)
Newark, St Leonard's, Nottinghamshire	1	1133–1642 century AD	South side of church General hospital	Bishop (1983); Manchester and Roberts (1986); Bayliss et al. (2004)
High Wycombe, St Margaret's, Buckinghamshire	1	13th–14th century AD	Leprosarium	Farley and Manchester (1989)
Oxford, St Bartholomew Hospital and Chapel, Oxfordshire	Disarticulated bone	Ca. AD 1126 onward	Charnel deposits Leprosarium	Levick (n.d.); Boylston (2012)
Abingdon Vineyard, Oxfordshire	1	1300–1540 AD	Abbey	Hacking and Wakely (n.d.)

Site	N	Date	Features	Reference
Ilford, St Mary and St Thomas, Essex	3	13th–14th century AD	Leprosarium	Manchester and Roberts (1986)
London, St Mary Spital, Spitalfields	2	1250–1400 AD	Marginal burial and one in a mass burial pit General hospital	Connell et al. (2012)
Stonar, Kent	2	—	—	Bayley n.d.; Manchester and Roberts (1986)
St Martin's Field, New Romney, Kent	1	13th–late 14th century AD	—	Clough and Loe (n.d.)
Lewes, St Nicholas, Sussex	2	12th–early 16th century AD	General hospital	Barber and Sibun (1998, 2010)
Cuddington, Nonesuch Palace, Ewell, Surrey	1	1100–1538	—	Møller-Christensen and Hughes (1962); Biddle (1959)
Chichester, St James and St Mary Magdalene, Sussex	75	12th–17th century AD	Leprosarium	Magilton and Boylston (2008)
Winchester, St Mary Magdalen, Hampshire	37	1070 AD–mid-12th century AD	Leprosarium	Roffey and Tucker (2012); Mendum et al. (2014); Schuenemann et al. (2013); Roffey et al. (2017)
Gloucester, St Oswald's Priory, Gloucestershire	1	1120–1230 AD	Priory	Mary Lewis, pers. comm., 2012
IRELAND (MODERN-DAY NORTHERN IRELAND AND REPUBLIC OF IRELAND)				
Armoy, St Patrick's Church, County Antrim (Northern Ireland)	1	1444–1636 AD	—	Murphy and Manchester (1998, 2002)
Dublin, St Michael le Pole, Leinster (Republic of Ireland)	3	11th century AD	Marginal	Buckley (2008a, 2008b)

(continued)

Site	N	Date	Features	Reference
Ardreigh, County Kildare (Republic of Ireland)	1	1264–1391 cal AD	—	Troy (2010)
CHANNEL ISLANDS				
St Tugal's Chapel, Herm	1	1251–1295 AD	—	Cataroche (2013); De Jersey and Cataroche (2012)
POST-MEDIEVAL (1550–1850 AD)				
London, St Marylebone Church	1	19th century AD	High status; amputation	Walker (2009)
Bristol, St Augustine the Less, Avon	1	"Post-medieval"	—	O'Connell (1998)

Appendix 4

Archaeological Hospitals, Including Leprosaria, Where Skeletons with or without Leprosy Have Been Excavated

Site name	Location	Reference
CARIBBEAN		
St. Eustatius	Caribbean Antilles	Gilmore (2008)
DENMARK		
Naestved	Zealand	Møller-Christensen (1953a–c, 1967, 1969, 1978)
St. Jørgensgård	Odense	Segal (2001); Matos (2009)
Stubbekøbing	Falster	Brander and Lynnerup (2002)
Svendborg	Funen	Møller-Christensen (1953c, 1967, 1978)
ENGLAND		
St Bartholomew	South Acre, Norfolk	Wells (1967)
St Bartholomew	Cowley, Oxford, Oxfordshire	Boylston (2012); Levick (n.d.)
St Giles	Brompton Bridge, North Yorkshire	Cardwell (1995)
St James's Hospital	Doncaster, South Yorkshire	Buckland and Magilton (1989)
St James and St Mary Magdalene	Chichester, Sussex	Magilton et al. (2008)
St John	Timberhill, Norfolk	Anderson (1996, 1998)
St John's church and hospital	New Romney, Kent	Clough and Loe (n.d.)
Leprosy hospital of St Stephen and St Thomas	New Romney, Kent	Clough and Loe (n.d.)
St Leonard	Grantham, Lincolnshire	Boulter (1992); Hadley (2001)
St Leonard	Newark, Nottinghamshire	Bishop (1983)

(continued)

Site name	Location	Reference
St Leonard	Hady Hill, Chesterfield, South Yorkshire	Magilton (2008e)
St Mary Magdalene	Bidlington, East Sussex	Magilton (2008e)
St Margaret's	High Wycombe, Buckinghamshire	Farley and Manchester (1989)
St Margaret's	London Road, Gloucester	Magilton (2008e)
St Margaret's	Spittal Link, Huntingdon, Huntingdonshire	Duhig (1993); Mitchell (1993)
St Mary Magdalene	Brook Street, Colchester, Essex	Crossan (2004)
St Mary Spital	London	Connell et al. (2012)
St Mary and St Thomas	Ilford, Essex	Manchester and Roberts (1986)
St Mary Magdalen	Winchester, Hampshire	Anderson (2001); Roffey and Tucker (2012)
St Nicholas	Lewes, Sussex	Barber and Sibun (1998)
St Saviour	Bury St Edmunds, Suffolk	Caruth and Anderson (1997)
St Stephen and St Thomas	New Romney, Kent	Magilton (2008e)
FRANCE		
St. Thomas d'Aizier	Calvados	Blondiaux et al. (2016)
Chapelle Saint-Lazare	Tours	Saint-Martin et al. (2006)
GERMANY		
Melaten	Aachen	Schmitz-Cliever (1971, 1972, 1973a, 1973b)
HUNGARY		
Szombathely Ferences	Szombathely	Marcsik et al. (2007)
NORWAY		
St Laurentius	Oslo	Per Holck (pers. comm., September 2011)
PORTUGAL		
Valle de Gafaria	Lagos, Algarve	Neves et al. (2012)
SWEDEN		
St. Jørgen	Malmo	Boldsen (2001)
St. Jørgen	Lund	Arentoft (1999)

Appendix 5

Useful Websites and Organizations

World Health Organization, Leprosy Elimination: http://www.who.int/topics/leprosy/en/

Centers for Disease Control and Prevention, Hansen's Disease (Leprosy): https://www.cdc.gov/leprosy/index.html

The Leprosy Mission International: https://www.leprosymission.org/

International Federation of Anti-Leprosy Associations: http://www.ilepfederation.org/

International Leprosy Association, History of Leprosy: http://www.leprosyhistory.org/

Damien Foundation, Brussels, Belgium: https://www.ilepfederation.org/member/damien-foundation-belgium/

National Hansen's Disease Museum, Carville, Louisiana, United States: http://www.hansen-dis.jp/english

American Leprosy Missions: http://www.leprosy.org/

Nepal Leprosy Trust: https://www.nlt.org.uk/

Leprosy Relief, Canada: http://www.slc-lr.ca/en/

National Leprosy Eradication Programme: http://nlep.nic.in/

Leprosy Mailing List: https://www.leprosy-information.org/resource/leprosy-mailing-list-lml

National Hansen's Disease (Leprosy) Clinical Center, Baton Rouge, Louisiana, United States: https://www.hrsa.gov/hansens-disease/index.html

Leprosy Information Services: https://www.leprosy-information.org/

Kalaupapa National Historical Park, Molokai, Hawaii, United States: http://www.gohawaii.com/molokai/regions-neighborhoods/central-molokai/kalaupapa-national-historical-park-molokai/

Robben Island, South Africa: http://www.robben-island.org.za/

Spinalonga Island, Crete: http://www.explorecrete.com/crete-east/EN-Spinalonga-leper-island.html

Father Damien Museum, Honolulu, Hawaii: http://www.bestplaceshawaii.com/tips/hidden_places/damien_museum.html#_=_

The Leprosy Museum, St. Jørgen Hospital , Bergen, Norway: http://www.bymuseet.no/en/museums/the-leprosy-museum-st-joergen-hospital/

National Hansen's Disease Museum, Tokyo, Japan: http://www.hansen-dis.jp/english

Culion Museum and Archives: https://www.thepoortraveler.net/2013/05/culion-museum-and-archives-palawan-philippines/

Notes

Introduction

1. "Neglected Tropical Diseases," World Health Assembly resolution WHA66.12, May 27, 2013, http://www.who.int/neglected_diseases/mediacentre/WHA_66.12_Eng.pdf?ua=1. The "roadmap" refers to World Health Organization, *Accelerating Work to Overcome the Global Impact of Neglected Tropical Diseases: A Roadmap for Implementation* (Geneva: World Health Organization, 2012), http://apps.who.int/iris/bitstream/10665/70809/1/WHO_HTM_NTD_2012.1_eng.pdf.

2. The BBC reporter used this term in the August 22, 2015, broadcast of *Today*. The *Times* Sunday magazine article appeared on July 5, 2015.

3. Ben Child, "Aardman Throws *Pirates*' Leprosy Gag Overboard," *The Guardian*, January 25, 2012, accessed August 15, 2016, https://www.theguardian.com/film/2012/jan/25/aardman-pirates-leprosy-gag.

4. Jiske Erlings, "FW: (LML) Leprosy Eponym: Need for a Medical Name," *Leprosy Mailing List*, January 30, 2020, http://leprosymailinglist.blogspot.com/search?q=naming+leprosy&max-results=20&by-date=true

5. Although I recognize that in the medical profession, "case" is used routinely to refer to patients, I have used quotation marks for that word in this book because I consider that using "case" to refer to a person objectifies a living or dead person.

Chapter 1. The Biology of Leprosy Bacteria and How They Are Transmitted to Humans

1. "About the Sustainable Development Goals," UN Sustainable Development Goals, accessed February 26, 2020, https://www.un.org/sustainabledevelopment/sustainable-development-goals/.

2. UNESCO, "Inscriptions of the Documentary Heritage in 2001," Memory of the World Register, accessed April 24, 2017, http://www.unesco.org/new/en/communication-and-information/memory-of-the-world/register/access-by-year/2001/#c184272.

3. Letter: Further Evidence of Leprosy in Isle of Wight Red Squirrels, *Veterinary Record* 180, no. 16 (2017): 407.

4. "National Hansen's Disease (Leprosy) Program: Caring and Curing since 1894," Health Resources & Service Administration, accessed July 2019, https://www.hrsa.gov/hansens-disease/index.html.

Chapter 2. How Leprosy Affects the Human Body

1. World Health Organization, "Leprosy," September 10, 2019, accessed January 27, 2020, https://www.who.int/en/news-room/fact-sheets/detail/leprosy.

Chapter 3. Past and Present Diagnosis, Treatment, and Prognosis

1. See World Health Organization, "Leprosy Elimination," accessed April 16, 2019, http://www.who.int/lep/en.

Chapter 5. The Bioarchaeological Evidence of Leprosy

1. For example, see the British Association for Biological Anthropology and Osteoarchaeology, "Code of Ethics," accessed February 2020, https://www.babao.org.uk/assets/Uploads/BABAO-Code-of-Ethics-2019.pdf; British Association for Biological Anthropology and Osteoarchaeology, "Code of Practice," accessed February 2020, https://www.babao.org.uk/assets/Uploads/BABAO-Code-of-Practice-2019.pdf; "Digital Imaging," accessed February 2020, https://www.babao.org.uk/assets/Uploads/BABAO-Digital-imaging-code-2019.pdf.

2. I have used quotation marks for the word "stress" because the word is frequently used in bioarchaeology but it can have many and varied definitions and causes. It can be defined as the body's physical, mental, or emotional reaction to pressure of whatever cause.

3. Because of these disagreements, I have not included this skeleton in the dataset I consider in subsequent discussions.

4. See "Camino de Santiago," accessed November 2018, http://santiago-compostela.net/. The cult of St James began in Spain in the sixth–seventh centuries AD; evidence in Britain in present the seventh and eighth centuries. The cult is evident in German-speaking countries by the ninth century. Documentary evidence suggests that interest increased in the twelfth century and declined in the sixteenth century.

5. "Leprosy Museum—Bergen City Museum," accessed February 2020, https://www.visitnorway.com/listings/leprosy-museum-bergen-city-museum/2585.

6. "Leprosarium valley."

7. De las Aguas (2006); Kalisch (1973); see also "The Leprosy of Tracadie," Virtual Museum.ca, accessed February 2020, http://www.virtualmuseum.ca/sgc-cms/histoires_de_chez_nous-community_memories/pm_v2.php?id=exhibit_home&fl=0&lg=English&ex=00000629.

Conclusions: A Future for Leprosy; Clinical and Bioarchaeological Perspectives

1. See the International Leprosy Association website about the history of leprosy at https://leprosyhistory.org/.

References

Abel, L., F. O. Sanchez, J. Oberti, N. V. Thuc, V. Hoa, V. D. Lap, E. Skamene, P. H. Lagrange, and E. Schurr. 1998. Susceptibility to Leprosy Is Linked to the Human NRAMP1 Gene. *Journal of Infectious Disease* 177: 133–145.

Abide, J. M., R. M. Webb, H. L. Jones, and L. Young. 2008. Three Indigenous Cases of Leprosy in the Mississippi Delta. *Southern Medical Journal* 101: 635–638.

Abraham, P. S. 1908. Leprosy. In *System of Medicine,* vol. 2, part 2, *Tropical Diseases and Animal Parasites,* edited by T. C. Albutt and H. D. Rolleston, 648–694. London: Macmillan.

Acevedo-Garcia, D., and J. Almeida. 2012. Introduction: Place, Migration and Health. *Social Science and Medicine* 75: 2055–2059.

Ackerknecht, E. H. 1965. *History and Geography of the Most Important Diseases.* New York: Hafner.

Adams, L. B. 2012. Insights from Animal Models on the Immunogenetics of Leprosy: A Review. *Memorias do Instituto Oswaldo Cruz* 107: 197–208.

Advisory Panel on the Archaeology of Burials in England. 2013. *Science and the Dead: A Guideline for the Destructive Sampling of Archaeological Human Remains for Scientific Analysis.* London: English Heritage and APABE.

Aftab, H., S. D. Nielsen, and I. C. Bygbjerg. 2016. Leprosy in Denmark, 1980-2010: A Review of 15 Cases. *BMC Research Notes* 9(1): 10.

Agarwal, S. C., B. B. Maheshwari, M. M. Mittal, and S. Kumar. 1973. A Histologic Study of Liver Lesions in Leprosy. *Indian Journal of Medical Research* 61: 389–395.

Ahlstrom, T., and C. Arcini. 2012. Swedish Paleopathology and Its Pioneers. In *The Global History of Paleopathology: Pioneers and Prospects,* edited by J. Buikstra and C. Roberts, 549–558. Oxford: Oxford University Press.

Albert, P., and P. D. O. Davies. 2008. Tuberculosis and Migration. In *Clinical Tuberculosis,* 4th ed., edited by P. D. O. Davies, P. F. Barnes, and S. B. Gordon, 367–381. London: Hodder Arnold.

Alfonso, J. L., F. A. Vich, J. J. Vilata, and J. Terencio de las Aguas. 2005. Factors Contributing to the Decline of Leprosy in Spain in the Second Half of the 20th Century. *International Journal of Leprosy and Other Mycobacterial Diseases* 73: 258–268.

Allen, M. 2018. Chapel Lane, Bingham, Nottinghamshire: Post-Excavation Assessment and Updated Project Design. Oxford Archaeology Client Report. Unpublished.

Almeida, Z. M., A. N. Ramos Jr., M. T. Raposo, F. R. Martins-Melo, and C. Vasconcellos. 2017. Oral Health Conditions in Leprosy Cases in Hyperendemic Area of the Brazilian Amazon. *Revista do Instituto de Medicina Tropical de Sao Paulo* 59: e50.

Alter, A., A. Grant, L. Abel, A. Alcaïs, and E. Schurr. 2011. Leprosy as a Genetic Disease. *Mammalian Genome* 22: 19–31.

Andersen, J. 1961. Plantar Ulcers in Leprosy. *Leprosy Review* 32: 16–27.

Andersen, J. G. 1969. *Studies in the Mediaeval Diagnosis of Leprosy in Denmark*. Copenhagen: Costers Bogtrykkeri.

Andersen, J. G., and J. W. Brandsma. 1984. *Management of Paralytic Deformities in Leprosy*. Addis Ababa, Ethiopia: All Africa Leprosy and Rehabilitation Training Centre.

Andersen, J. G., and K. Manchester. 1987. Grooving of the Proximal Phalanx in Leprosy: A Palaeopathological and Radiological Study. *Journal of Archaeological Science* 14: 77–82.

Andersen, J. G., and K. Manchester. 1988. Dorsal Tarsal Exostoses in Leprosy: A Palaeopathological and Radiological Study. *Journal of Archaeological Science* 15: 51–56.

———. 1992. The Rhinomaxillary Syndrome in Leprosy: A Palaeopathological and Radiological Study. *International Journal of Osteoarchaeology* 2: 121–129.

Andersen, J. G., K. Manchester, and R. S. Ali. 1992. Diaphyseal Remodelling in Leprosy: A Palaeopathological and Radiological Study. *International Journal of Osteoarchaeology* 2: 211–219.

Andersen, J. G., K. Manchester, and C. Roberts. 1994. Septic Bone Changes in Leprosy: A Palaeopathological and Radiological Review. *International Journal of Osteoarchaeology* 4: 21–30.

Anderson, H., B. Stryjewska, B. L. Boyanton, and M. R. Schwartz. 2007. Hansen Disease in the United States in the 21st Century. *Archives of Pathology and Laboratory Medicine* 131: 982–986.

Anderson, S. 1996. Human Skeletal Remains from Timberhill, Castle Mall, Norwich (Excavated 1989–1991). Ancient Monuments Laboratory Report 73/96. Unpublished.

———. 1998. Leprosy in a Medieval Churchyard in Norwich. In *Current and Recent Research in Osteoarchaeology: Proceedings of the 3rd Meeting of the Osteoarchaeology Research Group*, edited by S. Anderson, 31–37. Oxford: Oxbow Books.

———. 2001. Human Skeletal Remains from the Hospital of St Mary Magdalene, Winchester (AY30). Unpublished report.

———. 2009a. Cemeteries 1 and 4: St John at the Castle Gate (Later de Berstrete, Now St John the Baptist, Timberhill). In *Norwich Castle: Excavation and Historical Survey, 1987–98. Part 1, Anglo-Saxon to c. 1345*, edited by E. Shepherd Popescu, 215–237. East Anglian Archaeology Report no. 132. Norwich: Norfolk Museums and Archaeology Service.

———. 2009b. The Human Skeletal Remains. In *A Medieval Cemetery at Mill Lane, Ormesby, St Margaret, Norfolk*, edited by H. Wallis, 11–27. Dereham: East Anglian Archaeology, Norwich Archaeological Unit Archaeology and Historic Environment.

Angel, J. L. 1964. Osteoporosis: Thalassaemia? *American Journal of Physical Anthropology* 22: 369–374.

———. 1969. Human Skeletal Remains at Karataş. In *Excavations at Karataş-Semayük and Elmali, Lycia*, edited by M. J. Mellink, 253–258. *American Journal of Archaeology* 74: 245–259.

Antunes-Ferreira, N., V. M. J. Matos, and A. L. Santos. 2013. Leprosy in Individuals Unearthed near the Ermida de Santo André and Leprosarium of Beja, Portugal. *Anthropological Science* 121: 149–159.

Antunes-Ferreira, N., and A. F. Rodrigues. 2003. Intervençao Archeológica no Largo da Ermida Largo da Ermida de Santo André (Beja). *Al-Madan* 12 (2nd series): 193.

Araujo, S., L. O. Freitas, L. R. Goulart, and I. M. Goulart. 2016. Molecular Evidence for the Aerial Route of Infection of *Mycobacterium leprae* and the Role of Asymptomatic Carriers in the Persistence of Leprosy. *Clinical Infectious Diseases* 63: 1412–1420.

Arcini, C. 1999. *Health and Disease in Early Lund: Osteo-Pathologic Studies of 3,305 Individuals Buried in the First Cemetery Area of Lund, 990–1536.* Lund, Sweden: Department of Community Health Sciences, University of Lund.

———. 2018. *The Viking Age: A Time of Many Faces.* Oxford: Oxbow.

Arcini, C., and T. Artelius. 1993. Äldsta Fallet av Spetälska i Norden. *Arkeologi i Sverige* 2: 55–71.

Ardagna, Y., I. Bouchez, M. Maillot, M. Evina, and M. Baud. 2012. Osteoarchaeological Evidence of Leprosy from Medieval Sudanese Nubia. Poster presented at the 19th European meeting of the Paleopathology Association, Lille, France.

Ardagna, Y., I. Bouchez, and L. Vidal. 2012. Possible Evidence of Leprosy in Medieval "Gard Provencal": Burial SP 21373 from the Saint Jean de Todon Cemetery in Laudun l'Ardoise (Gard, France). Poster presented at the 19th European meeting of the Paleopathology Association, Lille, France.

Arentoft, E. 1999. *De Spedalskes Hospital Udgravning af Sankt Jørgensgården I. Odense Fynske Studier 18.* Odense, Denmark: Odense Universitetsforlag.

Arora, M., K. Katoch, M. Natrajan, R. Kamal, and V. S. Kamal. 2008. Changing Profile of Disease in Leprosy Patients Diagnosed in a Tertiary Care Centre during Years 1995–2000. *Indian Journal of Leprosy* 80: 257–265.

Arraes, M. L. B. M., M. V. Holanda, L. N. G. C. Lima, J. A. B. Sabadia, C. R. Duarte, R. L. F. Almeida, C. Kendall, L. R. S. Kerr, and C. C. Frota. 2017. Natural Environmental Water Sources in Endemic Regions of Northeastern Brazil Are Potential Reservoirs of Viable *Mycobacterium leprae*. *Memorias do Instituto Oswaldo Cruz* 112: 805–811.

Asano, M. 1958. Leprous Pink Spots of the Tooth. *Lepro* 27: 398–401.

Ashmead, A. S. 1895. Pre-Columbian Leprosy. *Journal of the American Medical Association* 24: 622–626, 669–672, 721–723, 753–754, 803–807, 850–853.

———. 1896. Introduction of Leprosy into Nova Scotia and the Province of New Brunswick: Micmacs Immune. *Journal of the American Medical Association* 26: 202–208.

Atkin, N. J. 2018. From Isolation to Prosperity: Rediscovering the Carville Leprosarium. *Clinical Dermatology* 36: 421–425.

Avanzi, C., J. Del-Pozo, A. Benjak, K. Stevenson, V. R. Simpson, P. Busso, J. McLuckie, C. Loiseau, C. Lawton, J. Schoening, et al. 2016. Red Squirrels in the British Isles Are Infected with Leprosy Bacilli. *Science* 354(6313): 744–747.

Avelleira, J. C. R., F. Bernardes Filho, M. V. Quaresma, and F. R. Vianna. 2014. History of Leprosy on Rio de Janeiro. *Anais Brasileiros de Dermatology* 89: 515–518.

Avi-Yonah, M. 1963. The Bath of the Lepers at Scythopolis. *Israel Exploration Journal* 13: 325–326.

Awasthi, S. K., G. Singh, R. K. Dutta, and O. P. Pahuja. 1990. Audiovestibular Involvement in Leprosy. *Indian Journal of Leprosy* 62: 429–434.

Aycock, W. L. 1940. Familial Susceptibility as a Factor in the Propagation of Leprosy in

North America. *International Journal of Leprosy and Other Mycobacterial Diseases* 8: 137–150.

Aye, K. S., Y. T. Oo, K. Kyaw, A. A. Win, and M. Matsuoka. 2012. Genotyping of *Mycobacterium leprae* in Myanamar and Supposed Transmission Mode. *Nihon Hansenbyo Gakkai Zasshi* 81: 191–8.

Azizi, M. H., and M. Bahadori. 2011. A History of Leprosy in Iran during the 19th and 20th Centuries. *Archives of Iranian Medicine* 14: 425–430.

Badger, L. F. 1959. Epidemiology. In *Leprosy in Theory and Practice*, edited by R. G. Cochrane, 69–97. Bristol: John Wright and Sons.

Baird, E. S. N.d. *Childhood in the Jungle*. Published privately.

Bakare, A. T., A. J. Yusuf, Z. G. Habib, and A. Obembe. 2015. Anxiety and Depression: A Study of People with Leprosy in Sokoto, North-Western Nigeria. *Journal of Psychiatry* S1: 004. https://www.omicsonline.org/open-access/anxiety-and-depression-a-study-of-people-with-leprosy-in-sokoto-north-western-nigeria-2378-5756-1000004.pdf.

Baker, B. J., and K. L. Bolhofner. 2014. Biological and Social Implications of a Medieval Burial from Cyprus for Understanding Leprosy in the Past. *International Journal of Paleopathology* 4: 17–24.

Baker, B. J., and M. A. Judd. 2012. Development of Palaeopathology in the Nile Valley. In *The Global History of Paleopathology: Pioneers and Prospects*, edited by J. Buikstra and C. Roberts, 209–234. Oxford: Oxford University Press.

Baker, B., G. Crane-Kramer, M. W. Dee, L. A. Gregoricka, M. Henneberg, C. Lee, S. A. Lukehart, D. C. Mabey, C. A. Roberts, A. L. W. Stodder, A. C. Stone, and S. Winingear. 2020. Advancing the Understanding of Treponemal Disease in the Past and Present. *Yearbook of Physical Anthropology*: https://doi.org/10.1002/ajpa.23988.

Bakija-Konsuo, A., and R. Mulić. 2011. Leprosy—Today Forgotten in Croatia? *Acta Medica Croatia* 65: 251–255.

Bakker, M. I., M. Hatta, A. Kwenang, P. R. Klatser, and L. Oskam. 2002. Epidemiology of Leprosy on Five Isolated Islands in the Flores Sea, Indonesia. *Tropical Medicine and International Health* 7: 780–187.

Balázs, J., Z. Rózsa, Z. Bereczki, A. Marsik, B. Tihanyi, K. Karlinger, G. Pölöskei, E. Molnár, H D. Donoghue, and G. Pálfi. 2019. Osteoarcheological and Biomolecular Evidence of Leprosy From an 11th–13th Century CE Muslim Cemetery in Europe (Orosháza, Southeast Hungary). *Homo* 70: 105–118.

Barber, L., and L. Sibun. 1998. The Medieval Hospital of St Nicholas, Lewes, East Sussex: Excavations 1994. Archaeology South-East Draft Report 1994/148.

———. 2010. The Medieval Hospital of St Nicholas, Lewes, East Sussex: Excavations 1994. *Sussex Archaeological Collections* 148: 79–109.

Barker, D. J. P. 1994. *Mothers, Babies, and Disease in Later Life*. London: BMJ Books.

Barnetson, J. 1951. Pathogenesis of Bone Changes in Neural Leprosy. *International Journal of Leprosy and Other Mycobacterial Diseases* 19: 297–307.

Barnetson, R., G. Bjune, J. M. H. Pearson, and G. Kronvall. 1976. Cell Mediated and Humoral Immunity in "Reversal Reactions." *International Journal of Leprosy* 44: 267–273.

Barrett, R., and G. J. Armelagos. 2013. *An Unnatural History of Emerging Infections*. Oxford: Oxford University Press.

Barrett, R., C. W. Kuzawa, T. McDade, and G. Armelagos. 1998. Emerging and Re-Emerging

Infectious Diseases: The Third Epidemiologic Transition. *Annual Review of Anthropology* 27: 247–271.

Barton, R. P. E. 1979. Radiological Changes in the Paranasal Sinuses in Lepromatous Leprosy. *Journal of Laryngology and Otology* 93: 597–600.

Baten, J., R. H. Steckel, C. S. Larsen, and C. A. Roberts. 2018. Multidimensional Patterns of European Health, Work, and Violence over the Past Two Millennia. In *The Backbone of Europe: Health, Diet, Work and Violence over Two Millennia*, edited by R. H. Steckel, C. S. Larsen, C. A. Roberts, and J. Baten, 380–396. Cambridge: Cambridge University Press.

Bayley, J. N.d. Skeletal Remains from Stonar, Kent. Unpublished manuscript.

Bayliss, A., E. Shepherd Popescu, N. Beavan-Athfield, and C. Bronk Ramsey. 2004. The Potential Significance of Dietary Offsets for the Interpretation of Radiocarbon Dates: An Archaeologically Significant Example from Medieval Norwich. *Journal of Archaeological Science* 31: 563–575.

Bayliss, J. 1979. Domus Leprosae—Community Care in Medieval England: Leper Houses in the Middle Ages. *Nursing Times* 75: 62–67.

Bedi, B. M., E. Narayanan, M. Sreevatsa, W. F. Kirchheimer, and M. Balasubrahmanyam. 1976. Dispersal of *Mycobacterium leprae* by Leprosy Patients While Breathing. *Annals of the Indian Academy of Medical Science* 12: 1–15.

Bedić, Z., M. Šlaus, and H. D. Donoghue. 2019. The Earliest Recorded Case of Lepromatous Leprosy in Continental Croatia. *Journal of Archaeological Science: Reports* 25: 47–55.

Behr, M. A., and E. Schurr. 2006. Mycobacteria in Crohn's Disease: A Persistent Hypothesis. *Inflammatory Bowel Disease* 12: 1000–1004.

Belcastro, M. G., V. Mariotti, F. Facchini, and O. Dutour. 2005. Leprosy in a Skeleton from the 7th Century Necropolis of Vicenne-Campochiaro (Molise, Italy). *International Journal of Osteoarchaeology* 15: 431–448.

Benediktsson, G., and O. Bjarnson. 1959. Leprosy in Iceland. *Nord Medicine* 62: 1225–1227.

Bennett, B. H., D. L. Parker, and M. Robson. 2008. Leprosy: Steps along the Journey of Eradication. *Public Health Reports* 123: 198–205.

Bennike, P. 2002. Vilhelm Møller-Christensen: His Work and Legacy. In *The Past and Present of Leprosy: Archaeological, Historical, Palaeopathological and Clinical Approaches*, edited by C. A. Roberts, M. E. Lewis, and K. Manchester, 135–144. British Archaeological Reports International Series 1054. Oxford: Archaeopress.

———. 2012. Paleopathology in Denmark: The Pioneers Vilhelm Møller-Christensen and Johannes G. Andersen. In *The Global History of Paleopathology: Pioneers and Prospects*, edited by J. Buikstra and C. Roberts, 361–374. Oxford: Oxford University Press.

Bennike, P., M. E. Lewis, H. Schutkowski, and F. Valentin. 2005. Comparisons of Child Morbidity in Two Contrasting Medieval Cemeteries from Denmark. *American Journal of Physical Anthropology* 128: 734–746.

Bergel, M. 1960. The Hutchinson Dietetic Hypothesis of Fish Eating as a Cause of Leprosy: A Reappraisal in Light of the Influence of Pro-Oxidant Nutritional Conditions. *Leprosy Review* 31: 302–304.

Berrington, W. R., M. Macdonald, S. Khadge, B. R. Sapkota, M. Janer, D. A. Hagge, G. Kaplan, and T. R. Hawn. 2010. Common Polymorphisms in the NOD2 Gene Region

Are Associated with Leprosy and Its Reactive States. *Journal of Infectious Diseases* 201: 1422–1435.

Bhattacharya, S. N., and V. N. Sehgal. 2002. Reappraisal of the Drifting Scenario of Leprosy Multi-Drug Therapy: New Approach Proposed for the New Millennium. *International Journal of Dermatology* 41: 321–326.

Bhoyroo, J. 1997. The British Leper: A Study of Leprosy in Britain, 1867–1951. BSc diss., Wellcome Institute for the History of Medicine.

Biddle, M. 1959. Nonesuch Palace, Ewell, Surrey. University of Cambridge Duckworth Laboratory Archives. Unpublished report.

Bieler-Gomez, L., E. Castoldi, E. Baldini, A. Cappella, and C. Cattaneo. 2018. Diabetic Bone Lesions: A Study on 38 Known Modern Skeletons and the Implications for Forensic Scenarios. *International Journal of Legal Medicine* 33: 1225–1239.

Binford, C. H., E. E. Storrs, and G. P. Walsh. 1976. Disseminated Infection in the Nine-Banded Armadillo (*Dasypus novemcinctus*) Resulting from Inoculation with *M. leprae*: Observations Made on 15 Animal Studies at Autopsy. *International Journal of Leprosy* 44: 80.

Birke, J. A., A. Novick, C. A. Patout, and W. C. Coleman. 1992. Healing Rates of Plantar Ulcers in Leprosy and Diabetes. *Leprosy Review* 63: 365–374.

Bishop, M. W. 1983. Burials from the Cemetery of the Hospital of St Leonard, Newark, Nottinghamshire. *Transactions of the Thoroton Society of Nottinghamshire* 87: 23–35.

Blau, S., and V. Yagodin. 2005. Osteoarchaeological Evidence for Leprosy from Central Asia. *American Journal of Physical Anthropology* 126: 150–158.

Blok, D. J., S. J. de Vlas, E. A. J. Fischer, and J. H. Richardus. 2015. Mathematical Modeling of Leprosy and Its Control. *Advances in Parasitology* 87: 33–51.

Blok, D. J., S. J. de Vlas, and J. H. Richardus. 2015. Global Elimination of Leprosy by 2020: Are We on Track? *Parasites and Vectors* 8: Article 548.

Blondiaux, J. 1995. DNA of *Mycobacterium leprae* Detected by PCR in Ancient Bone by Rafi et al (1). *International Journal of Osteoarchaeology* 5: 299.

Blondiaux, J., J. Durr, L. Khouchaf, and L. E. Eisenberg. 2002. Microscopic Study and X-Ray Analysis of Two 5th Century Cases of Leprosy: Palaeoepidemiological Inferences. In *The Past and Present of Leprosy: Archaeological, Historical, Palaeopathological and Clinical Approaches,* edited by C. A. Roberts, M. E. Lewis, and K. Manchester, 105–110. British Archaeological Reports International Series 1054. Oxford: Archaeopress.

Blondiaux, J., O. Dutour, and M. Sansilbano-Collilieux. 2012. The Pioneers of Palaeopathology in France. In *The Global History of Paleopathology: Pioneers and Prospects,* edited by J. Buikstra and C. Roberts, 375–386. Oxford: Oxford University Press.

Blondiaux, J., J.-F. Duvette, S. Vatteoni, and L. E. Eisenberg. 1994. Microradiographs of Leprosy from an Osteoarchaeological Context. *International Journal of Osteoarchaeology* 4: 13–20.

Blondiaux, J., S. Naji, J.-P. Bocquet-Appel, T. Colard, A. de Broucker, and C. de Seréville-Niel. 2016. The Leprosarium of Saint-Thomas d'Azier: The Cementochronological Proof of the Medieval Decline of Hansen's Disease in Europe? *International Journal of Paleopathology* 15: 140–151.

Bobosha, K., J. J. Van Der Ploeg-Van Schip, M. Zewdie, B. R. Sapkota, D. A. Hagge, K. L. M. C. Franken, W. Inbiale, A. Aseffa, T. H. M. Ottenhoff, and A. Geluk. 2011. Immunogenicity of *Mycobacterium leprae* Unique Antigens in Leprosy Endemic Populations in Asia and Africa. *Leprosy Review* 82: 445–58.

Bochud, P. Y., D. Sinsimer, A. Aderem, M. R. Siddiqui, P. Saunderson, S. Britton, I. Abraham, A. Tadess Argaw, M. Janer, T. R. Hawn, and G. Kaplan. 2009. Polymorphisms in Toll-Like Receptor 4 (TLR4) Are Associated with Protection against Leprosy. *European Journal of Clinical Infectious Diseases* 28: 1055–1065.

Boddington, A. 1996. *Raunds Furnells: The Anglo-Saxon Church and Churchyard*. English Heritage Archaeological Report 7. London: English Heritage.

Boeckl, C. M. 2011. *Images of Leprosy: Disease, Religion, and Politics in European Art*. Kirksville, MS: Truman State University Press.

Boel, L. W. T., and D. J. Ortner. 2013. Skeletal Manifestations of Skin Ulcer in the Lower Leg. *International Journal of Osteoarchaeology* 23: 303–309.

Boku, N., D. N. Lockwood, M. V. Balagon, F. E. Pardillo, A. A. Maghanoy, I. B. Mallari, and H. Cross. 2010. Impacts of the Diagnosis of Leprosy and of Visible Impairments amongst People Affected with Leprosy in Cebu, the Philippines. *Leprosy Review* 81: 111–120.

Boldsen, J. 2001. Epidemiological Approach to the Paleopathological Diagnosis of Leprosy. *American Journal of Physical Anthropology* 115: 380–387.

———. 2005. Leprosy and Mortality in the Medieval Danish Village of Tirup. *American Journal of Physical Anthropology* 126: 159–168.

Boldsen, J. L. 2008. Leprosy in the Early Medieval Lauchheim Community. *American Journal of Physical Anthropology* 135: 301–310.

———. 2009. Leprosy in Medieval Denmark—Osteological and Epidemiological Analyses. *Anthropologischer Anzeiger* 67: 407–425.

Boldsen, J. L., and L. Mollerup. 2006. Outside St. Jørgen: Leprosy in the Medieval Danish City of Odense. *American Journal of Physical Anthropology* 130: 344–351.

Bolek, E. C., A. Erden, C. Kulecki, U. Kalyoncu, and O. Karadag. 2017. Rare Occupational Cause of Nasal Septum Perforation: Nickel Exposure. *International Journal of Occupational Medicine and Environmental Health* 30: 963–967.

Bonnar, P. E., N. P. Cunningham, A. K. Boggild, N. M. Walsh, R. Sharma, and I. R. C. Davies. 2018. Leprosy in Non-Immigrant Canadian Man without Travel outside North America, 2014. *Emerging Infectious Diseases* 24: 165–166.

Boocock, P., K. Manchester, and C. A. Roberts. 1995. The Human Remains from Eccles, Kent. Calvin Wells Laboratory, University of Bradford. Unpublished manuscript.

Boocock, P., C. A. Roberts, and K. Manchester. 1995a. Prevalence of Maxillary Sinusitis in Leprous Individuals from a Medieval Leprosy Hospital. *International Journal of Leprosy and Other Mycobacterial Diseases* 63(2): 265–268.

———. 1995b. Maxillary Sinusitis in Medieval Chichester. *American Journal of Physical Anthropology* 98: 483–495.

Bosworth, C. E. 1976. *The Medieval Islamic Underworld*. Leiden: Brill Academic Publishers.

Bottene, I. M., and V. M. Reis. 2012. Quality of Life of Patients with Paucibacillary Leprosy. *Anais Brasileiros de Dermatology* 87: 408–411.

Boulter, S. 1992. Death and Disease in Medieval Grantham. Undergraduate diss., University of Sheffield.

Bourbou, C. 2003. The Leprosarium of Spinalonga (1903–1957) in Eastern Crete (Greece). *Eres Arqueologia/Bioantropologia* 14: 121–136.

Bouwman, A. S., and T. Brown. 2005. The Limits of Biomolecular Palaeopathology: Ancient DNA Cannot Be Used to Study Venereal Syphilis. *Journal of Archaeological Science* 32: 703–713.

Boyd, K. M. 2000. Disease, Illness, Sickness, Health, Healing and Wholeness: Exploring Some Elusive Concepts. *Journal of Medical Ethics: Medical Humanities* 26: 9–17.

Boyd, R. 2000. *The Coming of the Spirit of Pestilence: Introduced Diseases and Population Decline among Northwest Coast Indians, 774–1874*. Vancouver: University of British Columbia Press.

Boyle, A. 2017. The Human Bones. In *The Medieval Cemetery of St Leonard's Leper Hospital at Midland Road, Peterborough*, edited by J. M. McComish, G. Millward, and A. Boyle, 44–136. York Report Number YAT 11/2017. York: York Archaeological Trust. https://www.yorkarchaeology.co.uk/wp-content/uploads/2017/07/YAT-AY11-Midland-Road-Peterborough-1.pdf.

Boylston, A. 2012. Charnel Deposits from St Bartlemas' Chapel Excavation, Oxford 2011. Discussion of the Palaeopathology and Its Implications. Unpublished manuscript.

Brady, K. 2002. Brei Holm, Papa Stour: In the Footsteps of the Papar? In *The Papar in the North Atlantic: Environment and History*, edited by B. E. Crawford, 69–82. St Andrews: University of St Andrews.

Brand, P. W. 1959. Temperature Variation and Leprosy Deformities. *International Journal of Leprosy and Other Mycobacterial Diseases* 27: 1–7.

Brander, T., and N. Lynnerup. 2002. A Possible Leprosy Hospital in Stubbekøbing, Denmark. In *The Past and Present of Leprosy: Archaeological, Historical, Palaeopathological and Clinical Approaches*, edited by C. A. Roberts, M. E. Lewis, and K. Manchester, 145–148. British Archaeological Reports International Series 1054. Oxford: Archaeopress.

Bratschi, M. W., P. Steinmann, A. Wickenden, and T. P. Gillis. 2015. Current Knowledge on *Mycobacterium leprae* Transmission: A Systematic Literature Review. *Leprosy Review* 986: 142–155.

Brenner, E., and F.-O. Touati, eds. Forthcoming. *Leprosy and Identity in the Middle Ages: from England to the Mediterranean*. Manchester: Manchester University Press.

Breslow, L. 1968. Leprosy in California. *California Medicine* 108: 81–82.

Bret, S., B. Flageul, P. Y. Girault, E. Lightburne, and J. J. Morand. 2013. Epidemiological Survey of Leprosy Conducted in Metropolitan France between 2009 and 2010. *Annales de Dermatologie et de Venereologie* 140: 347–352.

Brickley, M., and R. Ives. 2008. *The Bioarchaeology of Metabolic Bone Disease*. London: Academic Press.

Bridgford, S., S. Desmond, V. Klontza-Jaklova, and I. Moutafi. 2014. Grave 1. An Important Early and Middle Byzantine Feature. In *A Cretan Landscape through Time: Priniatikos Pyrgos and Environs*, edited by B. P. C. Molloy and C. N. Duckworth, 54–60. British Archaeological Reports International Series 2634. Oxford: Archaeopress.

Brightbill, H. D., D. H. Libraty, S. R. Krutzik, R. B. Yang, J. T. Belisle, J. R. Bleharski, M. Maitland, M. V. Norgard, S. E. Plevy, S. T. Smale, et al. 1999. Host Defense Mechanisms Triggered by Microbial Lipoproteins through Toll-Like Receptors. *Science* 285: 732–736.

Britton, W. J. 1993. Leprosy 1962–1992. 3. Immunology of Leprosy. *Transactions of the Royal Society of Tropical Medicine and Hygiene* 87: 508–514.

Britton, W. J., and D. N. Lockwood. 2004. Leprosy. *Lancet* 363: 1209–1219.

Brødholt, E. T. 2007. *Skjelettene fra Mariakirken: Nytt Lys på Kongskirken Gjennom en Undersøkelse av Skjelettmaterialet*. Antropologiske skrifter nr. 7. Oslo: Avdeling for anatomi, Institutt for medisinske basalfag, Universitetet i Oslo.

Brody, S. N. 1974. *The Disease of the Soul: Leprosy in Medieval Literature*. Ithaca, NY: Cornell University Press.

Brosch, R., S. V. Gordon, K. Eiglmeier, T. Garnier, and S. T. Cole. 2000. Comparative Genomics of the Leprosy and Tubercle Bacilli. *Research in Microbiology* 151: 135–142.

Brothwell, D. R. 1958. Evidence of Leprosy in British Archaeological Material. *Medical History* 11: 287–291.

———. 1961. The Palaeopathology of Early British Man: An Essay on the Problems of Diagnosis and Analysis. *Journal of the Royal Anthropological Institute* 91: 318–344.

———. 1977. On a Mycoform Structure in Orkney, and Its Relevance to Possible Further Interpretations of So-Called Souterrains. *Bulletin of the Institute of Archaeology, London* 14: 179–190.

Brothwell, D. R., and W. Krzanowski. 1974. Evidence of Biological Difference between Early British Populations from Neolithic to Medieval Times, as Revealed by Eleven Commonly Available Cranial Vault Measurements. *Journal of Archaeological Science* 1: 249–260.

Brothwell, D. R., R. Powers, and S. Hirst. 2000. The Pathology. In *Cannington Cemetery: Excavations 1962–3 of Prehistoric, Roman, Post-Roman and Later Features at Cannington Park Quarry, Near Bridgwater, Somerset*, edited by P. Rahtz, S. Hirst, and S. M. Wright, 195–256. Britannia Monograph Series 17. London: Society for the Promotion of Roman Studies.

Brown, F., and C. Howard-Davis. 2008. *Norton Priory: Monastery to Museum: Excavations 1970–1987*. Lancaster: Oxford Archaeology North.

Brown, L. R., C. D. May, and S. E. Williams. 1962. A Non-Tuberculous Granuloma in Cats. *New Zealand Veterinary Journal* 10: 7–9.

Brown, P., T. Sutikna, M. J. Morwood, R. P. Soejono, E. Jatmiko, E. Wayhu Saptomo, and R. A. Due. 2004. A New Small-Bodied Hominin from the Late Pleistocene of Flores, Indonesia. *Nature* 431: 1055–1061.

Brown, R., S. Teague, L. Loe, B. Sudds, and E. Popescu. Forthcoming. *Excavations at Stoke Quay, Ipswich: Southern Gipeswic and the Parish of St Augustine*. Norwich, Norfolk: East Anglian Archaeology.

Brown, T., and K. Brown. 2011. *Biomolecular Archaeology: An Introduction*. Chichester: Wiley-Blackwell.

Browne, S. G. 1975a. *Leprosy in the Bible*. Christian Medical Fellowship Publication, London.

———. 1975b. Some Aspects of the History of Leprosy: The Leprosie of Yesterday. *Proceedings of the Royal Society of Medicine* 68: 485–493.

Brozou, A., N. Lynnerup, M. A. Mannino, A. R. Millard, and D. R. Grocke. 2019. Investigating Dietary Patterns and Organizational Structure by Using Stable Isotope Analysis: A Pilot Study of the Danish Medieval Hospital at Naestved. *Anthroplologia Anzeiger* 76: 167–168.

Brubaker, M. L., C. H. Binford, and J. R. Trautman. 1969. Occurrences of Leprosy in U. S. Veterans after Service in Endemic Areas Abroad. *Public Health Reports* 84: 1057–1058.

Brubaker, M. L., W. M. Meyers, and J. Bourland. 1985. Leprosy in Children One Year of Age and Under. *International Journal of Leprosy* 53: 517–523.

Bruintjes, T. D. 1990. The Auditory Ossicles in Human Skeletal Remains from a Leper Cemetery in Chichester, England. *Journal of Archaeological Science* 17: 627–633.

Bryceson, A., and R. E. Pfaltzgraff. 1990. *Leprosy*. 3rd ed. Edinburgh: Churchill Livingstone.

Buckingham, J. 2002. *Leprosy in Colonial South India: Medicine and Confinement*. Basingstoke: Palgrave.

Buckland, P. C., and J. R. Magilton. 1989. *The Archaeology of Doncaster*. Vol. 2, *The Medieval and Later Town*. British Archaeological Reports, British Series 202 (ii). Oxford: BAR.

Buckley, L. 2008a. Leprosy from Hiberno-Norse Dublin, Ireland c.1000AD. Poster presented at the 17th European meeting of the Paleopathology Association, Copenhagen, Denmark.

———. 2008b. Outcasts, or Care in the Community? *Archaeology Ireland* (Spring): 27–31.

———. 2011. Ireland/Éire. In *The Routledge Handbook of Archaeological Human Remains and Legislation*, edited by N. Márquez-Grant and L. Fibiger, 211–219. London: Routledge.

Buikstra, J. E. 2006. Repatriation and Bioarchaeology: Challenges and Opportunities. In *Bioarchaeology: The Contextual Analysis of Human Remains*, edited by J. E. Buikstra and L. A. Beck, 389–415. Amsterdam: Elsevier.

Buikstra, J. E., and C. A. Roberts, eds. 2012. *The Global History of Paleopathology: Pioneers and Prospects*. Oxford: Oxford University Press.

Buikstra, J. E., and C. C. Gordon. 1981. The Study and Restudy of Human Skeletal Series: The Importance of Long-Term Curation. *Annals of the New York Academy of Science* 376: 449–465.

Burki, T. 2010. Fight against Leprosy No Longer about the Numbers. *Lancet Infectious Diseases* 10: 74.

Butalia, U. 1992. *The Story within the Story: Women Fight against Leprosy*. New Delhi: Danlep.

Butlin, C. R., and P. Saunderson. 2014. Children with Leprosy. *Leprosy Review* 85: 69–73.

Buzhilova, A. 2012. Paleopathology in Russia. In *A Global History of Paleopathology: Pioneers and Prospects*, edited by J. Buikstra and C. Roberts, 519–527. Oxford: Oxford University Press.

Cabral-Miranda, W., F. Chiaravalloti Neto, and L. V. Barrozo. 2014. Socio-Economic and Environmental Effects Influencing the Development of Leprosy in Bahia, North-Eastern Brazil. *Tropical Medicine and Health* 19: 1504–1514.

Caffell, A. 2015. Assessment of Human Remains Recovered from Midland Road, Peterborough, Cambridgeshire (PMR 14). Report Prepared for Archaeological Services, Durham University. Unpublished.

Cairns Smith, W. 1996. Is There a Decline in Leprosy Publications and Research? *Leprosy Review* 67: 1–3.

Callender, G. R. 1925. The Leprosy Problem in the Philippine Islands. *American Journal of Tropical Medicine* 5: 351–358.

Cameron, A., and C. A. Roberts. 1984. Gambier-Parry Lodge, Gloucester. Human Skeletal Report. Unpublished.

Campbell, E., and J. Colton, eds. 1960. *The Surgery of Theodoric, ca. A. D. 1267*. 2 vols. New York: Appleton-Century-Crofts.

Carayan, A. 1977. Effets de la Malnutrition sur la Propagation la Gravite et les Infirmities de la Lèpre. *Journal of Medicine in the Tropics* 37: 393.

Cardona-Castro, N., J. C. Beltran-Alzate, I. M. Romero-Montoya, E. Melendez, F. Torres, R. M. Sakamuri, et al. 2009. Identification and Comparison of *Mycobacterium leprae* Genotypes in Two Geographical Regions of Colombia. *Leprosy Review* 80: 316–321.

Cardona-Castro, N., E. Cortés, C. Beltrán, M. Romero, J. E. Badel-Mogollón, and G. Bedoya. 2015. Human Genetic Ancestral Composition Correlaters with the Origin of *Mycobacterium leprae* Strains in a Leprosy Endemic Population. *PloS Neglected Tropical Diseases* 9(9): e0004045.

Cardoso, C. C., A. C. Pereira, C. deSales Marques, and M. O. Moraes. 2011. Leprosy Susceptibility: Genetic Variations Regulate Innate and Adaptive Immunity, and Disease Outcome. *Future Microbiology* 6: 533–549.

Cardwell, P. 1995. Excavation of the Hospital of St Giles by Brompton Bridge, North Yorkshire. *Archaeological Journal* 152: 109–245.

Cardy, A. 1996. *The Human Skeletal Remains from The Green, Aberdeen, 1994. (E38)*. Supplement to J. A. Stones, ed., *Three Scottish Carmelite Friaries: Excavations at Aberdeen, Linlithgow and Perth, 1980–1986*. Society of Antiquaries of Scotland Monograph Series Number 6. Edinburgh: Society of Antiquaries of Scotland.

Carnaud, C., S. T. Ishizaka, and O. Stutman. 1984. Early Loss of Precursors of CTL and IL-2-Producing Cells in the Development of Neonatal Tolerance to Alloantigens. *Journal of Immunology* 133: 45–51.

Carpintero-Benítez, P., and A. García-Frasquet. 1998. Bone Island and Leprosy. *Skeletal Radiology* 27: 330–333.

Carpintero-Benítez, P., C. Logroño, and E. Collantes-Estevez. 1996. Enthesopathy in Leprosy. *Journal of Rheumatology* 23: 1020–1021.

Caruth, I., and S. Anderson. 1997. St Saviour's Hospital, Bury St Edmunds (BSE 013): A Report on the Archaeological Excavations 1989–1994. Suffolk County Council Report 97/20. Unpublished.

Carvalho, A. S. 1932. *História da Lepra em Portugal*. Porto, Portugal: Oficinas Gráficas da Sociedade de Papelaria.

Cataroche, J. 2013. *Osteological Analysis of Human Skeletal Remains from St Tugal's Chapel, Herm*. St Peter Port, Guernsey: Guernsey Museum Archaeology Department, Guernsey Museum and Art Gallery.

Chakrabarty, A. N., and S. G. Dastidar. 1989. Correlation between Occurrence of Leprosy and Fossil Fuels: Role of Fossil Fuel Bacteria in the Origin and Global Epidemiology of Leprosy. *Indian Journal of Experimental Biology* 27: 483–496.

———. 2001–2002. Is Soil an Alternative Source of Leprosy Infection? *Acta Leprologia* 12: 79–84.

Chamberlain, W. E., N. E. Wayson, and N. H. Garland. 1931. Bone and Joint Changes of Leprosy: Roentgenologic Study. *Radiology* 17: 930–939.

Chambers, J. A., C. W. Baffi, and K. T. Nash. 2009. The Diagnostic Challenge of Hansen's Disease. *Military Medicine* 174: 652–656.

Chaudhury, D. S. 1970. Spina Bifida Occulta Causing Plantar Ulceration. *Leprosy Review* 41: 236–237.

Chauhan, S. L., A. Girdhar, B. Mishra, G. N. Malaviya, K. Venkatasen, and B. K. Girdhar. 1996. Calcification of Peripheral Nerves in Leprosy. *Acta Leprologia* 10: 51–56.

Chauhan, S., A. Wakhlu, and V. Agarwal. 2010. Arthritis in Leprosy. *Rheumatology (Oxon)* 49: 2237–2242.

Chaussinand, R. 1944. Tuberculose et Lèpre, Maladies Antagoniques: Éviction de la Lèpre par la Tuberculose. *Revue Médical Français d'Extrême-Orient* 22: 667.

———. 1948. Tuberculose et Lèpre, Maladies Antagoniques: Éviction de la Lèpre par la Tuberculose. *International Journal of Leprosy and Other Mycobacterial Diseases* 16: 431–438.

———. 1953. Tuberculosis and Leprosy: Mutually Antagonistic Diseases. *Leprosy Review* 24: 90–94.

———. 1955. *Leprosy.* Paris: Expansion Scientifique Française.

Cho, J. H. 2001. The Nod2 Gene in Crohn's Disease: Implications for Future Research into the Genetics and Immunology of Crohn's Disease. *Inflammatory Bowel Diseases* 7: 271–275.

Choi, S. M., B. C. Kim, S. S. Kweon, M. H. Shin, J. H. Park, H. S. Song, D. C. O. H. Kwon, M. H. Lee, L. Jong-Jun, et al. 2012. Restless Legs Syndrome in People Affected by Leprosy. *Leprosy Review* 83: 363–369.

Clark, B. M., C. K. Murray, L. L. Horvath, G. A. Deye, M. S. Rasnake, and R. N. Longfield. 2008. Case-Control Study of Armadillo Contact and Hansen's Disease. *American Journal of Tropical Medicine and Hygiene* 78: 962–967.

Clay, R. M. 1909. *The Mediaeval Hospitals of England.* London: Methuen.

Close-Brooks, J. 1979. Reports on the Body. Long Cists at Dryburn Bridge, Near Dunbar, East Lothian. *Transactions of the East Lothian Antiquarian and Field Naturalists Society* 16: 7–14.

Clough, S., and L. Loe (with a contribution by Helen Webb). N.d. St Martin's Field, New Romney and New Romney Pipeline Skeletal Report. Canterbury Archaeological Trust. Unpublished manuscript.

Cochrane, R. G. 1947. *A Practical Textbook of Leprosy.* Oxford: Oxford University Press.

Cochrane, R. G., and T. F. Davey. 1964. *Leprosy in Theory and Practice.* Bristol: John Wright and Sons.

Cole, G., and T. Waldron. 2011. Apple Down 152: A Putative Case of Syphilis from Sixth Century AD Anglo-Saxon England. *American Journal of Physical Anthropology* 144: 72–79.

———. 2012. Letter to the Editor: Syphilis Revisited. *American Journal of Physical Anthropology* 149: 149–150.

———. 2015. Letter to the Editor: Apple Down 152 Putative Syphilis: Pre-Columbian Date Confirmed. *American Journal of Physical Anthropology* 156: 489.

Cole, S. T., R. Brosch, J. Parkhill, T. Garnier, C. Churche, D. Harris, S. V. Gordon, K. Eiglmeier, S. Gas, C. E. Barry III, et al. 1998. Deciphering the Biology of *Mycobacterium tuberculosis* from the Complete Genome Sequence. *Nature* 396(6707): 537–544.

Cole, S. T., K. Eiglmeier, J. Parkhill, K. D. James, N. R. Thomson, P. R. Wheeler, N. Honoré, T. Garnier, C. Churcher, D. Harris, et al. 2001. Massive Gene Decay in Leprosy Bacillus. *Nature* 409: 1007–1011.

Cole, S. T., P. Supply, and N. Honoré. 2001. Repetitive Sequences in *Mycobacterium leprae* and Their Impact on Genome Plasticity. *Leprosy Review* 72: 449–461.

Connell, B., A. Gray Jones, R. Redfern, and D. Walker. 2012. *A Bioarchaeological Study of Medieval Burials on the Site of St Mary Spital: Excavations at Spitalfields Market, London E1, 1991–2007.* MOLA Monograph 60. London: Museum of London Archaeology.

Cook, D. C. 2002. Rhinomaxillary Syndrome in the Absence of Leprosy: An Exercise in Differential Diagnosis. In *The Past and Present of Leprosy: Archaeological, Historical, Palaeopathological and Clinical Approaches,* edited by C. A. Roberts, M. E. Lewis, and K. Manchester, 81–85. British Archaeological Reports International Series 1054. Oxford: Archaeopress.

Cooney, J. P., and E. H. Crosby. 1944. Absorptive Bone Changes of Leprosy. *International Journal of Leprosy and Other Mycobacterial Diseases* 42: 15–19.

Cooper, A., and H. N. Poinar. 2000. Ancient DNA: Do It Right or Not At All. *Science* 289: 1139.

Cottle, W. 1879. Chaulmoogra Oil in Leprosy. *British Medical Journal* 1(965): 968–969.

Courtenay, W. H. 2000. Constructions of Masculinity and Their Influence on Men's Well-Being. *Social Science and Medicine* 50: 1385–1401.

Courtin, F., M. Huerre, J. Fyfe, and P. Dumas. 2007. A Case of Feline Leprosy Caused by *Mycobacterium lepraemurium* Originating from the Island of Kythira (Greece): Diagnosis and Treatment. *Journal of Feline Medicine and Surgery* 9: 239–241.

Coutelier, L. 1971. The Bone Lesions in Leprosy: A Study Based on Microradiography and Fluorescence Microscopy. *International Journal of Leprosy* 39(2): 231–243.

Crane-Kramer, G. 2002. Was There a Medieval Diagnostic Confusion between Leprosy and Syphilis? In *The Past and Present of Leprosy: Archaeological, Historical, Palaeopathological and Clinical Approaches,* edited by C. A. Roberts, M. E. Lewis, and K. Manchester, 111–119. British Archaeological Reports International Series 1054. Oxford: Archaeopress.

Crespo, F. A., C. K. Klaes, A. E. Switala, and S. N. DeWitte. 2016. Do Leprosy and Tuberculosis Generate a Systemic Inflammatory Shift? Setting the Ground for a New Dialogue between Experimental Immunology and Bioarchaeology. *American Journal of Physical Anthropology* 162: 143–156.

Crespo, F., J. White, and C. Roberts. 2019. Revisiting the Tuberculosis and Leprosy Cross-Immunity Hypothesis: Expanding the Dialogue between Immunology and Paleopathology. *International Journal of Paleopathology* 26: 37–47.

Crook, N., R. Ramosubban, A. Samy, and B. Singh. 1991. An Educational Approach to Leprosy Control: An Evaluation of Knowledge, Attitude and Practice in 2 Poor Localities in Bombay, India. *Leprosy Review* 62: 395–401.

Cross, A. B. 1972. Foot Deformities in Leprosy: A Survey in the Solomon Islands. *Leprosy Review* 43: 45–52.

Cross, H., V. N. Kulkarni, A. Dey, and G. Rendall. 1996. Plantar Ulceration in Patients with Leprosy. *Journal of Wound Care* 5: 406–411.

Crossan, C. 2004. Excavations at St Mary Magdalen's Hospital, Brook Street, Colchester. *Essex Archaeology and History* 34: 91–154.

Cruz-Coke, R. 1988. Estudios Biomédicos Chilenos en Isla de Pascua. *Revista Médica de Chile* 116: 818–821.

Cule, J. 1970. The Diagnosis, Care and Treatment of Leprosy in Wales and the Border in the Middle Ages. *Transactions of the British Society for the History of Pharmacy* 1: 29–50.

Culpeper, N. 1805. *Culpeper's English Physician and Complete Herbal.* London: Lewis and Roden, Paternoster-Row (originally published 1653).

Cunha, C., V. L. Pedrosa, L. C. Dias, A. Braga, A. Chrusciak-Talhari, M. Santos, G. O. Penna, S. Talhari, and C. Talhari. 2015. A Historical Overview of Leprosy Epidemiology in Amazonas, Brazil. *Revista da Sociedade de Medicina Tropical* 48 supplement 1: 55–62.

Cunha, E., and S. Wasterlain. 2007. The Coimbra Identified Osteological Collections. *Documenta Archaeobiologiae* 5: 23–33.

Curtiss, R., III, S. Blower, K. Cooper, D. Russell, S. Silverstein, and L. Young. 2001. Leprosy Research in the Post-Genome Era. *Leprosy Review* 72: 8–22.

Cury, M. R., V. D. Paschoal, S. M. Nardi, A. P. Chierotti, A. L. Rodrigues Jr., and F. Chiaravalloti-Neto. 2012. Spatial Analysis of Leprosy Incidence and Associated Socioeconomic Factors. *Revista de Saúde Pública* 46: 110–114.

Cust, R. N. 1881. *Pictures of Indian Life: Sketched with a Pen from 1852–1881.* London: Trubner and Company.

Daffe, M., and P. Draper. 1998. The Envelope Layers of Mycobacteria with Reference to Their Pathogenicity. *Advances in Microbial Physiology* 39: 131–203.

Dalby, G. 1993. The Palaeopathology of Middle Ear and Mastoid Disease. PhD thesis, University of Bradford, England.

Daniel, E., P. S. Rao, and P. Courtright. 2013. Facial Sensory Loss in Multi-Bacillary Leprosy Patients. *Leprosy Review* 84: 194–198.

Danielsen, K. 1970. Odontodysplasia Leprosa in Danish Mediaeval Skeletons. *Særtryk af Tandlægebladet* 74: 603–625.

Danielssen, D.-C., and C. W. Boeck. 1847. *Atlas Colorié de Spedalskhed (Éléphantiasis des Grecs).* Bergen, Norway: Norwegian Government.

———. 1848. *Traité de la Spédalskhed ou Éléphantiasis des Grecs.* Paris: Baillière.

Da Silva, M. B., J. M. Portela, W. Li, M. Jackson, M. Gonzalez, A. S. Hidalgo, J. T. Belisle, R. C. Bouth, A. R. Gobbo, J. G. Barreto, A. H. H. Minervino, et al. 2018. Evidence of Zoonotic Leprosy in Pará, Brazilian Amazon, and Risks Associated with Human Contact or Consumption of Armadillos. *PLoS Neglected Tropical Diseases* 12(6): e0006532. https://journals.plos.org/plosntds/article?id=10.1371/journal.pntd.0006532.

Da Silva, S. A., P. S. Mazini, P. G. Reis, A. M. Sell, L. T. Tsuneto, P. R. Peixoto, and J. E. Visentainer. 2009. HLA-DR and HLA-DQ Alleles in Patients from the South of Brazil: Markers for Leprosy. *BMC Infectious Diseases* 9: 134.

Dave, D. S., and S. K. Agrawal. 1984. Prevalence of Leprosy in Children of Leprosy Patients. *Indian Journal of Leprosy* 56: 615–621.

Davies, R. P. O., K. Tocque, M. A. Bellis, T. Rimmington, and P. D. O. Davies. 1999. Historical Declines in Tuberculosis: Improving Social Conditions or Natural Selection? In *Tuberculosis: Past and Present,* edited by G. Pálfi, O. Dutour, J. Deák, and I. Hutás, 89–92. Szeged, Hungary: Golden Book Publisher and Budapest, Hungary: Tuberculosis Foundation.

Davison, A. R., R. Kooij, and J. Wainwright. 1960. Classification of Leprosy. 1. Application of

the Madrid Classification of Various Forms of Leprosy. *International Journal of Leprosy and Other Mycobacterial Diseases* 28: 113–125.
Dayal, R., N. A. Hashmi, P. P. Mathur, and R. Prasad. 1990. Leprosy in Childhood. *Indian Pediatrics* 27: 170–180.
Deacon, H. 1994. Leprosy and Racism on Robben Island. In *Studies in the History of Cape Town,* edited by E. van Heyningen, 45–83. Rondebosch: University of Cape Town Press.
———. 2003. Patterns of Exclusion on Robben Island, 1654–1992. In *Isolation: Places and Practices of Exclusion,* edited by C. Strange and A. Bashford, 153–172. London: Routledge.
Dean, G. 1905. Observations on a Leprosy-Like Disease of the Rat. *Journal of Hygiene* 5: 99–112.
Dean, L., R. Tolhurst, G. Gallo, K. Kollie, A. Betee, and S. Theobald. 2019. Neglected Tropical Disease as a "Biograophical Disruption": Listening to the Narratives of Affected Persons to Develop Integrated People Centred Care in Liberia. *PLoS Neglected Tropical Diseases:* doi.org/10.1371/journal.pntd.0007710.
De Jersey, P., and J. Cataroche. 2012. Excavation at St Tugal's Chapel, Herm, 2011. *Transactions de la Société Guernesiaise* 27: 381–431.
De las Aguas, J. T. 2006. Consideraciones Histórico-Epidemiológicas de la Lepra en América. *Medicina Cutánea Ibero-Latino-Americana* 34: 179–194.
Demaitre, L. 1985. The Description and Diagnosis of Leprosy by 14th Century Physicians. *Bulletin of the History of Medicine* 59: 327–344.
———. 2007. *Leprosy in Premodern Medicine: A Malady of the Whole Body.* Baltimore, MD: Johns Hopkins University Press.
De Miguel Ibáñez, M. P., J. M. Ballesteros Herrái, M. Unzu Urmeneta, J. A. Faro Carballa, and M. García-Barberena Unzu. 2011. Tres Posibles Casos de Lepra en La Plaza del Castillo (Pamplona, Navarra). In *Paleopatología: Ciencia Multidisciplinar: Proceedings of the 10th Meeting of the Spanish Society of Paleopathology,* edited by A. González Martín, O. Cambra-Moo, J. Rascón Pérez, M. Campo Martín, M. Robledo Acinas, E. Labajo González, and J. A. Sánchez Sánchez, 355–365. Madrid: Universidad Autónoma de Madrid.
De Miguel Ibáñez, M. P., M. Unzu Urmeneta, J. A. Faro Carballa, P. Prieto Sáez de Tejada, and M. García-Barberena Unzu. 2008. Evidencias de Ajusticiamiento: A Propósito de una Fosa Común de Época Romana (s. II–IV) (Plaza del Castillo, Pamploma, Navarra). In *Actas de las Jornadas de Antropología Física y Forense,* edited by C. Roca de Togores Muñoz and F. Rodes Lloret, 81–88. Alicante: Instituto de Cultura Juan Gil Albert, Diputación de Alicante.
Deps, P. D., B. L. Alves, C. G. Gripp, R. L. Aragao, B. Guedes, J. B. Filho, M. K. Andreatta, R. S. Marcari, I. Prates, and L. C. Rodrigues. 2008. Contact with Armadillos Increases the Risk of Leprosy in Brazil: A Case Control Study. *Indian Journal of Dermatology, Venereology and Leprology* 74: 338–342.
Deps, P. D., J. M. Antunes, and J. Tomimori-Tamashita. 2007. Detection of *Mycobacterium leprae* Infection in Wild Nine-Banded Armadillos (*Dasypus novemcinctus*) Using the Rapid ML Flow Test. *Revista da Sociedade Brasileira de Medicina Tropical* 40: 87–87.
Deps, P. D., S. M. Collins, S. Robin, and P. Charlier. 2020. Leprosy in Skulls from the Paris Catacombs. *Annals of Human Biology:* doi: 10.1080/03014460.2020.

Desikan, K. V. 1977. Viability of *Mycobacterium leprae* outside the Human Body. *Leprosy Review* 48: 231–235.

Desikan, K. V., and Sreevatsa. 1995. Extended Studies on the Viability of *M. leprae* outside the Human Body. *Leprosy Review* 66: 287–295.

DeWitte, S. N. 2014. Health in Post-Black Death London (1350–1538): Age Patterns of Periosteal New Bone Formation in a Post-Epidemic Population. *American Journal of Physical Anthropology* 155: 260–267.

———. 2015. Developmental Stress and Disease Susceptibility: The Association between Skeletal Indicators of Leprosy and Other Physiological Stressors. Poster presented at the 80th annual meeting of the Society for American Archaeology, March, San Francisco, CA.

De Witte, S. N., and C. Stojanowski. 2015. The Osteological Paradox 20 Years Later: Past Perspectives, Future Directions. *Journal of Archaeological Research* 23: 397–350.

Dharmendra. 1947. Leprosy in Ancient Indian Medicine. *International Journal of Leprosy and Other Mycobacterial Diseases* 15: 424–430.

———, ed. 1978. Deformities of the Face in Leprosy. In *Leprosy*, edited by Dharmendra, 237–241. Bombay: Kothari Medical Publishing House.

Dimitrov, D., and J. C. Szepietowski. 2017. Stigmatization in Dermatology with a Special Focus on Psoriatic Patients. *Postępy Higieny i Medycyny Doświadczalnej (Advances in Hygiene and Experimental Medicine)* 71: 1115–1122.

Dinwiddy, K. 2016. The Unburnt Human Bone. In *An Anglo-Saxon Cemetery at Collingbourne Ducis, Wiltshire*, edited by K. Egging Dinwiddy and N. Stoodley, 69–92. Wessex Archaeology Report 37. Salisbury: Wessex Archaeology.

Dokládal, M. 2002. The History of Leprosy in the Territory of the Czech Republic. In *The Past and Present of Leprosy: Archaeological, Historical, Palaeopathological and Clinical Approaches,* edited by C. A. Roberts, M. E. Lewis, and K. Manchester, 155–156. British Archaeological Reports International Series 1054. Oxford: Archaeopress.

Dols, M. W. 1979. Leprosy in Medieval Arabic Medicine. *Journal of Medical History* 34: 313–333.

Donham, K. J., and J. R. Leininger. 1977. Spontaneous Leprosy-Like Disease in a Chimpanzee. *Journal of Infectious Diseases* 136: 132.

Donoghue, H. D. 2008. Molecular Palaeopathology of Human Infectious Disease. In *Advances in Human Palaeopathology,* edited by R. Pinhasi and S. Mays, 147–176. New York: Wiley.

———. 2019. Tuberculosis and Leprosy Associated with Historical Human Population Movements in Europe and Beyond: An Overview Based on Mycobacterial Ancient DNA. *Annals of Human Biology:* doi.org/10.1080/03014460.2019.1624822.

Donoghue, H. D., Y. S. Erdal, R. Pinhasi, and M. Spigelman. 2005. A Possible Case of Congenital Leprosy in a Five-Month Old Child from Byzantine-Age Turkey. In *Proceedings of the 26th Congress of the European Society of Mycobacteriology, Istanbul, Turkey, June 6–29,* 177.

Donoghue, H. D., J. Gladykowska-Rzeczycka, A. Marcsik, J. Holton, and M. Spigelman. 2002. *Mycobacterium leprae* in Archaeological Samples. In *The Past and Present of Leprosy: Archaeological, Historical, Palaeopathological and Clinical Approaches,* edited by C.

A. Roberts, M. E. Lewis, and K. Manchester, 271–286. British Archaeological Reports International Series 1054. Oxford: Archaeopress.

Donoghue, H. D., J. Holton, and M. Spigelman. 2001. PCR Primers That Can Detect Low Levels of *Mycobacterium leprae* DNA. *Journal of Medical Microbiology* 50: 177–182.

Donoghue, H. D., A. Marcsik, C. Matheson, K. Vernon, E. Nuorala, J. E. Molto, C. L. Greenblatt, and M. Spigelman. 2005. Co-Infection of *Mycobacterium tuberculosis* and *Mycobacterium leprae* in Human Archaeological Samples: A Possible Explanation for the Historical Decline of Leprosy. *Proceedings of the Royal Society B* 272: 389–394.

Donoghue, H. D., and M. Spigelman. 2008. *Mycobacterium leprae és Mycobacterium tuberculosis* DNS-vizsgálata a Székesfehérvári I/11, VI/24, és VI/27, Jelzetű Mintákon. In *A székesfehérvári Királyi Bazilika embertani leletei: 1848–2002*, edited by K. Éry, 171–173. Ecclesia Beatae Mariae VirginisAlbaeregalis. Budapest: Balassi Kiadó.

Donoghue, H. D., G. M. Taylor, A. Marcsik, E. Molnár, G. Pálfi, I. Pap, M. Teschler-Nicola, R. Pinhasi, Y. S. Erdal, P. Velemínsky, et al. 2015. A Migration-Driven Model for the Historical Spread of Leprosy in Medieval Eastern and Central Europe. *Infection, Genetics and Evolution* 31: 250–256.

Donoghue, H. D., G. M. Taylor, R. Pinhasi, A. Marcsik, E. Molnár, M. Teschler-Nicola, Y. S. Erdal, and M. Spigelman. 2006. Who Brought Leprosy into Central Europe? Paper presented at the 16th European meeting of the Paleopathology Association, Santorini, Greece, August 28–September 1.

Donoghue, H. D., G. M. Taylor, G. R. Stewart, O. Y-C. Lee, H. H. T. Wu, G. S. Besra, and D. E. Minnikin. 2017. Positive Diagnosis of Ancient Leprosy and Tuberculosis Using Ancient DNA and Lipid Biomarkers. *Diversity* 9: Article 46.

Doss, C. G. P., N. Nagasundaram, J. S. Rajan, and C. Chiranjib. 2012. LSHGD: A Database for Human Leprosy Susceptible Genes. *Genomics* 1000: 162–166.

Doull, J. A., R. S. Guinto, and J. N. Rodriguez. 1942. The Incidence of Leprosy in Cordova and Talisay, Philippines. *International Journal of Leprosy and Other Mycobacterial Diseases* 10: 107–131.

Doyle, J. 1953. Case of Leprosy Seen in a Venereal Disease Clinic in Britain. *British Medical Journal* 2(4830): 261–262.

Drabik, R., and A. Drabik. 1992. Leprosy: One of the Scourges of Mankind. *Hamdard* 35: 25–51.

Drutz, D. J., T. S. N. Chen, and W.-H. Lu. 1972. The Continuous Bacteremia of Lepromatous Leprosy. *New England Journal of Medicine* 287: 159–164.

Duarte, V. A. 2015. De Scallabis a Chantirene: Aanálise Paleoantropológica de duas Aamostras Paleocristãs dos Séculos IV e VI da Necrópole da Avenida 5 de Outubro (Santarém). Dissertação de Mestrado em Evolução e Biologia Humanas, Universidade de Coimbra.

Duhig, C. 1993. Assessment of Skeletons from Pipe Trenches at Huntingdon. In *A Leper Cemetery at Spittal's Link, Huntingdon*, edited by D. Mitchell, 18–21. Archaeological Field Unit of Cambridgeshire County Council (Cambridgeshire Archaeology). Unpublished report.

———. 1998. The Human Skeletal Material. In *The Anglo-Saxon Cemetery at Edix Hill (Barrington A), Cambridgeshire*, edited by T. Malin and J. Hines, 154–199. Council for British Archaeology Research Report 112. York: Council for British Archaeology.

Dunbar, C. E., K. A. High, J. K. Joung, D. B. Kohn, K. Ozawa, and M. Sadelain. 2018. Gene Therapy Comes of Age. *Science* 359(6372). https://science.sciencemag.org/content/359/6372/eaan4672.

Duncan, K. 1994. Climate and the Decline of Leprosy in Britain. *Proceedings of the Royal College of Physicians of Edinburgh* 24: 114–120.

Duncan, M. E. 1980. Babies of Mothers with Leprosy Have Small Placentae, Low Birth Weights and Grow Slowly. *British Journal of Obstetrics and Gynaecology* 87: 471–479.

———. 1982. A Prospective Clinico-Pathological Study of Pregnancy and Leprosy in Ethiopia. MD thesis, University of Edinburgh.

———. 1985a. Leprosy and Procreation—A Historical Review of Social and Clinical Aspects. *Leprosy Review* 56: 153–162.

———. 1985b. Leprosy in Young Children—Past, Present and Future. *International Journal of Leprosy and Other Mycobacterial Diseases* 53: 468–473.

———. 1993. An Historical and Clinical View of the Interaction of Leprosy and Pregnancy: A Cycle to Be Broken. *Social Science and Medicine* 37: 457–472.

Duncan, M. E., R. Melsom, J. M. H. Pearson, S. Menzel, and R. St. C. Barnetson. 1983. A Clinical and Immunological Study of Four Babies of Mothers with Lepromatous Leprosy, Two of Whom Developed Leprosy in Infancy. *Indian Journal of Leprosy* 51: 7–17.

Duncan, M. E., R. Melsom, J. M. H. Pearson, and D. S. Ridley. 1981. The Association of Pregnancy and Leprosy. I. New Cases, Relapse of Cured Patients and Deterioration in Patients on Treatment during Pregnancy and Lactation—Results of a Prospective Study of 154 Pregnancies in 147 Ethiopian Women. *Leprosy Review* 52: 245–262.

Duncan, M. E., T. Miko, R. Howe, S. Hansen, S. Menzel, R. Melsom, D. Frommel, E. Bezuneh, M. Hunegnaw, G. Amare, et al. 2007. Growth and Development of Children of Mothers with Leprosy and Healthy Controls. *Ethiopian Medical Journal* 45: 9–23.

Duncan, M. E., and R. E. Oakey. 1982. Estrogen Excretion in Pregnant Women with Leprosy: Evidence of Diminished Fetoplacental Function. *Obstetrics and Gynaecology* 60: 82–86.

Duncan, M. E., J. M. H. Pearson, and R. J. W. Rees. 1981. The Association of Pregnancy and Leprosy II. Pregnancy and Dapsone-Resistant Leprosy. *Leprosy Review* 52: 263–270.

Duncan, M. E., J. M. H. Pearson, R. S. Ridley, R. Melsom, and G. Bjune. 1982. Pregnancy and Leprosy: The Consequences of Alterations of Cell Mediated Immunity during Pregnancy and Lactation. *Indian Journal of Leprosy* 50: 425–435.

Dupnik, K. M., M. M. Martins, A. T. Souza, S. M. Jeronimo, and M. L. Nobre. 2012. Nodular Secondary Syphilis Simulates Lepromatous Leprosy. *Leprosy Review* 83: 389–393.

Düppre, N. C., L. A. Camacho, S. S. da Cunha, C. J. Struchiner, A. M. Sales, J. A. Nery, and E. N. Sarno. 2008. Effectiveness of BCG Vaccination among Leprosy Contacts: A Cohort Study. *Transactions of the Royal Society of Tropical Medicine and Hygiene* 102: 631–638.

Dupras, T. L., and H. P. Schwarcz. 2001. Strangers in a Strange Land: Stable Isotope Evidence for Human Migration in the Dakhleh Oasis, Egypt. *Journal of Archaeological Science* 28: 1199–1208.

Dupras, T. L., J. Williams, H. Willems, and C. Peeters. 2010. Pathological Skeletal Remains from Ancient Egypt: The Earliest Case of Diabetes Mellitus? *Practical Diabetes International* 27: 358–363.

Duthie, M. S., T. P. Gillis, and S. G. Reed. 2011. Advances and Hurdles on the Way toward a Leprosy Vaccine. *Human Vaccines* 7: 1172–1183.

Duthie, M. S., M. T. Pena, G. J. Ebenezer, T. P. Gillis, R. Sharma, K. Cunningham, M. Polydefkis, Y. Maeda, M. Makino, R. W. Truman, and S. G. Reed. 2018. LepVax, a Defined Subunit Vaccine that Provides Effective Pre-Exposure and Post-Exposure Prophylaxis of *M. leprae* infection. *Vaccines* 3: doi.org/10.1038/s41541-018-0050-z.

Dwivedi, V. P., A. Banerjee, I. Das, A. Saha, M. Dutta, B. Bhardwaj, S. Biswas, and D. Chattopadhyay. 2019. Diet and Nutrition: An Important Risk Factor in Leprosy. *Microbial Pathogenesis* doi: 10.1016.

Dyer, I. 1921. Leprosy. In *The Principles and Practice of Medicine: Designed for the Use of Practitioners and Students of Medicine*, edited by W. Osler and T. McCrae, 152–155. London: D. Appleton and Company.

Dyer, I., and R. Hopkins. 1911. The Diagnosis of Leprosy. *Lepra* 12: 4–5.

Dzierzykray-Rogaliski, T. 1980. Palaeopathology of the Ptolemaic Inhabitants of Dahkleh Oasis (Egypt). *Journal of Human Evolution* 9: 71–74.

Ebenso, J., L. T. Muyiwa, and B. E. Ebenso. 2009. Self Care Groups and Ulcer Prevention in Okegbala, Nigeria. *Leprosy Review* 80: 187–196.

Economou, C., A. Kjellstrom, K. Lidén, and I. Panagopoulos. 2013a. Ancient-DNA Reveals an Asian Type of *Mycobacterium leprae* in Medieval Scandinavia. *Journal of Archaeological Science* 40: 465–470.

———. 2013b. Corrigendum to "Ancient-DNA Reveals an Asian Type of *Mycobacterium leprae* in Medieval Scandinavia" [J. Archaeol. Sci. 40(1): 465–470]. *Journal of Archaeological Science* 40: 2867.

Edelman, R. 1979. Malnutrition and Leprosy—An Analytical Review. *Leprosy in India* 51: 376–388.

Edmond, R. 2006. *Leprosy and Empire: A Medical and Cultural History*. Cambridge: Cambridge University Press.

Egnew, T. R. 2009. Suffering, Meaning, and Healing: Challenges of Contemporary Medicine. *Annals of Family Medicine* 7: 170–175.

Ehlers, E. 1895. On the Conditions under Which Leprosy Has Declined in Iceland. In *Prize Essays on Leprosy*, edited by G. Newman, 151–187. London: New Sydenham Society.

Eliopoulos, C., K. Moraitis, V. Vanna, and S. Manolis. 2011. Greece/ΛΛΑΔΑ. In *The Routledge Handbook of Archaeological Human Remains and Legislation*, edited by N. Márquez-Grant and L. Fibiger, 173–183. London: Routledge.

Ell, S. R. 1984. Blood and Sexuality in Medieval Leprosy. *Janus* 71: 153–164.

———. 1985. Diet and Leprosy in the Medieval West: The Noble Leper. *Janus* 72: 113–129.

———. 1986. Editorial: Leprosy and Social Class in the Middle Ages. *International Journal of Leprosy and Other Mycobacterial Diseases* 54: 300–305.

———. 1987. Plague and Leprosy in the Middle Ages: A Paradoxical Cross-Immunity. *International Journal of Leprosy and Other Mycobacterial Diseases* 55: 343–350.

———. 1988. Reconstructing the Epidemiology of Medieval Leprosy: Preliminary Efforts with Regard to Scandinavia. *Perspectives in Biology and Medicine* 31: 496–506.

Elton, S., and P. O'Higgins, eds. 2008. *Medicine and Evolution: Current Applications, Future Prospects*. Boca Raton, FL: CRC Press.

Enna, C. D., R. R. Jacobsen, and R. O. Rausch. 1971. Bone Changes in Leprosy: A Correlation of Clinical and Radiographic Features. *Radiology* 100: 295–306.

Erdal, Y. S. 2004. Kovuklukaya (Boyabat, Sinop) İnsanlarinin Sağlik ve Yaşam Biçimleriyle İlişkisi; Health Status of the Kovuklukaya (Boyabat/Sinop) People and Its Relations with Their Lifestyle. *Anadolu Arastirmalari* 17: 169–196.

Erickson, P. T., and F. A. Johansen. 1948. Bone Changes in Leprosy under Sulfone Treatment. *International Journal of Leprosy and Other Mycobacterial Diseases* 16: 147–156.

Erueti, C., P. Glasziou, C. D. Mar, and M. L. van Driel. 2012. Do You Think It's a Disease? A Survey of Medical Students. *BMC Medical Education* 12: Article 19.

Éry, K., A. Marcsik, and F. Szalai. 2008. A Földsírok Csontvázleletei (II–IX. csoport). In *A Székesfehérvári Királyi Bazilika Embertani Leletei, 1848–2002: Ecclesia Beatae Mariae Virginis Albaeregalis*, edited by K. Éry, 119–134. Budapest: Balassi Kiadó.

Esguerra-Gómez, G., and E. Acosta. 1948. Bone and Joint Lesions in Leprosy. *Radiology* 50: 619–631.

Etxeberria, F., L. Herrast, and M. A. Beguiristan. 1997. Signos de Lepro en un Individuo Altomedieval de Navarra. In *La Enfermedad en los Restos Humanios Arqueológicos*, edited by M. Macias Lopez and J. E. Pcazo Sánchez, 319–323. Cadiz: University of Cadiz, Servicio de Publicaciones.

Evison, V. I., and P. Hill, ed. 1996. *Two Anglo-Saxon Cemeteries at Hereford and Worcester.* Council for British Archaeology Research Report 103. York: Council for British Archaeology.

Fabrizii-Reuer, S., and S. Reuer. 2000. *Das Frühmittelalterliche Gräberfeld von Pottenbrunn.* Wein: Verlag der österreichischen Akademie der Wissenschaften.

Faerman, M. 2008. New Palaeopathological and Genetic Evidence of Leprosy in Israel: Research and Teaching Experience. Paper presented at the 9th International Conference on Ancient DNA and Associated Biomolecules, Pompeii, Italy, October.

Faget, G. H., and H. Mayoral. 1944. Bone Changes in Leprosy: A Clinical and Roentgenological Study of 505 Cases. *Radiology* 42: 1–13.

Faget, G. H., R. C. Pogge, F. A. Johansen, J. F. Dinan, B. M. Prejean, and C. G. Eccles. 1943. The Promin Treatment of Leprosy: A Progress Report. *Public Health Report* 58: 1728–1741.

Fairchild, A. L. 2008. Community and Confinement: The Evolving Experience of Isolation for Leprosy in Carville, Louisiana. *Public Health Report* 119:3 62–370.

Fairley, J. K., J. A. Ferreira, A. L. G. de Oliveira, T. de Filippis, M. A. de Faria Grossi, L. P. Chaves, L. N. Caldeira, P. S. Dos Santos, R. R. Costa, M. C. Diniz, C. S. Duarte, L. A. Bomjardim Pôrto, P. S. Suchdev, D. A. Negrão-Corrêa, F. do Carmo Magalhães, J. M. Peixoto Moreira, A. de Melo Freire Júnior, M. C. Cerqueira, U. Kitron, and S. Lyon. 2019. The Burden of Helminth Coinfections and Micronutrient Deficiencies in Patients with and without Leprosy Reactions: A Pilot Study in Minas Gerais, Brazil. *American Journal of Tropical Medicine and Hygiene* 101: 1058–1065.

Farkas, G., A. Marcsik, and F. Szalai. 1991. Békéscsaba Területének Embertani Leletei. In *Békéscsaba Története. Elsőkötet. Akezdetektől 1848–ig*, edited by D. Jankovich and G. Erdmann, 313–357. Békéscsaba, Hungary: Város Kiadványa.

Farley, M., and K. Manchester. 1989. The Cemetery of the Leper Hospital of St Margaret, High Wycombe, Buckinghamshire. *Medieval Archaeology* 33: 82–90.

Farwell, D. E., and T. Molleson. 1993. *Excavations at Poundbury 1966–80.* Vol. 2, *The Cemeteries.* Dorset Natural History and Archaeological Society Monograph Series 11. Dorset: Dorset Natural History and Archaeological Society.

Fava, V., M. Orlova, A. Cobat, A. Alcaí, M. Mira, and E. Schurr. 2012. Genetics of Leprosy Reactions: An Overview. *Memorias do Instituto Oswaldo Cruz* 107: 132–142.

Faye, L. 1882. Hospitaler og Milde Stiftelser i Norge i Middelalderen. *Norsk Mag. F. Lægev* 12: 93–125, 181–228.

Fava, V.M., M. Dallmann-Sauer, and E. Schurr. 2019. Genetics of Leprosy: Today and Beyond. *Human Genetics:* doi 10.1007/s00439-019-02087-5.

Fekete, L. 1874. *A Magyarországi Ragályos és Járványos Kórok Roved Történelme.* Debrecen: N.p.

Feldman, R. A., and M. Sturdivant. 1975. Leprosy in Louisiana, 1855–1970: An Epidemiologic Study of Long-Term Trends. *American Journal of Epidemiology* 102: 303–310.

———. 1976. Leprosy in the United States 1950–1969: An Epidemiologic Review. *Southern Medical Journal* 69: 970–979.

Fernandez, J. M. M. 1939. Estudio comparativo de la reacción de Mitsuda con las reacciones tuberculínicas (Comparative Study of the Mitsuda Reaction with the Tuberculin Reaction). *Revista Argentina Dermatosifilis* 23: 425–453.

———. 1957. Leprosy and Tuberculosis: Antagonistic Diseases. *Archives of Dermatology and Syphilis* 75: 101–106.

Fernando, S. I., and W. J. Britton. 2006. Genetic Susceptibility to Mycobacterial Disease in Humans. *Immunology and Cell Biology* 84: 125–137.

Ferreira, R. C., T. X. Gonçalves, A. R. D. S. Soares, L. R. A. Carvalho, F. L. Campos, M. T. F. Riberiro, A. M. E. B. L. Martins, and E. F. E. Ferreira. 2018. Dependence on Others for Oral Hygiene and Its Association with Hand Deformities and Functional Impairment in Elders with a History of Leprosy. *Gerodontology* 35(3): 237–245.

Ferreira, M. T., M. J. Neves, and S. N. Wasterlain. 2013. Lagos Leprosarium (Portugal): Evidence of Disease. *Journal of Archaeological Science* 40: 2298–2307.

Ferreira, J. D., D. A. Souza Oliveira, J. P. Santos, C. C. D. U. Ribeiro, B. A. Baêta, R. C. Teixeira, A. D. S. Neumann, P. S. Rosa, M. C. V. Pessolani, M. O. Moraes, G. H. Bechara, P. L. De Oliveira, M. H. F. Sorgine, P. N. Suffys, A. N. B. Fontes, L. Bell-Sakyi, A. H. Fonseca, and F. A. Lara. 2018. Ticks as Potential Vectors of *Mycobacterium leprae:* Use of Tick Cell Lines to Culture the Bacilli and Generate Transgenic Strains. *PLoS Neglected Tropical Diseases* 12(12): e0007001.

Field, V., P. Gautret, P. Schlagenhauf, G. D. Burchard, E. Caumes, M. Jensenius, F. Castelli, E. Gkrania-Klotsas, L. Weld, R. Lopez-Velez, et al. 2010. Travel and Migration Associated Infectious Diseases Morbidity in Europe, 2008. *BMC Infectious Diseases* 10: 330.

Fildes, V. A. 1988. *Wet Nursing: A History from Antiquity to the Present.* New York: Basil Blackwell.

Finch, R. G., P. Moss, D. J. Jeffries, and J. Anderson. 2002. Infectious Diseases, Tropical Medicine and Sexually Transmitted Diseases. In *Clinical Medicine,* 5th ed., edited by P. Kumar and M. Clark, 21–151. Edinburgh: W. B. Saunders.

Fine, P. E. M. 1981. Immunogenetics of Susceptibility to Leprosy, Tuberculosis and Leishmaniasis. An Epidemiological Perspective. *International Journal of Leprosy and Other Mycobacterial Diseases* 49: 437–454.

———. 1982. Leprosy: The Epidemiology of a Slow Bacterium. *Epidemiologic Reviews* 4: 161–188.

———. 1984. Leprosy and Tuberculosis—An Epidemiological Comparison. *Tubercle* 65: 137–153.

Fine, P. E. M., S. Floyd, J. L. Stanford, P. Nkosoa, A. Kasumga, S. Chaguluka, D. K. Warndorff, P. A. Jenkins, M. Yates, and J. M. Ponnighaus. 2001. Environmental Mycobacteria in Northern Malawi: Implications for the Epidemiology of Tuberculosis and Leprosy. *Epidemiology and Infection* 126: 379–387.

Fine, P. E. M., and L. C. Rodrigues. 1990. Modern Vaccines: Mycobacterial Diseases. *Lancet* 335: 1016–1020.

Fine, P. E. M., and D. K. Warndorff. 1997. Leprosy by the Year 2000—What Is Being Eliminated? *Leprosy Review* 68: 201–202.

Fite, G. L. 1941. Lesions of Leprosy. *International Journal of Leprosy and Other Mycobacterial Diseases* 9: 193–202.

Floyd-Richard, M., and S. Gurung. 2000. Stigma Reduction through Group Counselling of Persons Affected by Leprosy—A Pilot Study. *Leprosy Review* 71: 499–504.

Fornaciari, G., V. Giuffra, and S. Minozzi. 2012. Paleopathology in Italy. In *The Global History of Paleopathology: Pioneers and Prospects*, edited by J. Buikstra and C. Roberts, 416–425. Oxford: Oxford University Press.

Fornaciari, G., F. Mallegni, and P. De Leo. 1999. The Leprosy of Henry VII: Incarceration or Isolation? *Lancet* 353: 758.

Forson, R. B. 2012. *The Grateful Leper: Tales of Two Birds*. Bloomington, IN: Authorhouse.

Foster, R. L., A. L. Sanchez, W. Stuyvesant, F. N. Foster, C. Small, and B. H. S. Lau. 1988. Nutrition and Leprosy: A Review. *International Journal of Leprosy and Other Mycobacterial Diseases* 56: 66–81.

Fóthi, E., I. Pap, L. A. Kristóf, M. Barta, M. Maczel, and G. Pálfi. 2001. A Propos d'un Nouveau Cas Paléopathologique de Lépre en Hongrie. *Centre Archéologique du Var Revue* 6: 52–54.

Frontini, R., and R. Vecchi. 2014. Thermal Alteration of Small Mammal from El Guanaco 2 Site (Argentina): An Experimental Approach on Armadillo Bone Remains (*Cingulata, Dasypodidae*). *Journal of Archaeological Science* 44: 22–29.

Frota, C. C., L. N. Lima, S. Rocha Ada, P. N. Suffys, B. N. Rolim, L. C. Rodrigues, M. L. Barreto, C. Kendall, and L. R. Kerr. 2012. *Mycobacterium leprae* in Six-Banded (*Euphractus sexcinctus*) and Nine-Banded Armadillos (*Dasypus novemcinctus*) in Northeast Brazil. *Memorias do Instituto Oswaldo Cruz* 107 supplement 1: 209–213.

Fuk-Tan-Tang, S., C. P. C. Chen, S.-C. Lin, C.-K. Wu, C.-K. Chen, and S.-P. Cheng. 2015. Reduction of Plantar Pressures in Leprosy Patients by Using Custom Made Shoes and Total Contact Soles. *Clinical Neurology and Neurosurgery* 129: S12–S15.

Fulton, N., L. F. Anderson, J. M. Watson, and I. Abubakar. 2016. Leprosy in England and Wales 1953–2012: Surveillance and Challenges in Low Incidence Countries. *BMJ Open* 6(5). https://bmjopen.bmj.com/content/6/5/e010608.

Fuzikawa, P. L., F. A. De Acúrcio, J. P. Velema, and M. Cherchiglia. 2010a. Decentralisation of Leprosy Control Activities in the Municipality of Betim, Gerais State, Brazil. *Leprosy Review* 81: 184–195.

———. 2010b. Factors Which Influenced the Decentralisation of Leprosy Control Activities in the Municipality of Betim, Gerais State, Brazil. *Leprosy Review* 81: 196–205.

García-Moro, C., M. Hernández, P. Moral, and A. González-Martín. 2000. Epidemiological Transition in Easter Island (1914–1996). *American Journal of Human Biology* 12: 371–381.

Garrington, G. E., and M. C. Crump. 1968. Pulp Death in a Patient with Lepromatous Leprosy. *Oral Surgery* 25: 427–434.

Gausterer, C., C. Stein, and M. Teschler-Nicola. 2015. First Genetic Evidence of Leprosy in Early Medieval Austria. *Wiener Medizinische Wochenschrift* 165: 126–132.

Gawkrodger, D. J. 2004. Racial Influences on Skin Disease. In *Rook's Textbook of Dermatology*, 7th ed., vol. 1, edited by T. Burns, S. Breathnach, N. Cox, and C. Griffiths, 69.1–69.21. Oxford: Blackwell Science.

Geater, J. G. 1975. The Fly as Potential Vector in the Transmission of Leprosy. *Leprosy Review* 46: 279.

Gejvall, N.-G. 1960. *Westerhus: Medieval Population and Church in Light of Their Skeletal Remains*. Lund, Sweden: Hakak Ohlssons Boktryckeri.

Gernaey, A. M., D. E. Minnikin, M. Copley, R. Dixon, J. C. Middleton, and C. A. Roberts. 2001. Mycolic Acids and Ancient DNA Confirm an Osteological Diagnosis of Tuberculosis. *Tuberculosis* 81: 259–265.

Ghosh, K. K., Dharmendra, and N. C. Dey. 1951. Nose and Throat Lesions in Cases of Leprosy of the Lepromatous Type. *Indian Medical Gazette* 86: 400–403.

Giannakopoulou, N., and T. Suzuki. 2012. History of Palaeopathology in Japan. In *The Global History of Paleopathology: Pioneers and Prospects*, edited by J. Buikstra and C. Roberts, 439–550. Oxford: Oxford University Press.

Gibson, J. 1982–1983. Smallpox on the Northwest Coast, 1835–1838. *British Columbia Studies* 56: 61–81.

Gibson, S., C. Greenblatt, M. Spigelman, A. Gorski, H. D. Donoghue, K. Vernon, and C. D. Matheson. 2002. The Shroud Cave—A Unique Case Study Linking a Closed Loculus, a Shroud and Ancient Mycobacteria. *Ancient Biomolecules* 4: 134.

Gijsbers van Wijk, C. M. T., H. Huisman, and A. M. Holk. 1999. Gender Differences in Physical Symptoms and Illness Behaviour: A Health Diary Study. *Social Science and Medicine* 49: 1061–1074.

Gilbert, M. T. P., I. Barnes, M. J. Collins, C. Smith, J. Eklund, J. Goudsmit, H. Poinar, and A. Cooper. 2005. Notes and Comments: Long-Term Survival of Ancient DNA in Egypt: Response to Zink and Nerlich (2003). *American Journal of Physical Anthropology* 128: 10–114.

Gilchrist, R. 1995. *Contemplation and Action: The Other Monasticism*. London: Leicester University Press.

Gill, A. L., D. R. Bell, G. V. Gil, G. B. Wyatt, and N. J. Beeching. 2005. Leprosy in Britain: 50 Years Experience in Liverpool. *QJM: An International Journal of Medicine* 98: 505–511.

Gillis, T. P., D. M. Scollard, and D. N. Lockwood. 2011. What Is the Evidence That the Putative *Mycobacterium lepromatosis* Specifically Causes Diffuse Lepromatous Leprosy? *Leprosy Review* 82: 205–209.

Gillis, T. P., V. Vissa, M. Matsuoka, S. Young, J. H. Richardus, R. Truman, B. Hall, and P. Brennan. 2009. Characterisation of Short Tandem Repeats for Genotyping *Mycobacterium leprae*. *Leprosy Review* 80: 250–260.

Gilmore, J. K. 2008. Leprosy at the Lazaretto on St Eustatius, Netherlands Antilles. *International Journal of Osteoarchaeology* 18: 72–84.

Ginns, A. 2007. Preliminary Report on the Second Season of Excavations Conducted on Mis Island (AKSC). *Sudan & Nubia* 11: 20–25.

Gladykowska-Rzeczycka, J. 1976. A Case of Leprosy from a Medieval Burial Ground. *Folia Morphologica (Warszawa)* 35: 253–264.

———. 1982. Schorzenia Swoiste Ludności z Dawnych Cmentarzysk Polski. *Przegląd Antropologiczny* 48: 39–55. (In Polish with English abstract.)

Glofurushan, F., M. Sadechi, M. Goldust, and N. Yosefi. 2011. Leprosy in Iran: An Analysis of 195 Cases from 1994–2008. *Journal of Pakistan Medical Association* 61: 558–561.

Gluckman, L. K. 1962. Leprosy in New Zealand before the 20th Century. *New Zealand Medical Journal* 61: 404–409.

Gluckman, P. D., A. Beedle, and M. Hanson. 2009. *Principles of Evolutionary Medicine*. Oxford: Oxford University Press.

Gluckman, P. D., F. M. Low, T. Buklijas, M. A. Hanson, and A. S. Beedle. 2011. How Evolutionary Principles Improve the Understanding of Health and Disease. *Evolutionary Applications* 4: 249–263.

Godal, T. 1978. Immunological Aspects of Leprosy—Present Status. *Progress in Allergy* 25: 211–242.

Godal, T., B. Myrvang, D. R. Samuel, W. F. Ross, and M. Løfgren. 1973. Mechanism of "Reactions" in Borderline Tuberculoid (BT) Leprosy. *Acta Pathologica et Microbiologica Scandinavia* 236: 45–53.

Gokhale, S. K., and S. H. Godbole. 1957. Serum Lipolytic Enzyme Activity and Serum Lipid Partition in Leprosy and Tuberculosis. *Indian Journal of Medical Research* 45: 327–336.

Golfurushan, F., M. Sadeghi, M. Goldust, and N. Yosefi. 2011. Leprosy in Iran: An Analysis of 195 Cases from 1994–2009. *Journal of the Pakistan Medical Association* 61(6): 558–561.

Gonzalez Prendes, M. A. 1963. *Historia de la Lepra en Cuba*. La Habana: Academia de Ciencias de la Republica de Cuba.

Good, R. A., A. West, and G. Fernandes. 1980. Nutritional Modulation of Immune Responses. *Federation Proceedings* 39: 3098–3104.

Goode, J. L. 1993. A Consideration of the Effects of Nerve Thickening on the Skeleton in Individuals Suffering from Leprosy. BSc diss., University of Bradford.

Gormus, B. J., R. H. Wolf, G. B. Baskin, S. Ohkawa, P. J. Gerone, G. P. Walsh, W. M. Meyers, C. H. Binford, and W. E. Greer. 1988. A Second Sooty Mangabey Monkey with Naturally Acquired Leprosy: First Reported Possible Monkey-to-Monkey Transmission. *International Journal of Leprosy and Other Mycobacterial Diseases* 56: 61–65.

Goulart, I. M. B. 2013. Molecular Investigation of the Route of Infection and Transmission of *M. leprae*. Poster presented at the 18th International Leprosy Congress, Brussels, Belgium.

Govindharaj, P., S. Mani, J. Darlong, and A. S. John. 2017. Acceptance and Satisfaction of Micro-Cellular Rubber Ready-Made Footwear among Patients with Insensitive Feet Due to Leprosy. *Leprosy Review* 88: 381–390.

Graciano-Machuca, O., E. E. Velarde-de la Cruz, M. G. Ramirez-Duenas, and A. Alvarado-Navarro. 2013. University Students' Knowledge and Attitude towards Leprosy. *Journal of Infection in Developing Countries* 7: 658–664.

Grainger, C. R. 1980. Leprosy in the Seychelles. *Leprosy Review* 51: 43–59.
———. 1981. Leprosy and Curieuse Island. *Leprosy Review* 52: 151–154.
Grange, J. M. 2008. *Mycobacterium tuberculosis:* The Organism. In *Clinical Tuberculosis*, 4th ed., edited by P. D. O. Davies, P. F. Barnes, and S. B. Gordon, 65–78. London: Hodder Arnold.
Grange, J. M., and J. I. Lethaby. 2004. Leprosy of the Past and Today. *Seminars in Respiratory and Critical Care Medicine* 25: 271–281.
Grant, A. V., A. Alter, N. T. Huong, M. Orlova, N. Van Thuc, N. N. Ba, V. H. Thai, L. Abel, E. Schurr, and A. Alcais. 2012. Crohn's Disease Susceptibility Genes Are Associated with Leprosy in the Vietnamese Population. *Journal of Infectious Diseases* 206: 1763–1767.
Graves, C. P. 2007. Sensing and Believing: Exploring Worlds of Difference in Pre-Modern England: A Contribution to the Debate Opened by Kate Giles. *World Archaeology* 39: 515–531.
Green, M. 2012. The Value of Historical Perspective. In *The Ashgate Research Companion to the Globalization of Health*, edited by T. Schrecker, 17–37. Farnham, Surrey: Ashgate Publishing Limited.
Grieshaber, B. M., B. L. Osborne, A. F. Doubleday, and F. A. Kaestle. 2007. A Pilot Study into the Effects of X-Ray and Computed Tomography Exposure on the Amplification of DNA from Bone. *Journal of Archaeological Science* 35: 681–687.
Grmek, M. 1989. *Diseases in the Ancient Greek World.* Translated by M. Muellner and L. Muellner. Baltimore, MD: Johns Hopkins University Press.
Grossi, M. A., M. A. Leboeuf, A. R. Andrade, S. Lyon, C. M. Antunes, and S. Bührer-Sékula. 2008. The Influence of ML Flow Test in Leprosy Classification. *Revista da Sociedade Brasileira de Medicina Tropical* 41 supplement 2: 34–38.
Grzybowski, A., J. Sak, J. Pawlikowski, and M. Nita. 2016. Leprosy: Social Implications from Antiquity to the Present. *Clinical Dermatology* 34: 8–10.
Guha, P. K., S. S. Pandey, G. Singh, and P. Kaur. 1981. Age of Onset of Leprosy. *Leprosy in India* 53: 83–87.
Guijo, J. M., R. LaCalle, and I. Lopez. 1999. Evidencias de Lepra en una Communidad Islámica Medieval de Sevilla. In *Sistematización Metodológica Paleopatología*, edited by J. A. Sánchez, 113–138. Actas del V. Congreso Nacional de Paleopatología. Jaén: Alcalá La Real.
Gupta, S. C., M. Tiwari, and K. G. Singh. 2004. Radiological and Antroscopic Study of Maxillary Antrum in Multibacillary Leprosy Patients. *Indian Journal of Leprosy* 76: 305–309.
Haas, C. J., A. Zink, G. Pálfi, U. Szeimies, and A. G. Nerlich. 2000. Detection of Leprosy in Ancient Human Skeletal Remains by Molecular Identification of *Mycobacterium leprae*. *American Journal of Clinical Pathology* 114: 428–436.
Haas, C. J., A. Zink, U. Szeimies, and A. G. Nerlich. 2002. Molecular Evidence of *Mycobacterium leprae* in Historic Bone Samples from South Germany. In *The Past and Present of Leprosy: Archaeological, Historical, Palaeopathological and Clinical Approaches*, edited by C. A. Roberts, M. E. Lewis, and K. Manchester, 287–292. British Archaeological Reports International Series 1054. Oxford: Archaeopress.
Hacking, P., and J. Wakely. N.d. The Human Skeletal Remains from Abingdon Abbey. Unpublished report.
Hadju, T., L. Márk, H. D. Donoghue, S. Fábián, T. Marton, G. Serlegi, A. Marcsik, and K.

Koehler. 2010. Preliminary Study of Biochemical Analysis of Osseous Leprosy Dated to the Late Copper Age (3700–3600 BC) Mass Grave in Hungary. Paper presented at the 18th European Meeting of the Paleopathology Association, Vienna, Austria.

Hadley, D. 2001. *Death in Medieval England: An Archaeology.* Oxford: Tempus.

Hamilton, H. K., W. R. Levis, F. Martiniuk, A. Cabrera, and J. Wolf. 2008. The Role of the Armadillo and Sooty Mangabey Monkey in Human Leprosy. *International Society of Dermatology* 47: 545–550.

Han, X. Y., F. M. Aung, S. E. Choon, and B. Werner. 2014. Analysis of the Leprosy Agents *Mycobacterium leprae* and *Mycobacterium lepromatosis* in Four Countries. *American Journal of Clinical Pathology* 142: 524–532.

Han, X. Y., and J. Jessurun. 2013. Severe Leprosy Reactions due to *Mycobacterium lepromatosis*. *American Journal of Medical Science* 345: 65–69.

Han, X. Y., N. A. Mistry, E. J. Thompson, H.-L. Tang, K. Khanna, and L. Zharng. 2015. Draft Genome of New Leprosy Agent *Mycobacterium lepromatosis*. *Genome Announcements* 3(3): e00513–e00515.

Han, X. Y., Y. H. Seo, K. C. Sizer, T. Schoberle, G. S. May, W. Li, and R. G. Nair. 2008. A New *Mycobacterium* Species Causing Diffuse Lepromatous Leprosy. *American Journal of Clinical Pathology* 130: 856–864.

Han, X. Y., and F. J. Silva. 2014. On the Age of Leprosy. *PLoS Neglected Tropical Diseases* 8(2): e2544.

Han, X. Y., K. C. Sizer, and H. H. Tan. 2012. Identification of the Leprosy Agent *Mycobacterium lepromatosis* in Singapore. *Journal of Drugs in Dermatology* 11: 168–172.

Han, X. Y., K. C. Sizer, E. J. Thompson, J. Kabanja, J. Li, P. Hu, L. Gómez-Valero, and F. J. Silva. 2009. Comparative Sequence Analysis of *Mycobacterium leprae* and the New Leprosy-Causing *Mycobacterium lepromatosis*. *Journal of Bacteriology* 191: 6067–6074.

Han, X. Y., K. C. Sizer, J. S. Velarde-Félix, L. O. Frias-Castro, and F. Vargas-Ocampo. 2012. The Leprosy Agents *Mycobacterium lepromatosis* and *Mycobacterium leprae* in Mexico. *International Journal of Dermatology* 51: 952–959.

Hansen, G. A. 1880. Studien uber *Bacillus leprae*. *Virchows Archiv* 79: 32–42.

Hansen, G. A., and C. Looft. 1895. *Leprosy in Its Clinical and Pathological Aspects.* Bristol: John Wright.

Harbitz, F. 1910. Trophoneurotic Changes in Bone and Joint in Leprosy. *Archives of Internal Medicine* 6: 147–169.

Harboe, M. 1983. The Work and Concepts of Armauer Hansen: How Do They Stand Today? *Ethiopian Medical Journal* 21: 123–126.

Hardas, U., R. Survey, and D. Chakrawarti. 1972. Leprosy in Gynecology and Obstetrics. *International Journal of Leprosy and Other Mycobacterial Diseases* 40: 398–401.

Harman, M. 1993. General Description of the Skeletons, Explanation of the Catalogue Entries, the Human Burials: Discussion. In A. G. Kinsley, *Broughton Lodge: Excavations on the Romano-British Settlement and Anglo-Saxon Cemetery at Broughton Lodge, Willoughby-on-the-Wolds, Nottinghamshire 1964-8*, 55–58. Nottingham Archaeological Monographs 4. Nottingham: Department of Classical and Archaeological Studies, University of Nottingham.

Harper, K., and G. J. Armelagos. 2010. The Changing Disease-Scape in the Third Epidemio-

logical Transition. *International Journal of Environmental Research and Public Health* 7: 675–697.
Harris, J. R., and P. W. Brand. 1966. Patterns of Disintegration of the Tarsus of the Anaesthetic Foot. *Journal of Bone and Joint Surgery* 48B: 4–16.
Hart, J. 1989. Leprosy in Cornwall and Devon. In *From Cornwall to Caithness: Some Aspects of British Field Archaeology. Papers Presented to Norman V. Quinnell,* edited by M. Bowden, D. Mackay, and P. Topping, 261–269. British Archaeological Reports British Series 209. Oxford: BAR.
Hasan, S. 1974. A Survey of Plantar Ulcers among the Inmates of a Leprosy Home and Rehabilitation Centre, Moula Ali, Hyderabad—A. P. *Leprosy in India* 46: 99–102.
Hauhnar, C. Z., S. Kaur, V. K. Sharma, and S. B. S. Mann. 1992. A Clinical and Radiological Study of the Maxillary Antrum in Lepromatous Leprosy. *Indian Journal of Leprosy* 64: 487–494.
Hauhnar, C. Z., S. B. S. Mann, V. K. Sharma, S. Kaur, S. Mehta, and B. D. Radotra. 1992. Maxillary Antrum Involvement in Multibacillary Leprosy: A Radiologic, Sinuscopic, and Histologic Assessment. *International Journal of Leprosy and Other Mycobacterial Diseases* 60: 390–395.
Heagerty, J. J. 1933. Leprosy in Canada. *International Journal of Leprosy and Other Mycobacterial Diseases* 1: 466–468.
Heiberg, H. 1886. Om Lepra Mutilans. *Klin Aarbog* 3: 301–319.
Heijinders, M. L. 2004. Experiencing Leprosy: Perceiving and Coping with Leprosy and Its Treatment: A Qualitative Study Conducted in Nepal. *Leprosy Review* 75: 327–337.
Herrick, A. B., and T. W. Earhart. 1911. The Value of Trophic Bone Changes in the Diagnosis of Leprosy. *Archives of Internal Medicine* 7: 801–811.
Hershkovitz, I. 2012. Paleopathology in Israel: Nicu Haas and His Successors. In *The Global History of Paleopathology: Pioneers and Prospects,* edited by J. Buikstra and C. Roberts, 426–438. Oxford: Oxford University Press.
Hershkovitz, I., M. Spiers, and B. Arensberg. 1993. Leprosy or Madura Foot? The Ambiguous Nature of Infectious Disease in Palaeopathology: Reply to Dr Manchester. *American Journal of Physical Anthropology* 91: 251–253.
Hershkovitz, I., M. Spiers, A. Katznelson, and B. Arensberg. 1992. Unusual Pathological Conditions in the Lower Extremities of a Skeleton from Ancient Israel. *American Journal of Physical Anthropology* 88: 23–26.
Hershkovitz, I., and M. Spigelman. 2007. Bones, Teeth and DNA Unravel Major Issues in Levantine Bio-History. In *Faces from the Past: Diachronic Patterns in the Biology of Human Populations from the Eastern Mediterranean: Papers in Honour of Patricia Smith,* edited by M. Faerman, L. K. Horwitz, T. Kahana, and U. Zilberman, 233–245. British Archaeological Reports International Series 1603. Oxford: Archaeopress.
Hillson, S. 1996. *Dental Anthropology.* Cambridge: Cambridge University Press.
Hinkle, C. A. 2007. *The Thankful Leper.* St. Louis, MO: Concordia Publishing House.
Hirata, K., C. Oku, and I. Morimoto. 2000. A Case of Leprosy in a Medieval Japanese. *Anthropological Science* 108: 114.
Hirschberg, M., and R. Biehler. 1909. Lepra der Knochen. *Dermatologische Zeitschrift* 16: 415–438.
Hogeweg, M., K. U. Kiran, and S. Suneetha. 1991. The Significance of Facial Patches and

Type I Reaction for the Development of Facial Nerve Damage in Leprosy: A Retrospective Study among 1226 Paucibacillary Patients. *Leprosy Review* 62: 143–149.

Hohler, C. 1957. The Badge of St James. In *The Scallop: Studies of a Shell and Its Influences on Humankind*, edited by I. Cox, 49–70. London: Shell Transport and Trading Company.

Hohmann, N., and A. Voss-Böhme. 2013. The Epidemiological Consequences of Leprosy-Tuberculosis Co-Infection. *Mathematical Biosciences* 241: 225–237.

Holck, P. 2012. Paleopathology in Norway. In *The Global History of Paleopathology: Pioneers and Prospects*, edited by J. Buikstra and C. Roberts, 487–489. Oxford: Oxford University Press.

Holst, M. 2005. Artefacts & Environmental Evidence: The Human Bone. In Blue Bridge Lane and Fishergate House, York: Report on Excavations, July 2000–July 2002, edited by C. A. Spall and N. J. Toop. http://www.mgassocs.com/mono/001/rep_bone_hum1a.html.

———. 2012. Osteological Analysis Dixon Lane and George Street York. York Osteoarchaeology. Unpublished report. https://www.yorkarchaeology.co.uk/wp-content/uploads/2015/05/Dixons-Lane-Osteological-Report.pdf.

Honap, T. P., L. A. Pfister, G. Housman, S. Mills, R. P. Tarara, K. Suzuki, F. P. Cuozzo, M. L. Sauther, M. S. Rosenberg, and A. C. Stone. 2018. *Mycobacterium leprae* Genomes from Naturally Infected Nonhuman Primates. *PLoS Neglected Tropical Diseases* 12(1): e0006190.

Hong, K. W., and B. Oh. 2010. Overview of Personalized Medicine in the Disease Genomic Era. *Biochemistry and Molecular Biology Reports* 43: 643–648.

Hotez, P. J., L. Savioloi, and A. Fenwick. 2012. Neglected Tropical Diseases of the Middle East and North Africa: Review of Their Prevalence, Distribution, and Opportunities for Control. *PLoS Neglected Tropical Diseases* 6(2): e1475. https://journals.plos.org/plosntds/article?id=10.1371/journal.pntd.0001475.

Hsieh, N. K., C. C. Chu, N. S. Lee, H. L. Lee, and M. Lin. 2010. Association of HLA-DRB1*0405 with Resistance to Multibacillary Leprosy in Taiwanese. *Human Immunology* 71: 712–716.

Huang, C. L. 1980. The Transmission of Leprosy in Man. *International Journal of Leprosy and Other Mycobacterial Diseases* 48: 309–318.

Hudson, T., and J. Genesse. 1982. Hansen's Disease in the United States. *Social Science and Medicine* 16: 997–1004.

Hughes, M. S., N. W. Ball, L. A. Beck, G. W. de Lisle, R. A. Skuce, and S. D. Neill. 1997. Determination of the Etiology of Presumptive Feline Leprosy by 16S rRNA Gene Analysis. *Journal of Clinical Microbiology* 35: 2464–2471.

Hukelova, Z. 2018. Nitra-Selenac II (Slovakia): An 11th Century Cemetery for the Sick and Dying? Paper presented at the 20th annual meeting of the British Association of Biological Anthropology and Osteoarchaeology, Cranfield University, September.

Humphry, A. H. 1952. Leprosy amongst Full-Blooded Aborigines in the Northern Territory. *Medical Journal of Australia* 39: 570–573.

Hunter, J. M. 1986. Hansen's Disease and Towns in Africa Part I. *The Star* (May–June): 4–7.

Hunter, J. M., and M. O. Thomas. 1984. Hypothesis of Leprosy, Tuberculosis and Urbanization in Africa. *Social Science and Medicine* 19: 27–57.

Hutchinson, D. L., and D. S. Weaver. 1998. Two Cases of Facial Involvement in Probable

Treponemal Infection from Late Prehistoric Coastal North Carolina. *International Journal of Osteoarchaeology* 8: 444–453.

Hutchinson, J. 1906. *On Leprosy and Fish Eating*. London: Constable.

Immel, A., A. Le Cabec, M. Bonazzi, A. Herbig, H. Temming, V. J. Schuenemann, K. I. Bos, F. Langbein, K. Harvati, A. Bridault, et al. 2016. Effect of X-Ray Irradiation on Ancient DNA in Sub-Fossil Bones—Guidelines for Safe X-Ray Imaging. *Scientific Reports* 6. https://www.ncbi.nlm.nih.gov/pmc/articles/PMC5018823/.

Inskip, S. 2008. Great Chesterford: A Catalogue of Burials. In *Proceedings of the British Association of Biological Anthropology and Osteoarchaeology 8th Annual Conference 2006, University of Bradford,* edited by M. Brickley and M. Smith, 57–66. British Archaeological Reports International Series 1743. Oxford: Archaeopress.

Inskip, S., G. M. Taylor, and G. Stewart. 2017. Leprosy in Pre-Norman Suffolk, UK: Biomolecular and Geochemical Analysis of the Woman from Hoxne. *Journal of Medical Microbiology* 66(11): 1640–1649.

Inskip, S. A., G. M. Taylor, S. R. Zakrzewski, S. A. Mays, A. W. G. Pike, G. Llewellyn, C. M. Williams, O. Y.-C. Lee, H. H. T. Wu, D. E. Minnikin, et al. 2015. Osteological, Biomolecular and Geochemical Examination of an Early Anglo-Saxon Case of Lepromatous Leprosy. *PLoS ONE* 10(5): e0124282.https://journals.plos.org/plosone/article?id=10.1371/journal.pone.0124282.

International Leprosy Association and the International Council on Archives. 2000. *Leprosy Archives: Preserve Them!* N.p.: International Leprosy Association and the International Council on Archives.

International Leprosy Association and Netherlands Leprosy Relief Association. 2011a. *Guidelines to Reduce Stigma: What Is Health Related Stigma?* Amsterdam and London: ILEP and Netherlands Leprosy Relief Association.

———. 2011b. *Guidelines to Reduce Stigma: How to Assess Health-Related Stigma*. Amsterdam and London: ILEP and Netherlands Leprosy Relief Association.

———. 2011c. *Guidelines to Reduce Stigma. A Roadmap to Stigma Reduction: An Empowerment Intervention*. Amsterdam and London: ILEP and Netherlands Leprosy Relief Association.

———. 2011d. *Guidelines to Reduce Stigma: Counselling to Reduce Stigma*. Amsterdam and London: ILEP and Netherlands Leprosy Relief Association.

Irgens, L. M. 1973. Leprosy in Norway: An Interplay of Research and Public Health. *International Journal of Leprosy and Other Mycobacterial Diseases* 41: 189–198.

———. 1980. Leprosy in Norway. *Leprosy Review* 51 supplement 1: 1–130.

———. 1981. Epidemiological Aspects and Implications of the Disappearance of Leprosy from Norway: Some Factors Contributing to the Decline. *Leprosy Review* 52 supplement 1: 147–165.

Irgens, L. M., and T. Bjerkedal. 2006. Epidemiology of Leprosy in Norway: The History of the National Leprosy Registry from 1856 until Today. In *Leprosy*, edited by L. M. Irgens, Y. Nedrebø, S. Sandmo, and A. Skivenes, 25–31. Førde, Norway: Selja Forlag.

Irgens, L. M., J. Kazda, K. Müller, and G. E. Eide. 1981. Conditions Relevant to the Occurrence of Acid-Fast Bacilli in Sphagnum Vegetation. *Acta Pathologica Microbiologica Scandinavica. Section B, Microbiology* 89: 41–47.

Isager, K. 1936. *Skeletfundene ved Øm Kloster: Til Belysning af Middelalderlig Patologi og Klosteret som Hospital.* København: Lewin and Munksgaard.

Ishikawa, S., A. Ishikawa, J. Yoh, H. Tanaka, and M. Fujiwara. 1999. Osteoporosis in Male and Female Leprosy Patients. *Calcified Tissue International* 64: 144–147.

Ishikawa, S., M. Mizushima, M. Furuta, A. Ishikawa, and K. Kawamura. 2000. Leydig Cell Hyperplasia and the Maintenance of Bone Volume: Bone Histomorphometry and Testicular Histopathology in 29 Male Leprosy Autopsy Cases. *International Journal of Leprosy and Other Mycobacterial Diseases* 68: 258–266.

Itakura, T. 1940. The Histo-Pathological Studies on Teeth of the Lepers, Especially on Its Pulp-Tissues. *Transactions of the Society of Pathology Japan* 30: 357–367.

Iyere, B. B. 1990. Leprosy Deformities: Experience in Molai Leprosy Hospital, Maiduguri, Nigeria. *Leprosy Review* 61: 171–179.

Jacob Raja, S. A., J. J. Raja, R. Vijayashree, B. M. Priva, G. S. Anusuya, and P. Ravishankar. 2016. Evaluation of Oral and Periodontal Status of Leprosy Patients in Dindigul District. *Journal of Pharmacy & Bioallied Sciences* 8 supplement 1: S119–S121.

Jaiswal, A. K., and N. T. Subbarao. 2010. Bilateral Lagopthalmos in Leprosy: Is It a Rare Phenomenon? *Indian Journal of Leprosy* 82: 201–203.

Jana, S., and G. S. Shekhawat. 2011. Critical Review on Medicinally Potent Plant Species: *Gloriosa superba*. *Fitoterapia* 82: 293–301.

Jaskulska, E. 2014. Adaptation to Cold Climate in the Nasal Cavity Skeleton: A Comparison of Archaeological Crania from Different Climatic Zones. PhD diss., University of Warsaw.

Jensen, A. G., I. C. Bygbjerg, H. Jensen, and S. Ullman. 1992. Leprosy—Still a Possibility in Denmark. *Ugeskrift for Laeger* 154: 1712–1713.

Jessamine, P. G., M. Desjardins, T. Gillis, D. Scollard, F. Jamieson, G. Broukhanski, P. Chedore, and A. McCarthy. 2012. Leprosy-Like Illness in a Patient with *Mycobacterium lepromatosis* from Ontario, Canada. *Journal of Drugs in Dermatology* 11: 229–233.

Jim, R., E. Johnson, and B. I. Pavlin. 2010. Role of GIS Technology during Leprosy Elimination Efforts in Pohnpei. *Pacific Health Dialogue* 16: 109–114.

Jindal, K. C., G. P. Singh, V. Mohan, and B. B. Mahajan. 2013. Psychiatric Morbidity among Inmates of Leprosy Homes. *Indian Journal of Psychological Medicine* 35: 335–340.

Job, C. K., A. B. A. Karat, and S. Karat. 1966. The Histopathological Appearance of Leprous Rhinitis and Pathogenesis of Septal Perforation in Leprosy. *Journal of Laryngology and Otology* 80: 718–732.

Job, C. K., R. M. Sanchez, and R. C. Hastings. 1987. Lepromatous Placentitis and Intrauterine Fetal Infection in Lepromatous Nine-Banded Armadillos. *Laboratory Investigations* 56: 44.

John, A. S., P. S. Rao, and S. Das. 2010. Assessment of Needs and Quality Care Issues of Women with Leprosy. *Leprosy Review* 81: 34–40.

Johnston, W. 1874. Observations on Leprosy and Its Treatment by Means of Vaporized Carbolic Acid in Union with Watery Vapour. *Indian Medical Gazette* 9: 319–322.

———. 1875. Observations on Leprosy and Its Treatment by Means of Vaporized Carbolic Acid in Union with Watery Vapour. *Indian Medical Gazette* 10: 9–12.

Jonquieres, E. 1977. An Ancient Briton Adds to the Story of Leprosy. *International Journal of Leprosy and Other Mycobacterial Diseases* 45: 66.

Jopling, W. H. 1982. Clinical Aspects of Leprosy. *Tubercle* 63: 295–305.
———. 1985. Environmental Mycobacteria and the Skin: Clinical Aspects. *The Star* (November–December): N.p.
———. 1991. Leprosy Stigma. *Leprosy Review* 62: 1–12.
———. 1992. Recollections and Reflections. *The Star* (March–April): 5–10.
Jopling, W. H., and A. C. McDougall. 1988. *Handbook of Leprosy*. 4th ed. London: Heinemann.
Judd, M. 2008. Trauma. In *"Lepers outside the Gate": Excavations at the Cemetery of the Hospital of St James and St Mary Magdalene, Chichester, 1986–87 and 1993*, edited by J. R. Magilton, F. Lee, and A. Boylston, 229–238. Council for British Archaeology Research Report 158 and Chichester Excavations vol. 10. Bootham: Council for British Archaeology.
Judd, M., and C. A. Roberts. 1988. Fracture Patterns at the Medieval Leper Hospital in Chichester. *American Journal of Physical Anthropology* 105: 43–55.
Justus, H. M., and A. M. Agnew. 2012. Two Possible Cases of Leprosy in Medieval Poland. Poster presented at the 19th European meeting of the Palaeopathology Association, Lille, France.
Jütte, R. 2005. *A History of the Senses: From Antiquity to Cyberspace*. Oxford: Polity.
Kaimal, S., and D. M. Thappa. 2009. Relapse in Leprosy. *Indian Journal of Dermatology, Venereology and Leprology* 75: 126–135.
Kakar, S. 1996. Leprosy in British India, 1860–1940: Colonial Politics and Missionary Medicine. *Medical History* 40: 215–230.
Kalisch, P. A. 1973. Tracadie and Penikese Leprosaria: A Comparative Analysis of Societal Response to Leprosy in New Brunswick, 1844–1880, and Massachusetts, 1904–1921. *Bulletin of the History of Medicine* 47: 480–512.
Kang, T. J., and G. T. Chae. 2001. Detection of Toll-Like Receptor 2 (TLR2) Mutation in the Lepromatous Leprosy Patients. *FEMS Immunology and Medical Microbiology* 31: 53–58.
Kang, T. J., S. B. Lee, and G. T. Chae. 2002. A Polymorphism in the Toll-Receptor 2 Is Associated with IL-12 Production from Monocytes in Lepromatous Leprosy. *Cytokine* 20: 56–62.
Kant, L., and D. Mukherji. 1987. Editorial: Childhood Leprosy. *Indian Pediatrics* 24: 105–107.
Kaplan, M., and R. H. Belber. 1988. Care of Plantar Ulcerations: Comparing Applications, Materials and Non-Casting. *Leprosy Review* 59: 66.
Karat, S., A. B. Karat, and R. Foster. 1968. Radiological Changes in Bones of the Limbs in Leprosy. *Leprosy Review* 39: 147–169.
Karat, S., P. S. S. Rao, and A. B. A. Karat. 1972. Prevalence of Deformities and Disabilities among Leprosy Patients in an Endemic Area. Part 2: Nerve Involvement in the Limbs. *International Journal of Leprosy and Other Mycobacterial Diseases* 40: 265–270.
Kasai, N., O. Kondo, K. Suzuki, Y. Aoki, N. Ishii, and M. Goto. 2018. Quantitative Evaluation of Maxillary Bone Deformation by Computed Tomography in Patients with Leprosy. *PLoS Neglected Tropical Diseases* 12(3): e0006341. https://journals.plos.org/plosntds/article?id=10.1371/journal.pntd.0006341.
Kato, S. 2000. The Function of Vitamin D Receptor in Vitamin D Action. *Journal of Biochemistry* 127: 717–722.

Katoch, V. M. 1999. Molecular Techniques for Leprosy: Present Applications and Future Perspectives. *Indian Journal of Leprosy* 71: 45–59.

Kaufman, M. H., and W. J. MacLennan. 2000. Robert the Bruce and Leprosy. *Proceedings of the Royal College of Physicians Edinburgh* 30: 75–80.

Kaur, H., and V. Ramesh. 1994. Social Problems of Women Leprosy Patients—A Study Conducted at Two Urban Leprosy Centres in Delhi. *Leprosy Review* 65: 361–375.

Kaur, S., S. K. Malik, B. Kumar, M. P. Singh, and R. N. Chakravarty. 1979. Respiratory System Involvement in Leprosy. *International Journal of Leprosy* 47: 18–25.

Kazda, J. 1981. Occurrence of Non-Cultivable Acid-Fast Bacilli in the Environment and Their Relationship to *M. leprae*. *Leprosy Review* 52 supplement: 85–91.

———. 2000. *The Ecology of* Mycobacterium leprae. London: Kluwer Academic Publishers.

———. 2006. Occurrence of Non-Cultivable Acid-Fast Bacilli in the Environment and Their Relationship to M. leprae. *International Journal of Epidemiology* 35: 994–1000.

Kazda, J., R. Ganapati, C. Revankar, T. M. Buchanan, D. B. Young, and L. M. Irgens. 1986. Isolation of Environment-Derived *Mycobacterium leprae* from Soil in Bombay. *Leprosy Review* 57: 201–208.

Kazda, J., L. M. Irgens, and A. H. Kolk. 1990. Acid-Fast Bacilli Found in Sphagnum Vegetation of Coastal Norway Containing *Mycobacterium leprae*-Specific Phenolic Glyolipid-1. *International Journal of Leprosy* 58: 353–357.

Kazda, J., L. M. Irgens, and K. Muller. 1980. Isolation of Noncultivable Acid-Fast Bacilli in Sphagnum and Moss Vegetation by Footpad Technique in Mice. *International Journal of Leprosy and Other Mycobacterial Diseases* 48: 1–6.

Kazen, R. O. 1993. Management of Plantar Ulcers—Theory or Practice? *Leprosy Review* 64: 188–189.

Kieffer-Olsen, J. 2001. *Grav og Graviskik I det Middelalderlige Danmark*. Åarhus, Denmark: Department of Medieval Archaeology, University of Åarhus.

Kellersberger, E. R. 1951. The Social Stigma of Leprosy. *Annals of the New York Academy of Science* 54: 126–133.

Kelmelis, K. S., and D. D. Pedersen. 2019. Impact of Urbanization on Tuberculosis and Leprosy Prevalence in Medieval Denmark. *Anthropologia Anzeiger* 76: 149–166.

Kelmelis, K. S., M. H. Price, and J. Wood. 2017. The Effect of Leprotic Infection on the Risk of Death in Medieval Rural Denmark. *American Journal of Physical Anthropology* 164: 763–775.

Kerr, L., C. Kendall, C. A. Sousa, C. C. Frota, J. Graham, L. Rodrigues, R. L. Fernandes, and M. L. Barreto. 2015. Human-Armadillo Interaction in Ceará, Brazil: Potential for Transmission of *Mycobacterium leprae*. *Acta Tropica* 152: 74–79.

Kerr-Pontes, L. R. S., M. L. Barreto, C. M. N. Ebangelista, L. C. Rodrigues, J. Heukelbach, and H. Feldmeier. 2006. Socioeconomic, Environmental, and Behavioural Risk Factors for Leprosy in North-East Brazil: Results of a Case-Control Study. *International Journal of Epidemiology* 35: 994–1000.

Kerudin, A. 2019. Genotyping of *Mycobacterium tuberculosis* and *Mycobacterium leprae* ancient DNA. PhD thesis, University of Manchester.

Kerudin, A., R. Müller, J. Buckberry, C. Knüsel, and T. Brown. 2019. Ancient *Mycobacterium Leprae* Genomes from Mediaeval England. *Journal of Archaeological Science* 112: doi.org/10.1016/j.jas.2019.105035.

King, R. 2010. *People on the Move: An Atlas of Migration.* Berkeley: University of California Press.

King, S. E. 1994. The Human Skeletal Remains from Glasgow Cathedral. Unpublished report.

———. 2002. The Skeletal Remains from the Nave Crypt, Choir and Treasury. In *Excavations at Glasgow Cathedral: 1988-1997,* edited by S. T. Driscoll, 134-152. Society for Medieval Archaeology Monograph 18. Durham: Society for Medieval Archaeology.

Kirchheimer, W. F. 1976. The Role of Arthropods in the Transmission of Leprosy. *International Journal of Leprosy and Other Mycobacterial Diseases* 44: 104-107.

Kirchheimer, W. F., and E. E. Storrs. 1971. Attempts to Establish the Armadillo (*Dasypus novemcinctus* Linn.) as a Model for the Study of Leprosy. 1. Report of Lepromatoid Leprosy in an Experimentally Infected Armadillo. *International Journal of Leprosy and Other Mycobacterial Diseases* 39: 693-702.

Kirchheimer, W. F., E. E. Storrs, and C. H. Binford. 1972. Attempts to Establish the Armadillo (*Dasypus novemcinctus* Linn.) as a Model for the Study of Leprosy. II. Histopathologic and Bacteriologic Post-mortem Findings in Lepromatoid Leprosy in the Armadillo. *International Journal of Leprosy and Other Mycobacterial Diseases* 40: 229-242.

Kiris, A., T. Karlidag, E. Kocakoc, Z. Bozgeyik, and M. Sarsilmaz. 2007. Paranasal Sinus Computed Tomography Findings in Patients Treated for Lepromatous Leprosy. *Journal of Laryngology and Otology* 121: 15-18.

Kishmore, B. N., G. Kamath, and H. D'Silva. 1985. A Study of Leprosy in Mangalore City through School Surveys. *Indian Journal of Leprosy* 57: 588-592.

Kjellstrom, A. 2012. Possible Cases of Leprosy and Tuberculosis in Medieval Sigtuna, Sweden. *International Journal of Osteoarchaeology* 22: 261-283.

Kjellstrom, A., and A. Linderholm. 2008. Stable Isotope Analysis of a Medieval Skeletal Sample with Signs of Leprosy from Sigtuna, Sweden. Poster presented at the 17th European meeting of the Paleopathology Association, Copenhagen, Denmark.

Kjellstrom, A., and A. Wikstrom. 2008. Kyrkan och det Sakrala Stadsrummet. In *På Vag mot Paradiset: Arkeologisk Undersökning i Kvarteret Humlegården 3 i Sigtuna 2006,* edited by A. J. Wikstrom, 165-216. Meddelanden och Rapporter från Sigtuna Museum 33. Sigtuna: Sigtuna Museum.

Kluger, M. J. 1978. The History of Bloodletting. *Natural History* 87: 78-83.

Knüsel, C., and S. Göggel. 1993. A Cripple from the Medieval Hospital of Sts James and Mary Magdalen, Chichester. *International Journal of Osteoarchaeology* 3: 155-165.

Köhler, K., S. Fábián, T. Marton, G. Serlegi, A. Marcsik, H. D. Donoghue, L. Márk, and T. Hajdu. 2010. Paleopathological Analysis of Possible Cases of Leprosy Dated to a Late Copper Age (3700-360 BC) Mass Grave in Hungary. Poster presented at the 18th European meeting of the Paleopathology Association, Vienna, Austria.

Köhler, K., A. Marcsik, P. Zádori, G. Biro, T. Szeniczey, S. Fábiá, G. Serlegi, T. Marton, H. D. Donoghue, and T. Hajdu. 2017. Possible Cases of Leprosy from the Late Copper Age (3780-3650 cal BC) in Hungary. *PLoS ONE* 12(10): e0185966. https://journals.plos.org/plosone/article?id=10.1371/journal.pone.0185966.

Kohout, E., T. Hushangi, and B. Azadeh. 1973. Leprosy in Iran. *International Journal of Leprosy and Other Mycobacterial Diseases* 41: 102-111.

Kozak, A. D., J. Gresky, N. Roumelis, and M. Schultz. 2008. Leprosy in the Old Kingdom

Elephantine (Egyptian Nubia). Poster presented at the 17th European meeting of the Paleopathology Association, Copenhagen, Denmark.

Kozak, A. D., and M. Schultz. 2006a. New Data on Leprosy in Medieval Kiev, Ukraine. Paper presented at the 16th European meeting of the Paleopathology Association, Santorini, Greece.

———. 2006b. Prokaza v drevnerusskom Kieve po dannym paleopatologii. *Vestnik Antropologii* 14: 34–40.

Kozlowski, T. 2012a. History of Paleopathology in Poland. In *The Global History of Paleopathology: Pioneers and Prospects*, edited by J. Buikstra and C. Roberts, 490–502. Oxford: Oxford University Press.

———. 2012b. *Stan Biologiczny i Warunki Życia Ludności in Culmine na Pomorzu Nadwiślańskim (X–XIII wiek): Studium Antropologiczne* (Biological State and Life Conditions of the Population Living in Culmine, Pomerania Vistula (10th–13th century): An Anthropological Study). Mons Sancti Laurentii vol. 7. Toruń: Wydawnictwo Naukowe Uniwersytetu Mikołaja Kopernika. (In Polish with English abstract).

Kozłowski, T., and A. Drozd. 2008. Występowanie Trądu na Obszarze Mezoregionu Kałdus-Gruczno w Okresie od XI do XIV Wieku. In *Epidemie, klęski wojny*, edited by W. Dzieduszycki and J. Wrzesiński, 147–155. Funeralia Lednickie Spotkanie 10. Poznań: SNA. (In Polish.)

Krahenbuhl, J. L., and L. B. Adams. 2000. Exploitation of Gene Knockout Mice Models to Study the Pathogenesis of Leprosy. *Leprosy Review* 71: S170–S175.

Krause-Kyora, B., M. Nutsoa, L. Boehme, F. Pierini, D. Dangvard Pedersen, S.-C. Kornell, D. Drichel, M. Nonazzi, L. Möbus, P. Tarp, et al. 2018. Ancient DNA Study Reveals HLA Susceptibility Locus for Leprosy in Medieval Europeans. *Nature Communications* 9: Article 1569. https://www.nature.com/articles/s41467-018-03857-x.

Krishnamoorthy, K. V. 1994. Protective Footwear for Leprosy Patients with Sole Sensory Loss or Ulceration of the Foot. *Leprosy Review* 65: 400–402.

Kula, B., and J. L. Robinson. 2013. Mistreatment of Immigrants: The History of Leprosy in Canada. *Leprosy Review* 84: 322–324.

Kulkarni, V. N., N. H. Antia, and J. M. Mehta. 1990. Newer Designs in Footwear for Leprosy Patients. *Indian Journal of Leprosy* 62: 483–487.

Kulkarni, V. N., and J. M. Mehta. 1983. Tarsal Disintegration in Leprosy. *Leprosy in India* 55: 338–370.

Kumar, A. S., S. Kumar, S. Abraham, and P. S. Rao. 2011. Leprosy among Tribal Populations of Chhattisgarh State, India. *Indian Journal of Leprosy* 83: 23–29.

Kumar, A., S. Lambert, and D. N. J. Lockwood. 2019. Picturing Health: A New Face for Leprosy. *Lancet* 393: 629–638.

Kumar, A., Y. C. Mathur, and P. Rao. 1973. Leprosy in Childhood in Rural and Urban Areas of Hyderabad. *Indian Pediatrics* 9: 337–338.

Kumar, B., S. Dogra, and I. Kaur. 2004. Epidemiological Characteristics of Leprosy Reactions: 15 Years Experience from North India. *International Journal of Leprosy and Other Mycobacterial Diseases* 72: 125–133.

Kumar, M., M. A. Sheikh, and R. W. Bussmann. 2011. Ethnomedicinal and Ecological Status of Plants in Garhwal Himalaya, India. *Journal of Ethnobiology and Ethnomedicine* 7: 1–13.

Kumar, R., P. Singhasivanon, J. B. Sherchand, P. Mahaisavariya, J. Kaewkungwal, S. Peerapakorn, and K. Mahotarn. 2004. Gender Differences in Socio-Epidemiological Factors for Leprosy in the Most Hyper-Endemic District of Nepal. *Nepal Medical College Journal* 6: 98–105.

Kumaresan, J. A., and E. T. Maganu. 1994. Socio-Cultural Dimensions of Leprosy in North-Western Botswana. *Social Science and Medicine* 39: 537–541.

Kundacki, N., and C. Erdem. 2019. Leprosy: A Great Imitator. *Clinical Dermatology* 37: 200–212.

Kuruwa, S., V. Vissa, and N. Mistry. 2012. A Study of the Distribution of *Mycobacterium leprae* Strains among Cases in Rural and Urban Populations of Maharashtra. *Journal of Clinical Microbiology* 50: 1406–1411.

Kusaka, T., K. Kohsaka, Y. Fukunishi, and H. Akimori. 1981. Isolation and Identification of Mycolic Acids in *Mycobacterium leprae* and *Mycobacterium lepraemurium*. *International Journal of Leprosy and Other Mycobacterial Diseases* 49: 406–416.

Kwan, Z., J. Pailoor, L. L. Tan, S. Robinson, S.-M. Wong, and R. Ismail. 2014. Leprosy—An Imported Disease. *Leprosy Review* 85: 170–176.

Kwiatkowska, B. 2005. *Mieszkańcy Średniowiecznego Wrocławia. Ocena Warunków Życia i Stanu Zdrowia w Ujęciu Antropologicznym*. Wrocław: Wydawn. Uniwersytetu Wrocławskiego. (In Polish with English abstract.)

Kyriakos, K. P. 2003. Leprosy in Greece at the End of the 20th Century. *International Journal of Leprosy and Other Mycobacterial Diseases* 71: 357–360.

———. 2010. Active Leprosy in Greece: A 20 Year Survey (1988–2007). *Scandinavian Journal of Infectious Diseases* 42: 594–597.

L'Abbé, E. N., and M. Steyn. 2007. Health Status of the Venda, a Post-Antibiotic Community in Rural South Africa. *International Journal of Osteoarchaeology* 17: 492–503.

Lafferty, K. D. 2009. The Ecology of Climate Change and Infectious Diseases. *Ecology* 90: 888–900.

L'Africain, J.-L. 1956. *Description de l'Afrique*. Edited and translated by A. Épaulard. Paris: Adrien-Maisonneuve.

Lagia, A., I. Petroutsa, O. Dretaki, and S. K. Manolis. 2006. Demographic and Paleopathologic Analysis of an Early Christian Era Population from the Kos Island (Aegean Archipelago, Greece). Paper presented at the 16th European meeting of the Paleopathology Association, Santorini, Greece, August 28–September 1.

Lagia, A., I. Petroutsa, and S. K. Manolis. 2002. A Demographic and Paleopathological Assessment of a Paleochristian Population from the Island of Kos. In *Proceedings of the 24th Annual Conference of the Hellenic Society for Biological Sciences, Eretria, 2002*, 163–164.

Lahiri, R., and J. L. Krahenbuhl. 2008. The Role of Free-Living Pathogenic Amoeba in the Transmission of Leprosy: A Proof of Principle. *Leprosy Review* 79: 401–409.

Lai, P. C., C. T. Low, W. S. Tse, C. K. Tsui, H. Lee, and P. C. Hui. 2013. Risk of Tuberculosis in High-Rise and High-Density Dwellings: An Exploratory Spatial Analysis. *Environmental Pollution* 183: 40–45.

Lane, J. E., D. S. Walsh, and W. M. Meyers. 2006. Borderline Tuberculoid Leprosy in a Woman from the State of Georgia with Armadillo Exposure. *Journal of the American Academy of Dermatology* 55: 714–716.

Lapinsky, S. E. 1992. Anaemia, Iron-Related Measurements and Erythropoietin Levels in Untreated Patients with Active Leprosy. *Journal of Internal Medicine* 232: 273–278.

Larson, L. M. 1917. *The King's Mirror (Speculum Regale-Konungs Skuggsjá)*. Translated from Old Norwegian. New York: American-Scandinavian Foundation.

Lavado-Valenzueala, R., M. José Bravo, A. P. Junqueira-Kipnis, M. Ramos de Souza, C. Moreno, A. Alonso, T. Liberman-Kipnis, W. D. da Silva, and A. Caballero. 2011. Distribution of the HLA Class II Frequency Alleles in Patients with Leprosy from the Mid-West of Brazil. *International Journal of Immunogenetics* 38: 255–258.

Lavania, M., R. S. Jadhav, R. P. Turankar, V. S. Chaitanya, M. Singh, and U. Sengupta. 2013. Single Nucleotide Polymorphisms Typing of *Mycobacterium leprae* Reveals Focal Transmission of Leprosy in High Endemic Regions of India. *Clinical Microbiology and Infection* 19: 1058–1062.

Lavania, M., K. Katoch, V. M. Katoch, A. K. Gupta, D. S. Chauhan, R. Sharma, R. Gandhi, V. Chauhan, G. Bansai, P. Sachan, et al. 2008. Detection of Viable *Mycobacterium leprae* in Soil Samples: Insights into Possible Sources of Transmission of Leprosy. *Infection Genetics and Evolution* 8: 627–631.

Lavania, M., K. Katoch, P. Sachan, A. Dubey, S. Kapoor, M. Kashyap, D. S. Chauhan, H. B. Sungh, V. D. Sharma, R. S. Jadhav, and V. M. Katoch. 2006. Detection of *Mycobacterium leprae* DNA from Soil Samples by PCR Targeting RLEP Sequences. *Journal of Communicable Diseases* 38: 269–273.

Lavania, M., K. Katoch, R. Sharma, P. Sharma, R. Das, A. K. Gupta, D. S. Chauhan, and V. M. Katoch. 2011. Molecular Typing of *Mycobacterium leprae* Strains from Northern India Using Short Tandem Repeats. *Indian Journal of Medical Research* 133: 618–626.

Lavania, M., R. P. Turankar, S. Karri, V. S. Chaitanya, U. Sengupta, and R. S. Jadhav. 2013. Cohort Study of the Seasonal Effect on Nasal Carriage and the Presence of *Mycobacterium leprae* in an Endemic Area in the General Population. *Clinical Microbiology and Infection* 19: 970–974.

Lavania, M., R. Turankar, I. Singh, A. Nigam, and U. Sengupta. 2014. Detection of *Mycobaterium gilvum* First Time from the Bathing Water of Leprosy Patients from Purulia, West Bengal. *International Journal of Mycobacteriology* 3: 286–289.

Law, A. S. 2012. *Kalaupapa: A Collective Memory*. Honolulu: University of Hawaii Press.

Lechat, M. F. 1961. Mutilations in Leprosy. *Tropical and Geographical Medicine* 13: 99–103.

———. 1981. The International Leprosy Association at 50 Years. *International Journal of Leprosy and Other Mycobacterial Diseases* 49: 60–64.

———. 1999. The Paleoepidemiology of Leprosy: An Overview. *International Journal of Leprosy and Other Mycobacterial Diseases* 67: 460–470.

Lechat, M. F., and J. Chardome. 1955. Altérations Radiologiques des Os de la Face chez le Lépreux Congolais. *Annales Société Belge Medicine Tropicale* 35: 603–611.

Lee, C. 2006. Changing Faces: Leprosy in Anglo-Saxon England. In *Conversion and Colonization in Anglo-Saxon England*, edited by C. E. Kartov and N. Howe, 59–81. Essays in Anglo-Saxon Studies vol. 2. Tempe: Center for Medieval and Renaissance Studies, Arizona State University.

Lee, F., and A. Boylston. 2008a. Infection: Tuberculosis and Other Infectious Diseases. In *"Lepers outside the Gate": Excavations at the Cemetery of the Hospital of St James and St Mary Magdalene, Chichester, 1986–87 and 1993*, edited by J. R. Magilton, F. Lee, and A.

Boylston, 218–228. Council for British Archaeology Research Report 158 and Chichester Excavations vol. 10. Bootham: Council for British Archaeology.

———. 2008b. Other Pathological Conditions. In *"Lepers outside the Gate": Excavations at the Cemetery of the Hospital of St James and St Mary Magdalene, Chichester, 1986–87 and 1993*, edited by J. R. Magilton, F. Lee, and A. Boylston, 252–259. Council for British Archaeology Research Report 158 and Chichester Excavations vol. 10. Bootham: Council for British Archaeology.

Lee, F., and J. Magilton. 1989. The Cemetery of the Hospital of St James and St Mary Magdalene, Chichester—A Case Study. *World Archaeology* 21: 273–282.

———. 2008a. Discussion. In *"Lepers outside the Gate": Excavations at the Cemetery of the Hospital of St James and St Mary Magdalene, Chichester, 1986–87 and 1993*, edited by J. R. Magilton, F. Lee, and A. Boylston, 263–269. Council for British Archaeology Research Report 158 and Chichester Excavations vol. 10. Bootham: Council for British Archaeology.

———. 2008b. Physical Anthropology. In *"Lepers outside the Gate": Excavations at the Cemetery of the Hospital of St James and St Mary Magdalene, Chichester, 1986–87 and 1993*, edited by J. R. Magilton, F. Lee, and A. Boylston, 150–156. Council for British Archaeology Research Report 158 and Chichester Excavations vol. 10. Bootham: Council for British Archaeology.

Lee, F., and K. Manchester. 2008. Leprosy: A Review of the Evidence on the Chichester Sample. In *"Lepers outside the Gate": Excavations at the Cemetery of the Hospital of St James and St Mary Magdalene, Chichester, 1986–87 and 1993*, edited by J. R. Magilton, F. Lee, and A. Boylston, 208–217. Council for British Archaeology Research Report 158 and Chichester Excavations vol. 10. Bootham: Council for British Archaeology.

Lee, O. Y.-C. 2008. Lipid Biomarkers in the Diagnosis of Ancient and Modern Tuberculosis and Leprosy. PhD thesis, University of Birmingham.

Lee, O. Y.-C., I. D. Bull, E. Molnár, A. Marcsi, G. Pálfi, H. D. Donoghue, G. S. Besra, and D. E. Minnikin. 2012. Integrated Strategies for the Use of Lipid Biomarkers in the Diagnosis of Ancient Mycobacterial Disease. In *Proceedings of the 12th Annual Conference of the British Association of Biological Anthropology and Osteoarchaeology. Department of Archaeology and Anthropology, University of Cambridge 2010*, edited by P. D. Mitchell and J. Buckberry, 63–69. British Archaeological Reports International Series 2380. Oxford: Archaeopress.

Le Grand, A. 1997. Women and Leprosy: A Review. *Leprosy Review* 68: 203–211.

Leiker, D. L. 1977. On the Mode of Transmission of *Mycobacterium leprae*. *Leprosy Review* 48: 9–16.

———. 1980. Epidemiology of Leprosy in the Netherlands. *Quaderni di Cooperazione Sanitaria* 1: 60–64.

Leiker, D. L., Y. Otsyula, and M. Ziedses des Plantes. 1968. Leprosy and Tuberculosis in Kenya. *Leprosy Review* 39: 79–83.

Leininger, J. R., K. J. Donham, and W. M. Meyers. 1980. Leprosy in a Chimpanzee: Postmortem Lesions. *International Journal of Leprosy and Other Mycobacterial Diseases* 48: 414–421.

Leitman, T., T. Porco, and S. Blower. 1997. Leprosy and Tuberculosis: The Epidemiological Consequences of Cross-Immunity. *American Journal of Public Health* 87: 1923–1927.

Lejbkowicz, F., B. Tsilman, R. Wexler, and H. I. Cohen. 2001. Leprosy in Israel: An Imported Disease—The Support of Histopathological Examination for Its Detection. *Acta Histochemica* 103: 433–436.

Leloir, H. 1886. *Traité Pratique et Théorétique de la Lêpra*. Paris.

Leprosy Museum. 2003. *The Leprosy Museum, St Jørgens Hospital*. Bergen: The Leprosy Museum.

Lethbridge, T. C. 1931. The Anglo-Saxon Cemetery at Burwell, Cambridgeshire. *Proceedings of the Cambridge Antiquarian Society Quarto Publications*3: 47–70.

Leung, A. K. C. 2009. *Leprosy in China*. New York: Columbia University Press.

Levick, P. N.d. Bartlemas Chapel Excavations 22 September–4 November 2011. Human Remains—Preliminary Assessment. Unpublished manuscript.

Levy, L., C. C. Shepard, and P. Fasal. 1976. The Bactericidal Effect of Rifampicin on *M. leprae* in Man: (a) Single Doses of 600, 9900 and 1200mg; and (b) Daily Doses of 300mg. *International Journal of Leprosy and Other Mycobacterial Diseases* 44: 183–187.

Lewis, G. 1987. A Lesson from Leviticus: Leprosy. *Journal of the Royal Anthropological Institute* 22: 593–612.

Lewis, M. E. 2002. Infant and Childhood Leprosy: Present and Past. In *The Past and Present of Leprosy: Archaeological, Historical, Palaeopathological and Clinical Approaches*, edited by C. A. Roberts, M. E. Lewis, and K. Manchester, 163–170. British Archaeological Reports International Series 1054. Oxford: Archaeopress.

———. 2004. Endocranial Lesions in Non-Adult Skeletons: Understanding Their Aetiology. *International Journal of Osteoarchaeology* 14: 82–87.

———. 2008. The Children. In *"Lepers outside the Gate": Excavations at the Cemetery of the Hospital of St James and St Mary Magdalene, Chichester, 1986–87 and 1993*, edited by J. R. Magilton, F. Lee, and A. Boylston, 174–187. Council for British Archaeology Research Report 158 and Chichester Excavations vol. 10. Bootham: Council for British Archaeology.

———. 2018. *The Paleopathology of Children: Identification of Pathological Conditions in the Human Skeletal Remains of Non-Adults*. London: Academic Press.

Lewis, M. E., C. A. Roberts, and K. Manchester. 1995a. A Comparative Study of the Prevalence of Maxillary Sinusitis in Medieval Urban and Rural Populations in Northern England. *American Journal of Physical Anthropology* 98(4): 497–506.

———. 1995b. Inflammatory Bone Changes in Leprous Skeletons from the Medieval Hospital of St. James and St. Mary Magdalene, Chichester, England. *International Journal of Leprosy and Other Mycobacterial Diseases* 63(1): 77–85.

Lie, H. P. 1929. Why Is Leprosy Declining in Norway? *Transactions of the Royal Society of Tropical Medicine and Hygiene* 22: 357–366.

———. 1938. On Leprosy in the Bible. *Leprosy Review* 9: 25–67.

Likovsky, J., M. Urbanová, M. Hájek, V. Černy, and P. Čech. 2006. Two Cases of Leprosy from Žatec (Bohemia) Dated to the Turn of the 12th Century and Confirmed by DNA Analysis for *Mycobacterium leprae*. *Journal of Archaeological Science* 33: 1276–1283.

Lima, M. C. V., F. R. Barbosa, D. C. M. D. Santos, R. D. D. Nascimento, and S. S. P. D'Azevedo. 2018. Practices for Self-Care in Hansen's Disease: Face, Hands and Feet. *Revista Gaúcha Enfermagem* 39: e2018045.

Linderholm, A., and A. Kjellstrom. 2011. Stable Isotope Analysis of a Medieval Skeletal

Sample Indicative of Systemic Disease from Sigtuna, Sweden. *Journal of Archaeological Science* 38: 925–933.

Lira, K. B., J. J. Leite, D. Castelo Branco de Souza Collares Maia, M. Freitas Rde, and A. R. Feijão. 2012. Knowledge of the Patients Regarding Leprosy and Adherence to Treatment. *Brazilian Journal of Infectious Diseases* 16: 472–5.

Little, K. L. 1943. A Study of a Series of Human Skulls from Castle Hill, Scarborough. *Biometrika* 33: 25–35.

Lockwood, D. N. J. 2004. Leprosy. In *Rook's Textbook of Dermatology*, 7th ed., vol. 1, edited by T. Burns, S. Breathnach, N. Cox, and C. Griffiths, 29.1–29.21. Oxford: Blackwell Science Ltd.

Lockwood, D. N. J., and S. N. Lambert. 2011. Human Immunodeficiency Virus and Leprosy: An Update. *Dermatologic Clinic* 29: 125–128.

López Flores, I., and F. Barrionuevo Contreras. 2009. Dos Nuevos Casos de Lepra Procedentes de la Necrópolis Islámica de Jerez de La Frontera (Cádiz). In *Investigaciones Histórico-Médicas Sobre Salud y Enfermedad en el Pasado: Actas del IX Congreso Nacional de Paleopatologia Morella (Casetlló), 26–29 septiembre de 2007*, edited by I. López Flores and F. Barrionuevo Contreras 575–584. Valencia: Sociedad Española de Paleopatología Valenciana.

Lorentz, K. O. 2011. Cyprus. In *The Routledge Handbook of Archaeological Human Remains and Legislation*, edited by N. Márquez-Grant and L. Fibiger, 99–111. London: Routledge.

Lowe, J. 1947. Comments on the History of Leprosy. *Leprosy Review* 18: 54–55.

Lowe, J., and F. McNulty. 1953a. Tuberculosis and Leprosy: Immunological Studies. *Leprosy Review* 24: 61–90.

———. 1953b. Tuberculosis and Leprosy: Immunological Studies in Healthy Persons. *British Medical Journal* 2(4836): 579–584.

Lumpkin, L. R., G. F. Cox, and J. E. Wolf Jr. 1983. Leprosy in Five Armadillo Handlers. *Journal of the American Academy of Dermatology* 9: 899–903.

Lunt, D. A. 1996. Dental and Anatomical Reports: The Dentitions. In E. Proudfoot, ed., Excavations at the Long Cist Cemetery on the Hallow Hill, St Andrews, Fife 1975–1977. *Proceedings of the Society of Antiquaries of Scotland* 126: 424–429.

———. 1997. The Dentitions. In *St Mary's Church, Kirk Hill, St Andrews*, edited by M. J. Rains and D. W. Hall, 131–137. Glenrothes: Tayside and Fife Archaeological Committee.

———. 2013. The First Evidence for Leprosy in Early Mediaeval Scotland: Two Individuals from Cemeteries in St Andrews, Fife, Scotland with Evidence for Normal Burial Treatment. *International Journal of Osteoarchaeology* 23: 310–318.

Lu'ong, K. V. Q., and L. T. H. Nguyễn. 2012. Role of Vitamin D in Leprosy. *American Journal of Medical Sciences* 343: 471–482.

Lustosa, A. A., L. T. Noqueirs, J. I. Pedrosa, J. B. Teles, and V. Campelo. 2011. The Impact of Leprosy on Health-Related Quality of Life. *Revista da Sociedade Brasileira de Medicina Tropical* 44: 621–626.

Lynnerup, N., and J. Boldsen. 2012. Leprosy (Hansen's Disease). In *A Companion to Paleopathology*, edited by A. L. Grauer, 458–471. Oxford: Wiley-Blackwell.

MacArthur, W. P. 1925. Some Notes on Old Time Leprosy. *Journal of the Royal Army Medical Corps* 45: 410–422.

Magilton, J. 2008a. Leprosy, Lepers and Their Hospitals. In *"Lepers outside the Gate": Exca-*

vations at the Cemetery of the Hospital of St James and St Mary Magdalene, Chichester, 1986-87 and 1993,* edited by J. R. Magilton, F. Lee, and A. Boylston, 9-26. Council for British Archaeology, Research Report 158 and Chichester Excavations vol. 10. Bootham: Council for British Archaeology.

———. 2008b. The Cemetery. In *"Lepers outside the Gate": Excavations at the Cemetery of the Hospital of St James and St Mary Magdalene, Chichester, 1986-87 and 1993,* edited by J. R. Magilton, F. Lee, and A. Boylston, 84-132. Council for British Archaeology Research Report 158 and Chichester Excavations vol. 10. Bootham: Council for British Archaeology.

———. 2008c. A. History of the Hospital. In *"Lepers outside the Gate": Excavations at the Cemetery of the Hospital of St James and St Mary Magdalene, Chichester, 1986-87 and 1993,* edited by J. R. Magilton, F. Lee, and A. Boylston, 57-68. Council for British Archaeology Research Report 158 and Chichester Excavations vol. 10. Bootham: Council for British Archaeology.

———. 2008d. Medieval and Early Modern Cemeteries in England: An Introduction. In *"Lepers outside the Gate": Excavations at the Cemetery of the Hospital of St James and St Mary Magdalene, Chichester, 1986-87 and 1993,* edited by J. R. Magilton, F. Lee, and A. Boylston, 27-48. Council for British Archaeology Research Report 158 and Chichester Excavations vol. 10. Bootham: Council for British Archaeology.

Magilton, J., and F. Lee. 1989. Leper Hospital of St James and St Mary Magdalene, Chichester. In *Burial Archaeology: Current Research, Methods and Developments,* edited by C. A. Roberts, F. Lee, and J. Bintliff, 249-265. British Archaeological Reports British Series 211. Oxford: BAR.

Magilton, J. R., F. Lee, and A. Boylston, eds. 2008. *"Lepers outside the Gate": Excavations at the Cemetery of the Hospital of St James and St Mary Magdalene, Chichester, 1986-87 and 1993.* Council for British Archaeology Research Report 158 and Chichester Excavations vol. 10. Bootham: Council for British Archaeology.

Magnússon, Þ. 1970. Skýrsla um Þjóðminjasafnið 1969. *Árbók Hins Íslenzka Fornleifafélags* 1969: 127-146.

Major, R. H., ed. 1949. A Thirteenth Century Clinical Description of Leprosy. *Journal of the History of Medicine and Allied Science*s 4: 237-239.

Malik, A. N., R. W. Morris, and T. J. Ffytche. 2011. The Prevalence of Ocular Complications in Leprosy Patients Seen in the United Kingdom over a Period of 21 Years. *Eye (London)* 25: 740-745.

Malin, T., and J. Hines, eds. 1998. *The Anglo-Saxon Cemetery at Edix Hill (Barrington A), Cambridgeshire,.* Council for British Archaeology Research Report 112. York: Council for British Archaeology.

Manchester, K. 1981. Study Groups: Leprosy. *Paleopathology Association Newsletter* 35: 4-5.

———. 1983. *The Archaeology of Disease.* Bradford: University of Bradford.

———. 1984. Tuberculosis and Leprosy in Antiquity: An Interpretation. *Medical History* 28: 162-173.

———. 1989. Bone Changes in Leprosy: Pathogenesis and Palaeopathological Diagnostic Criteria. University of Bradford, Department of Archaeological Sciences. Unpublished manuscript.

———. 1991. Tuberculosis and Leprosy: Evidence for Interaction of Disease. In *Human*

Paleopathology: Current Syntheses and Future Options, edited by D. J. Ortner and A. C. Aufderheide, 25–35. Washington, DC: Smithsonian Institution Press.

———. 1993. Unusual Pathological Condition in the Lower Extremities of a Skeleton from Ancient Israel. *American Journal of Physical Anthropology* 91: 249–250.

———. 2002. Infective Bone Changes of Leprosy. In *The Past and Present of Leprosy: Archaeological, Historical, Palaeopathological and Clinical Approaches,* edited by C. A. Roberts, M. E. Lewis, and K. Manchester, 69–72. British Archaeological Reports International Series 1054. Oxford: Archaeopress.

———. 2014. Bone Changes in Leprosy: Pathogenesis and Palaeopathological Diagnostic Criteria. University of Bradford, Department of Archaeological Sciences. Unpublished manuscript.

Manchester, K., and C. A. Roberts. 1986. Palaeopathological Evidence of Leprosy and Tuberculosis in Britain. SERC Report for Grant Number 337.367. Unpublished.

Manchester, K., R. Storm, and A. Ogden. 2012. Workshop on Leprosy. European Meeting of the Paleopathology Association, Lille, France, August 27–29.

Mandal, D., A. H. Reja, N. Biswas, P. Bhattacharyya, P. K. Patra, and B. Bhattacharyya. 2015. Vitamin D Receptor Expression Levels Determine the Severity and Complexity of Disease Progression among Leprosy Reaction Patients. *New Microbes and New Infections* 6: 35–39.

Manes, C. 2013. *Out of the Shadow of Leprosy: The Carville Letters and Stories of the Landry Family.* Jackson: University Press of Mississippi.

Marchoux, N. E. 1912. The Problem of Leprosy. *British Medical Journal* 2: 1191M.

Marcombe, D. 2003. *Leper Knights: The Order of St Lazarus of Jerusalem in England, c. 1150–1544.* Woodbridge, Suffolk: The Boydell Press.

Marcsik, A. 1998. Az Ópusztaszeri Csontvázanyag Paleopatológiás Elváltozásai. In Ópusztaszer-Monostor *Lelőhely Antropológiai Leletei,* edited by G. L. Farkas, 97–105. Szeged: JATE Embertani Tanszéke Kiadványa.

———. 2001. A Csengelei Sírok Embertani Vizsgálata. In *A Csengelei Kunok ura és Népe,* edited by F. Horváth, 326–330. Budapest: Archaolingua.

———. 2003. Ibrány-Esbo Halom 10–11. Századi Humán Csontvázanyagának Paleopatológiai Jellegzetességei. In *A Rétköz Honfoglalás* és *Árpád-kori Emlékanyaga,* edited by E. Istvánovits, 392–399. Nyíregyháza, Hungary: Jósa András Múzeum, Magyar NemzetiMúőzeum, MTA Régészeti Intézet kiadványa.

Marcsik, A., E. Fóthi, and H. D. Donoghue. 2006. Leprosy or Syphilis—A Differential Diagnostic Palaeopathological Problem. Paper presented at the 16th European meeting of the Paleopathology Association, Santorini, Greece,.

Marcsik, A., E. Molnár, and B. Ősz. 2007. *Specifikus Fertőző Megbetegedések Csontelváltozásai Történeti Népesség Körében.* Szeged, Hungary: JETE Press.

Marcsik, A., G. Pálfi, L. Márk, and E. Molnár. 2010. Cases of Leprosy and Tuberculosis in an 8th–9th Century Cemetery from Hungary. Poster presented at the 18th European meeting of the Paleopathology Association, Vienna, Austria.

Mariotti, V., O. Dutour, M. G. Belcastro, F. Facchini, and P. Brasili. 2005. Probable Early Presence of Leprosy in Europe in a Celtic Skeleton of the 4th–3rd Century BC. *International Journal of Osteoarchaeology* 15: 311–325.

Mark, S. 2002. Alexander the Great, Seafaring, and the Spread of Leprosy. *Journal of the History of Medicine* 57: 285–311.

———. 2017. Early Human Migrations (ca. 13,000 Years Ago) or Postcontact Europeans for the Earliest Spread of *Mycobacterium leprae* and *Mycobacterium lepromatosis* to the Americas. *Interdisciplinary Perspectives on Infectious Diseases* 2017: Article 6491606. https://www.hindawi.com/journals/ipid/2017/6491606/.

Marks, S. C., and G. Grossetete. 1988. *Facies Leprosa:* Resorption of Maxillary Anterior Alveolar Bone and the Anterior Nasal Spine in Patients with Lepromatous Leprosy in Mali. *International Journal of Leprosy and Other Mycobacterial Diseases* 56: 21–26.

Márquez-Grant, N., and L. Fibiger, eds. 2011. *The Routledge Handbook of Archaeological Human Remains and Legislation: An International Guide to Laws and Practice in the Excavation and Treatment of Archaeological Human Remains.* London: Routledge.

Marrakchi, H., M. A. Lanéelle, and M. Daffé. 2014. Mycolic Acids: Structures, Biosynthesis, and Beyond. *Chemistry and Biology* 21: 67–85.

Martin, A. 1921. The Representation of Leprosy and of Lepers in Minor Art, Particularly in Germany. *Urologic and Cutaneous Review* 25: 445–453.

Martin, B. 1963. *Miracle at Carville: The Story of One Girl's Triumph over the World's Most Feared Malady.* New York: Image Books.

Martinez, T. S., M. M. Figueira, A. V. Costa, M. A. Gonçalves, L. R. Goulart, and I. M. Goulart. 2011. Oral Mucosa as a Source of Mycobacterium leprae Infection and Transmission, and Implications of Bacterial DNA Detection and the Immunological Status. *Clinical Microbiology and Infection* 17: 1653–1658.

Masaki, T., J. Qu, J. Cholewa-Waclaw, K. Burr, A. Rambukkana, and R. Raaum. 2013. Reprogramming Adult Schwann Cells to Stem Cell-Like Cells by Leprosy Bacilli Promotes Dissemination of Infection. *Cell* 152: 51–67.

Masland, A. L., E. A. Bachen, S. Cogen, B. Robin, and S. B. Manuck. 2002. Stress, Immune Reactivity and Susceptibility to Infectious Disease. *Physiology and Behaviour* 77: 711–716.

Massone, C., A. M. Brunasso, S. Noto, T. M. Campbell, A. Clapasson, and E. Nunzi. 2012. Imported Leprosy in Italy. *Journal of the European Academy of Dermatology and Venereology* 26: 999–1006.

Matos, V. M. J. 2009. Odiagnóstico Retrospective da Lepra: Complimentaridade Clínica e Paleopatológica no Arquivo Médico do Hospital-Colónia Rovisco Pais (Século X. X., Tocha, Portugal) e na Colecção de Esqueletos da Leprosaria Medieval de St Jørgen's (Odense, Dinamarca) [The Retrospective Diagnosis of Leprosy: Clinical and Paleopathological Complementarities in the Medical Files from the Rovisco Pais Hospital-Colony (20th Century, Tocha, Portugal) and in the Skeletal Collection from the Medieval Leprosarium of St. Jorgen's (Odense, Denmark)]. PhD thesis, Universidade de Coimbra, Faculdade de Ciências e Technologia, Coimbra.

Matos, V. M. J., G. Magno, A. Amoroso, and S. Garcia. 2017. The Case of a Portuguese Postman Who Died from Leprosy in 1931 and the Paleopathological Analysis of His Skeleton. Poster presented at the 44th annual North American meeting of the Paleopathology Association, New Orleans, LA, April 17–19.

Matos, V. M. J., and A. L. Santos. 2013a. Diagnóstico, Terapêutica Investigãço Cientifica. In *Leprosaría Nasional: Modernidade e Ruína no Hospital-Colónia Rovisco Pais,* edited by

P. Providência, V. M. J. Matos, A. L. Santos, S. Xavier, L. Quintais, and E. Brás, Porto: Dafne Editoria. *Coleçes de Arquitectua* 24: 99–123.

———. 2013b. Leprogenic Odontodysplasia: New Evidence from the St Jørgen's Medieval Leprosarium Cemetery (Odense, Denmark). *Anthropological Science* 121: 43–47.

———. 2013c. Some Reflections Considering the Paleopathological Diagnosis of Lepromatous and Tuberculoid Leprosy on Human Skeletal Remains. Poster presented at the 18th International Leprosy Congress, Brussels, Belgium.

Matsuoka, M. 2009. Recent Advances in the Molecular Epidemiology of Leprosy. *Nihon Hansenbyo Gakkai Zasshi* 78: 67–73.

Matsuoka, M., T. Budiawan, T. Mukai, and S. Izumi. 2013. *Mycobacterium leprae* in Daily Used Water in Endemic Area: Its Quantification and Evaluation of the Viability. Poster presented at the 18th International Leprosy Congress, Brussels, Belgium.

Matsuoka, M., S. Izumi, T. Budiawan, N. Natak, and K. Saeki. 1999. *M. leprae* DNA in Daily Using Water as a Possible Source of Leprosy Infection. *Indian Journal of Leprosy* 71: 61–67.

Matsuoka, M., L. Zhang, T. Budiawan, K. Saeki, and S. Izumi. 2004. Genotyping of *Mycobacterium leprae* on the Basis of the Polymorphisms of TTC Repeats for Analysis of Leprosy Transmission. *Journal of Clinical Microbiology* 42: 741–745.

Mays, S. A. 1989. The Anglo-Saxon Human Bone from School Street, Ipswich, Suffolk. English Heritage, Ancient Monuments Laboratory Report 115/89. Unpublished.

———. 1991. The Medieval Burials from the Blackfriars Friary, School Street, Ipswich, Suffolk. English Heritage, Ancient Monuments Laboratory Report 16/91. Unpublished.

———. 2007. The Human Remains. In *Wharram: A Study of Settlement on the Yorkshire Wolds*, Vol. 11, *The Churchyard,* edited by S. A. Mays, C. Harding, and C. Heighway, 77–192. York University Archaeological Publications 13. York: University of York.

———. 2012. The Relationship between Paleopathology and the Clinical Sciences. In *A Companion to Paleopathology,* edited by A Grauer, 285–309. Oxford: Wiley-Blackwell.

Mays, S. A., C. Harding, and C. Heighway, eds. 2007. *Wharram: A Study of Settlement on the Yorkshire Wolds*. Vol. 11, *The Churchyard*. York University Archaeological Publications 13. York: University of York.

Mays, S. A., and G. M. Taylor. 2003. The First Prehistoric Case of Tuberculosis from Britain. *International Journal of Osteoarchaeology* 13: 189–196.

Mazini, P. S., H. V. Alves, P. G. Reis, A. P. Lopes, A. M. Sell, M. Santos-Rosa, J. E. Visentainer, and P. Rodrigues-Santos. 2016. Gene Association with Leprosy: A Review of Published Data. *Frontiers of Immunology* 6(January 12): 658.

McCarthy, D. D., and J. Numa. 1962. Leprosy in the Cook Islands. *New Zealand Medical Journal* 61: 77–85.

McComish, J. 2015. *Roman, Anglian and Anglo-Scandinavian Activity and a Medieval Cemetery on Land at the Junction of Dixon Lane and George Street, York.* Report no. AYW9. York: York Archaeological Trust. Accessed November 2018. https://www.yorkarchaeology.co.uk/wp-content/uploads/2015/05/AYW9-Dixon-Lane-and-George-Street1.pdf.

McComish, J. M., G. Millward, and A. Boyle. 2017. *The Medieval Cemetery of St Leonard's Leper Hospital at Midland Road, Peterborough.* Report Number YAT 11/2017. York: York Archaeological Trust. Accessed November 2019. https://www.yorkarchaeology.co.uk/wp-content/uploads/2017/07/YAT-AY11-Midland-Road-Peterborough-1.pdf.

McDonough, A. 2008. *The Man with Leprosy*. Unley, Australia: Lost Sheep Resources Pty.

McDougall, A. C., R. J. W. Rees, A. G. M. Weddell, and M. W. Kanan. 1975. The Histopathology of Lepromatous Leprosy in the Nose. *Journal of Pathology* 115: 215–126.

McElroy, A., and P. K. Townsend. 2009. *Medical Anthropology in Ecological Perspective*. Boulder: Westview Press.

McIntyre, L. M. 2016. The York 113: Osteological Analysis of the 10 Mass Graves from Fishergate, York. *Journal of Conflict Archaeology* 11: 115–134.

McIntyre, L., and A. T. Chamberlain. 2009. Osteological Analysis of the Multi-Period Human Skeletal Assemblage from All Saint's, Fishergate, York. Unpublished Report for On Site Archaeology.

McKinley, J. I. Forthcoming. Human Bone from Horton Kingsmead Quarry, Berkshire (54635, 54636, 54637, 71800). In *Horton Kingsmead Quarry*. Vol. 1, *2003-2009 Excavations*, edited by G. Chaffey, A. J. Barclay, and R. Pelling. Wessex Archaeology Report 32. Salisbury: Wessex Archaeology.

McVaugh, M. R. 1993. *Medicine before the Plague: Practitioners and Their Patients in the Crown of Aragon, 1285-1345*. Cambridge: Cambridge University Press.

Medeiros, S., M. G. Catorze, and M. R. Vieira. 2009. Hansen's Disease in Portugal: Multibacillary Patients Treated between 1988 and 2003. *Journal of the European Academy of Dermatology and Venereology* 23: 29–35.

Medley, G. F., R. E. Crump, and D. N. J. Lockwood. 2017. Interpreting Data in Policy & Control: The Case of Leprosy. *Indian Journal of Medical Research* 145: 1–3.

Meima, A., L. M. Irgens, G. J. van Oortmarssen, J. H. Richardus, and J. D. K. Habbema. 2002. Disappearance of Leprosy from Norway: An Exploration of Critical Factors Using an Epidemiological Modelling Approach. *International Journal of Epidemiology* 31: 991–1000.

Meima, A., W. C. Smith, G. J. van Oortmarssen, J. H. Richardus, and J. D. Hebbema. 2004. The Future Incidence of Leprosy: A Scenario Analysis. *Bulletin of the World Health Organization* 82: 373–380.

Melo, L., A. M. Silva, V. Matos, A. L. Santos, and C. Ferreira. 2018. Lepra no Norte de Portugal: Evidências num Indivíduo do Sexo Masculino Exumado do Cemitério da Igreja Paroquial de Travanca (Santa Maria da Feira). Poster Presented at the 6th Jornadas Portuguesas de Paleopatologia (30 November–1 December, University of Coimbra, Portugal.

Melsom, R., M. E. Duncan, and G. Bjune. 1980. Immunoglobulin Concentration in Mothers with Leprosy and in Healthy Controls and Their Babies at the Time of Birth. *Leprosy Review* 51: 19–28.

Melsom, R., M. Harboe, and M. E. Duncan. 1982. IgA and IgM Anti-*M. leprae* Antibodies in Babies of Leprosy Mothers during the First Two Years of Life. *Clinical and Experimental Immunology* 49: 532–542.

Melsom, R., M. Harboe, M. E. Duncan, and H. Bergsvik. 1981. IgA and IgM Antibodies against *Mycobacterium leprae* in Cord Sera and in Patients with Leprosy: An Indication of Intrauterine Infection in Leprosy. *Scandinavian Journal of Immunology* 14: 343–352.

Mendum, T. A., V. J. Schuenemann, S. Roffey, G. M. Taylor, H. Wu, P. Singh, K. Tucker, J. Hinds, S. T. Cole, A. M. Kierzek, et al. 2014. *Mycobacterium leprae* Genomes from a British Medieval Leprosy Hospital: Towards Understanding an Ancient Epidemic. *BMC Genomics* 15(April 8): Article 270.

Meredith, A., J. Del Pozo, S. Smith, E. Milne, K. Stevenson, and J. McLuckie. 2014. Leprosy in Red Squirrels in Scotland. *Veterinary Record* 175: 285–286.

Metzl, J. M., and A. Kirkland, eds. 2011. *Against Health: How Health Became the New Morality.* London: New York University Press.

Meyer, W. H. 1955. History of Leprosy in Louisiana. *Journal of the Louisiana State Medical Society* 107: 359–366.

Meyers, W. M., C. H. Binford, G. P. Walsh, R. H. Wolf, B. J. Gormus, L. N. Martin, and P. J. Gerone. 1984. Animal Models in Leprosy. In *Microbiology—1984,* edited by L. Leive and D. Schlessinger, 307–311. Washington DC: American Society for Microbiology.

Meyers, W. M., G. P. Walsh, H. L. Brown, C. H. Binford, G. D. Imes Jr., T. L. Hadfield, C. J. Schlagel, Y. Fukunishi, P. J. Gerone, R. H. Wolf, et al. 1985. Leprosy in a Mangabey Monkey—Naturally Acquired Infection. *International Journal of Leprosy and Other Mycobacterial Diseases* 53: 1–14.

Michman, J., and F. Sager. 1957. Changes in the Anterior Nasal Spine and the Alveolar Process of the Maxillary Bone in Leprosy. *International Journal of Leprosy and Other Mycobacterial Diseases* 25: 217–222.

Miles, J. 1997. *Infectious Diseases: Colonising the Pacific?* Dunedin: University of Otago Press.

Millard, L. G., and J. A. Cotterill. 2004. Psychocutaneous disorders. In *Rook's Textbook of Dermatology,* 7th ed., vol. 1, edited by T. Burns, S. Breathnach, N. Cox, and C. Griffiths, 61.1–61.41. Oxford: Blackwell Science Ltd.

Miller, T. 1913. Early Diagnosis of a Case of Leprosy Much Assisted by the X-Rays. *Lancet* (July 26): 219.

Mira, M. T., A. Alcais, N. van Thuc, V. H. Thai, N. T. Huong, N. N. Ba, A. Verner, T. J. Hudson, L. Abel, and E. Schurr. 2003. Chromosome 6q25 Is Linked to Susceptibility to Leprosy in a Vietnamese Population. *Nature Genetics* 33: 412–415.

Mitchell, P. D. 1993. Leprosy and the Case of King Baldwin IV of Jerusalem: Mycobacterial Disease in the Crusader States of the 12th and 13th Centuries. *International Journal of Leprosy and Other Mycobacterial Diseases* 61: 283–291.

———. 2000a. An Evaluation of the Leprosy of King Baldwin IV of Jerusalem in the Context of the Medieval World. In *The Leper King and His Heirs: Baldwin IV and the Crusader Kingdom of Jerusalem,* edited by B. Hamilton, 245–258. Cambridge: Cambridge University Press.

———. 2000b. The Evolution of Social Attitudes to the Medical Care of Those with Leprosy within the Crusader States. In *Histoire et Archéologique de la Lèpre et des Lépreux en Europe et en Méditerranée de l'Antiquité aux Temps Modernes,* edited by B. Tabuteau, 21–28. Rouen: University of Rouen.

———. 2002. The Myth of the Spread of Leprosy with the Crusades. In *The Past and Present of Leprosy: Archaeological, Historical, Palaeopathological and Clinical Approaches,* edited by C. A. Roberts, M. E. Lewis, and K. Manchester, 171–178. British Archaeological Reports International Series 1054. Oxford: Archaeopress.

Mitsuda, K. 1952. *Atlas of Leprosy.* Okayama, Japan: Chōtōkai Foundation (Chojyo Kwai) and Nankodo Company.

Mitsuda, K., and M. A. Ogawa. 1937. A Study of 150 Autopsies of Cases of Leprosy. *International Journal of Leprosy and Other Mycobacterial Diseases* 5: 53–60.

Modlin, R. L., and T. H. Rea. 1987. Leprosy: New Insight into an Ancient Disease. *Journal of the American Academy of Dermatology* 17: 1–13.

Mogren, M. 1984. *Spetälska och Spetälskehospital i Norden under Medeltiden.* Lund, Sweden: Institutionen för Medeltidsarkeologi, Lunds Universitet.

Mohanty, P. S., F. Naaz, D. Katara, L. Misba, D. Kumar, D. K. Dwivedi, A. K. Tiwari, D. S. Chauhan, A. K. Bansal, S. P. Tripathy, and K. Katoch. 2016. Viability of *Mycobacterium leprae* in the Environment and Its Role in Leprosy Dissemination. *Indian Journal of Dermatology, Venereology and Leprology* 82: 23–27.

Molesworth, E. H. 1933. The Influence of Natural Selection on the Incidence of Leprosy. *International Journal of Leprosy and Other Mycobacterial Diseases* 1: 265–282.

Møller-Christensen, V. 1952. Case of Leprosy from the Middle Ages of Denmark. *Acta Medica Scandinavica* 142 supplement 266: 101–108.

———. 1953a. Changes in the Maxillary Bone with Leprosy. *International Journal of Leprosy and Other Mycobacterial Diseases* 21: 616–617.

———. 1953b. Location and Excavation of the First Danish Leper Graveyard from the Middle Ages. *Bulletin of the History of Medicine* 17: 112–123.

———. 1953c. *Ten Lepers from Naestved in Denmark.* Copenhagen: Danish Science Press.

———. 1958. *Bogen om Æbelholt Kloster.* Copenhagen: Nationalmuseet.

———. 1961. *Bone Changes in Leprosy.* Copenhagen: Munksgaard.

———. 1963. Skeletfundene fra St Jørgens Kirke i Svendborg. *Fynske Minder* 5: 35–49.

———. 1965. New Knowledge of Leprosy through Palaeopathology. *International Journal of Leprosy and Other Mycobacterial Diseases* 33: 603–610.

———. 1967. Evidence of Leprosy in Earlier Peoples. In *Diseases in Antiquity,* edited by D. R. Brothwell and A. T. Sandison, 295–307. Springfield, IL: Charles Thomas.

———. 1969. Provisional Results of the Examination of the Whole Naestved Leprosy Hospital Churchyard—ab. 1250–1550 A. D. *Nordisk Medicinhistorik Årsbok* Part 4: 29–36.

———. 1974. Changes in the Anterior Nasal Spine and the Alveolar Process of the Maxilla in Leprosy. *International Journal of Leprosy and Other Mycobacterial Diseases* 42: 431–435.

———. 1976. Leprosy and Its Way from the Old to the New World. *Dansk Medicinsk Historisk Årbog:* 107–115.

———. 1978. *Leprosy Changes of the Skull.* Odense: Odense University Press.

Møller-Christensen, V., S. N. Bakke, R. S. Melsom, and E. Waaler. 1952. Changes in the Anterior Nasal Spine and the Alveolar Process of the Maxillary Bone in Leprosy. *International Journal of Leprosy and Other Mycobacterial Diseases* 20: 335–340.

Møller-Christensen, V., and B. Faber. 1952. Leprous Changes in a Material of Medieval Skeletons from the St George's Court, Næstved. *Acta Radiologica* 37: 308–317.

Møller-Christensen, V., and D. R. Hughes. 1962. Two Early Cases of Leprosy in Great Britain. *Man* 287: 177–179.

———. 1966. An Early Case of Leprosy from Nubia. *Man* 62: 177–179.

Møller-Christensen, V., and R. G. Inkster. 1965. Cases of Leprosy and Syphilis in the Osteological Collection of the Department of Anatomy, University of Edinburgh with a Note on the Skull of Robert the Bruce. *Danish Medical Bulletin* 12: 11–18.

Møller-Christensen, V., and W. H. Jopling. 1964. An Examination of the Skulls in the Catacombs of Paris. *Medical History* 8: 187–188.

Molnár, E., H. D. Donoghue, O. Y. Lee, H. H. Wu, G. S. Besra, D. E. Minniki, I. D. Bul,

G. Llewellyn, C. M. Williams, O. Spekker, and G. Pálfi. 2015. Morphological and Biomolecular Evidence for Tuberculosis in 8th Century A. D. Skeletons from Bélmegyer-Csömöki Domb, Hungary. *Tuberculosis* 95: S35–S41.

Molnár, E., A. Marcsik, Z. Bereczki, and H. D. Donoghue. 2006. Pathological Cases from the 7th Century in Hungary. Paper presented at the 16th European meeting of the Paleopathology Association, Santorini, Greece.

Molto, J. E. 2002. Leprosy in Roman Period Skeletons from Kellis 2, Dakhleh, Egypt. In *The Past and Present of Leprosy: Archaeological, Historical, Palaeopathological and Clinical Approaches,* edited by C. A. Roberts, M. E. Lewis, and K. Manchester, 179–192. British Archaeological Reports International Series 1054. Oxford: Archaeopress.

Monot, M., N. Honoré, M. T. Garnier, R. Arao, J.-Y. Coppé, C. Lacroix, S. Sow, J. S. Spencer, R. W. Truman, D. L. Williams, et al. 2005. On the Origin of Leprosy. *Science* 308(5724): 1040–1042.

Monot, M., N. Honoré, T. Garnier, N. Zidane, D. Sherafi, A. Paniz-Mondolfi, M. Matsuoka, G. M. Taylor, H. D. Donoghue, A. Bouwman, et al. 2009. Comparative Genomic and Phylogeographic Analysis of *Mycobacterium leprae. Nature Genetics* 41: 1282–1289.

Montestruce, E., and R. Berdoinneau. 1954. Two New Cases of Leprosy in Infants in Martinique. *Bulletin de la Societé de Pathologic Exotique et de ses Filiales* 47: 781–783. (In French.)

Montgomerie, J. Z. 1988. Leprosy in New Zealand. J. *Polynesian Society* 97: 115–152.

Montiel, R., C. Garcia, M. P. Cañadas, A. Isidro, J. M. Guijo, and A. Malgosa. 2003. DNA Sequences of *Mycobacterium leprae* Recovered from Ancient Bones. *FEMS Microbiology Letters* 226: 413–414.

Morgado de Abreu, M. A., A. M. Roselino, M. Enokihara, S. Nonogaki, L. E. Prestes-Careiro, L. L. Weckx, and M. M. Alchorne. 2014. *Mycobacterium leprae* Is Identified in the Oral Mucosa from Paucibacillary and Multibacillary Leprosy Patients. *Clinical Microbiology and Infection* 20: 59–64.

Morimoto, I. 1995. Paleopathological Aspects of Specific Inflammations. *Japanese Red Cross College of Nursing* 9: 1–7. (In Japanese.)

Morris, A. G., and M. Steyn. 2012. Paleopathological Studies in South Africa. In *The Global History of Paleopathology: Pioneers and Prospects,* edited by J. Buikstra and C. Roberts, 235–242. Oxford: Oxford University Press.

Morrison, A. 2000. A Woman with Leprosy Is in Double Jeopardy. *Leprosy Review* 71: 128–143.

Morrow, H., E. L. Walker, and H. E. Miller. 1922. Experience with Chaulmoogra Oil Derivatives in Treatment of Leprosy. *Journal of the American Medical Association* 79: 434–440.

Moschella, S. L. 2004. An Update on the Diagnosis and Treatment of Leprosy. *Journal of the American Academy of Dermatology* 51: 417–26.

Motta, A. C., K. J. Pereira, D. C. Tarquínio, M. B. Vieira, K. Miyake, and N. T. Foss. 2012. Leprosy Reactions: Coinfections as a Possible Risk Factor. *Clinics (São Paulo)* 67: 1145–1148.

Motta, C. P. 1981. The Epidemiological Situation in the Americas. *Leprosy Review* 52 supplement 1: 61–68.

Mouat, F. J. 1854. Notes on Native Remedies. *Indian Annals of Medical Science* 1: 646–652.

Muir, E. 1957. Relationship of Leprosy to Tuberculosis. *Leprosy Review* 28: 11–19.

Mukherjee, R. 1989. *Vaccines for Leprosy: Present Status and Future Prospects.* Calcutta: Greater Calcutta Leprosy Treatment and Health Education Scheme.

Mull, J. D., C. S. Wood, L. P. Gans, and D. S. Mull. 1989. Culture and Compliance among Leprosy Patients in Pakistan. *Social Science and Medicine* 29: 799–811.

Murdock, J. R., and H. J. Hutter. 1932. Leprosy: A Roentgenological Survey. *American Journal of Roentgenology Radium Therapy and Nuclear Medicine* 28: 598–621.

Murenius, B. 1908. *Boëtius Murenius Acta Visitatoria 1637–1666 Utgitna av Kaarol Österbladh.* Borgå: Suomen Kirkkohistoriallisen Seuran Toimituksia VI.

Murphy, E., and K. Manchester. 1998. Be Thou Dead to the World: Leprosy in Ireland, Evidence from Armoy, Co. Antrim. *Archaeology Ireland* 12: 12–14.

———. 2002. Evidence for Leprosy in Medieval Ireland. In *The Past and Present of Leprosy: Archaeological, Historical, Palaeopathological and Clinical Approaches,* edited by C. A. Roberts, M. E. Lewis, and K. Manchester, 193–200. British Archaeological Reports International Series 1054. Oxford: Archaeopress.

Murto, C., F. Chammartin, K. Schwarz, L. M. da Costa, C. Kaplan, and J. Heukelbach. 2013. Patterns of Migration and Risk Associated with Leprosy among Migrants in Maranhao, Brazil. *PLoS Neglected Tropical Diseases* 7(9): e2422. https://journals.plos.org/plosntds/article?id=10.1371/journal.pntd.0002422.

Naafs, B. 2006. Treatment of Leprosy: Science or Politics? *Tropical Medicine and International Health* 11: 268–278.

Nadkarni, N. J., A. Grugni, M. S. Kini, and M. Balakrishnan. 1988. Childhood Leprosy in Bombay: A Clinico-Epidemiological Study. *Indian Journal of Leprosy* 60: 173–188.

Naik, S. S., U. H. Thakar, A. M. Phrande, and R. Ganapati. 1999. Survey of Leprosy in Unapproachable and Uncovered Areas. *Indian Journal of Leprosy* 71: 333–335.

Nakajo, S. 1914. Über Die Primare Lepra der Neugeborenen (Primary Leprosy in the Newborn). *Japanische Zeitschrift für Dermatologie und Urologie* 14: 1026–1034.

Narayanan, E., K. S. Manja, and B. M. S. Bedi. 1972. Arthropod Feeding Experiments in Lepromatous Leprosy. *Leprosy Review* 43: 188–193.

Narayanan, E., K. S. Manja, and W. F. Kirchheimer. 1972. Occurrence of *Mycobacterium leprae* in Arthropods. *Leprosy Review* 43: 194–198.

Nerlich, A., S. Marlow, and A. Zink. 2006. Molecular Analysis of Leprosy and Tuberculosis in a Skeletal Series from a South Germany Ossuary. Paper presented at the 16th European meeting of the Paleopathology Association, Santorini, Greece.

Nesse, R. M., and G. C. Williams. 1994. *Why We Get Sick: The New Science of Darwinian Medicine.* New York: Vintage Books.

Neves, M. J., M. T. Ferreira, and S. N. Wasterlain. 2012. *Lagos Leprosarium (Portugal): Direct and Indirect Evidence of Disease.* Poster presented at the 19th European meeting of the Paleopathology Association, Lille, France.

Newell, K. W. 1966. An Epidemiologist's View of Leprosy. *Bulletin of the World Health Organization* 34: 827.

Newman, G. 1895. *Prize Essays on Leprosy.* London: New Sydenham Society.

Nicolle, C. 1905. Reproduction Expérimentale de la Lèpre Chez le Singe. *Comptes Rendus de l'Académie des Sciences* 140: 539–542.

Nolan, L., D. Haberling, D. Scollard, R. Truman, A. Rodriguez-Lainz, L. Blum, and D.

Blaney. 2014. Incidence of Hansen's Disease—United States, 1994–2011. *Morbidity and Mortality Monthly Report* 63: 969–972.

Noordeen, S. K. 1993. Leprosy 1962–1902. 4. Epidemiology and Control of Leprosy—A Review of Progress over the Last 30 Years. *Transactions of the Royal Society of Tropical Medicine and Hygiene* 87: 515–517.

Norton, S. A. 1994. Useful Plants of Dermatology I: Hydnocarpus and Chaulmoogra. *Journal of the American Academy of Dermatology* 31: 683–686.

———. 1998. Herbal Medicines in Hawaii from Tradition to Convention. *Hawaii Medical Journal* 57: 382–386.

Noussitou, F. M., H. Sansarricq, and J. Walter. 1976. *Leprosy in Children.* Geneva: World Health Organization.

Nsagha, D. S., E. A. Bamgyove, J. C. Assob, A. L. Njunda, H. L. Kamga, A. C. Bissek Zoung-Kanyi, E. N. Tabah, A. B. Oyediran, and A. K. Njamnshi. 2011. Elimination of Leprosy as a Public Health Problem by 2000 AD: An Epidemiological Perspective. *Pan African Medical Journal* 9: 4.

Núñez, M., M. M.-L. Niskanen, J.-A. Kortelainen, K. Junno, S. Paavola, S. Niinimäki, and M. Modarress. 2011. Finland/Suomi. In *The Routledge Handbook of Archaeological Human Remains and Legislation,* edited by N. Márquez-Grant and L. Fibiger, 139–149. London: Routledge.

Nunn, C. L., S. C. Alberts, C. R. McClain, S. R. Meshnick, T. J. Vision, B. M. Wiegmann, and A. G. Rodrigo. 2015. Linking Evolution, Ecology, and Health: TriCEM. *Bioscience* 65: 748–749.

Nuorala, E., H. D. Donoghue, M. Spigelman, A. Götherström, B. Hårding, L. Grundberg, V. Alexanderson, I. Leden, and K. Lidén. 2004. Diet and Disease in Björned, a Viking-Early Medieval Site in Northern Sweden. In E. Nuorala, *Molecular Palaeopathology: Ancient DNA Analyses of the Bacterial Diseases Tuberculosis and Leprosy.* Theses and Papers in Scientific Archaeology Volume 6. Stockholm University, Archaeological Research Laboratory, Stockholm, Sweden.

Ober, W. B. 1983. Can the Leper Change His Spots? The Iconography of Leprosy. Part II. *American Journal of Dermatology* 5: 173–186.

O'Connell, L. 1998. The Articulated Human Skeletal Remains from St Augustine the Less, Bristol. Bournemouth University. Unpublished report.

Ogden, A., and F. Lee. 2008. Dental Health and Disease. In *"Lepers outside the Gate": Excavations at the Cemetery of the Hospital of St James and St Mary Magdalene, Chichester, 1986–87 and 1993,* edited by J. R. Magilton, F. Lee, and A. Boylston, 188–197. Council for British Archaeology Research Report 158 and Chichester Excavations vol. 10. Bootham: Council for British Archaeology.

Ohyama, H., H. Hongyo, N. Shimizu, Y. Shimizu, F. Nishimura, M. Nakagawa, H. Arai, N. Kato-Kogoe, N. Terada, A. Nagai, et al. 2010. Clinical and Immunological Assessment of Periodontal Disease in Japanese Leprosy Patients. *Japan Journal of Infectious Disease* 63: 427–432.

Ojha, K. S., R. C. Chaudhury, and S. K. Choudhary. 1984. Socio-Environmental Factors in Relation to Leprosy at Jaipur. *Indian Journal of Leprosy* 56: 884–888.

O'Keefe, S. F. 2000. An Overview of Oils and Fats, with a Special Emphasis on Olive Oil.

In *The Cambridge World History of Food*, vol. 1, edited by K. F. Kiple and K. C. Ornelas, 375–377. Cambridge: Cambridge University Press.

Oktaria, S., N. S. Hurif, W. Naim, H. B. Thio, T. E. C. Nijsten, and J. H. Richardus. 2018. Dietary Diversity and Poverty as Risk Factors for Leprosy in Indonesia: A Case-Control Study. *PLoS Neglected Tropical Diseases* 12(3): e0006317.

Oliveira, I., P. Deps, and J. Antunes. 2019. Armadillos and Leprosy: From Infection to Biological Model. *Revista do Instituto de Medicina Tropical de São Paulo* 61: doi 10.1590/s1678-9946201961044.

Oommen, S. T. 2002. The History of the Treatment of Leprosy and the Use of Hydnocarpus Oil. In *The Past and Present of Leprosy: Archaeological, Historical, Palaeopathological and Clinical Approaches,* edited by C. A. Roberts, M. E. Lewis, and K. Manchester, 201–204. British Archaeological Reports International Series 1054. Oxford: Archaeopress.

Ortner, D. 1998. Male-Female Immune Reactivity and Its Implications for Interpreting Evidence in Human Skeletal Palaeopathology. In *Sex and Gender in Palaeopathological Perspective,* edited by A. L. Grauer and P. Stuart-Macadam, 79–92. Cambridge: Cambridge University Press.

———. 2002. Observations on the Pathogenesis of Skeletal Disease in Leprosy. In *The Past and Present of Leprosy: Archaeological, Historical, Palaeopathological and Clinical Approaches,* edited by C. A. Roberts, M. E. Lewis, and K. Manchester, 73–80. British Archaeological Reports International Series 1054. Oxford: Archaeopress.

———. 2003. *Identification of Pathological Conditions in Human Skeletal Remains.* 3rd ed. London: Academic Press.

———. 2006. Differential Diagnosis of Skeletal Lesions in Infectious Disease. In *Advances in Human Palaeopathology,* edited by R. Pinhasi and S. Mays, 191–214. Chichester: John Wiley and Sons.

———. 2008. Skeletal Manifestations of Leprosy. In *"Lepers outside the Gate": Excavations at the Cemetery of the Hospital of St James and St Mary Magdalene, Chichester, 1986–87 and 1993,* edited by J. R. Magilton, F. Lee, and A. Boylston, 198–207. Council for British Archaeology Research Report 158 and Chichester Excavations vol. 10. Bootham: Council for British Archaeology.

Ortner, D. J., K. Manchester, and F. Lee. 1991. Metastatic Carcinoma in a Leper Skeleton from a Medieval Cemetery in Chichester, England. *International Journal of Osteoarchaeology* 1: 91–98.

Padhi, T., and S. Pradhan. 2014. Prevalence of Restless Legs Syndrome among Leprosy Patients: A Hospital Based Study. *Leprosy Review* 85: 218–223.

Pálfi, G. 1991. The First Osteoarchaeological Evidence of Leprosy in Hungary. *International Journal of Osteoarchaeology* 1: 99–192.

Pálfi, G., A. Marcsik, and I. Pap. 2012. A Short History of Paleopathological Research in Hungary. In *The Global History of Paleopathology: Pioneers and Prospects,* edited by J. Buikstra and C. Roberts, 405–415. Oxford: Oxford University Press.

Pálfi, G., and E. Molnár. 2009. The Paleopathology of Specific Infectious Diseases from Southeastern Hungary: A Brief Overview. *Acta Biologica Szegediensis* 53: 111–116.

Pálfi, G., E. Molnár, A. Marcsik, Z. Bereczki, I. Pap, E. Fóthi, Á. Kustár, B. G. Mende, D. E. Minnikin, O. Y.-C. Lee, et al. 2012. Mycobacterium tuberculosis—Mycobacterium leprae Coinfections from Hungary: Osteological and Biomolecular Findings. In *ICEPT-2: The*

Past and Present of Tuberculosis: A Multidisciplinary Overview on the Origin and Evolution of TB, edited by G. Pálfi, Z. Bereczki, E. Molnár, and O. Dutour. Szeged: JATE Press.

Pálfi, G., E. Molnár, I. Pap, E. Fóthi, Á. Kustár, D. E. Minnikin, O.-Y. Lee, G. S. Besra, M. Spigelman, and H. D. Donoghue. 2010. Visual and Molecular Biological Evidence of Leprosy-TB Co-Infection in a Medieval Skeleton from Hungary. Paper presented at the 18th European meeting of the Paleopathology Association, Vienna, Austria.

Pálfi, G., I. Pap, and E. Fóthi. 2001. Mycobacterialis Fertőzések új Paleopatológia Esetei. In *II. Kárpát-medencei Biológiai Szimpozium*, edited by I. Isépy, Z. Korsós, and I. Pap, 325–331. Budapest: Biológiai Társaság és a Magyar Természettudományi Múzeum Kiadványa.

Pálfi, G., A. Zink, C. J. Haas, A. Marcsik, O. Dutour, and A. G. Nerlich. 2002. Historical and Palaeopathological Evidence of Leprosy in Hungary. In *The Past and Present of Leprosy: Archaeological, Historical, Palaeopathological and Clinical Approaches*, edited by C. A. Roberts, M. E. Lewis, and K. Manchester, 205–212. British Archaeological Reports International Series 1054. Oxford: Archaeopress.

Palheta Neto, F. X., M. Silva Filho, J. M. Pantoja Jr., L. L. Teixeira, R. V. Miranda, and A. C. Palheta. 2010. Main Vocal Complaints of Elderly Patients after Leprosy Treatment. *Brazilian Journal of Otorhinolaryngology* 76: 156–163.

Palit, A., and A. C. Inamadar. 2014. Childhood Leprosy in India over the Past Two Decades. *Leprosy Review* 85: 93–99.

Pandey, A., and H. Rathod. 2010. Integration of Leprosy in General Health System vis-à-vis Leprosy Endemicity, Health Situation and Socioeconomic Development: Observations from Chhattisgarh & Kerala. *Leprosy Review* 81: 121–128.

Parascandola, J. 2003. Chaulmoogra Oil and the Treatment of Leprosy. *Pharmacy in History* 45: 47–57.

Parkash, O., and B. P. Singh. 2012. Advances in Proteomics of *Mycobacterium leprae*. *Scandinavian Journal of Immunology* 75: 369–378.

Paschoal, J. A., V. D. Paschoal, S. M. Nardi, P. S. Rosa, M. G. Ismael, and E. P. Sichieri. 2013. Identification of Urban Leprosy Clusters. *Scientific World Journal* 2013: Article 219143. https://www.hindawi.com/journals/tswj/2013/219143/.

Passos Vázquez, C. M., R. S. Mendes Netto, K. B. Ferreira Barbosa, T. Rodrigues de Moura, R. P. de Almeida, M. S. Duthie, and A. Ribeiro de Jesus. 2014. Micronutrients Influencing the Immune Response in Leprosy. *Nutricion Hospitalaria* 29: 26–36.

Paterson, D. E., and C. K. Job. 1964. Bone Changes and Absorption in Leprosy: A Radiological, Pathological and Clinical Study. In *Leprosy in Theory and Practice*, edited by R. G. Cochrane and T. F. Davey, 425–446. Bristol: John Wright and Sons.

Paterson, D. E., and M. Rad. 1961. Bone Changes of Leprosy: Their Incidence, Progress, Prevention and Arrest. *International Journal of Leprosy and Other Mycobacterial Diseases* 29: 393–422.

Patil, K. M., and S. Jacob. 2000. Mechanics of Tarsal Disintegration and Plantar Ulcers in Leprosy by Stress Analysis in Three Dimensional Foot Models. *Indian Journal of Leprosy* 72: 69–88.

Pearson, M. 2011. Antarctica. In *The Routledge Handbook of Archaeological Human Remains and Legislation*, edited by N. Márquez-Grant and L. Fibiger, 673–688. London: Routledge.

Pechenkina, E. A. 2012. From Morphometrics to Holistics: The Emergence of Paleopathology in China. In *The Global History of Paleopathology: Pioneers and Prospects*, edited by J. Buikstra and C. Roberts, 345–360. Oxford: Oxford University Press.

Pedley, J. C. 1967. The Presence of *M. leprae* in Human Milk. *Leprosy Review* 38: 239–242.

———. 1968. The Presence of *M. leprae* in the Lumina of the Female Mammary Gland. *Leprosy Review* 39: 201–202.

Pena, M., A. Geluk, J. J. Van der Ploeg-Van Schip, K. L. Franken, R. Sharma, and R. Truman. 2011. Cytokine Responses to *Mycobacterium leprae* Unique Proteins Differentiate between *Mycobacterium leprae* Infected and Naïve Armadillos. *Leprosy Review* 82: 422–431.

Peters, E. S., and A. L. Eshiet. 2002. Male-Female (Sex) Differences in Leprosy Patients in South Eastern Nigeria: Females Present Late for Diagnosis and Treatment and Have Higher Rates of Deformity. *Leprosy Review* 73: 262–267.

Petersone-Gordina, E., C. Roberts, and G. Gerhards. 2016. A Skeleton with Probable Leprosy from a Post-Medieval Cemetery in Riga, Latvia. Poster presented at the 43rd annual North American meeting of the Paleopathology Association, Atlanta, GA, April.

Pfeiffer, D. U. 2008. Animal Tuberculosis. In *Clinical Tuberculosis*, 4th ed., edited by P. D. O. Davies, P. F. Barnes, and S. B. Gordon, 519–528. London: Hodder Arnold.

Pfrengle, S., J. Neukamm, S. Inskip, R. I. Tukhbatova, N. Y. Berezina, A. P. Buzhilova, S. S. Hamre, V. Matos, M. T. Ferreira, E. Reiter, et al. 2018. Reconstruction of New Ancient *Mycobacterium leprae* Genomes from Europe. Poster presented at the 22nd European meeting of the Paleopathology Association, Zagreb, Croatia, August 28–September 1.

Phetsuksiri, B., S. Srisungngam, J. Rudeeaneksin, S. Bunchoo, A. Lukebua, R. Wongtrungkapun, S. Paitoon, R. M. Sakamuri, P. J. Brennan, and V. Vissa. 2012. SNP Genotypes of *Mycobacterium leprae* Isolates in Thailand and Their Combination with rpoT and TTC Genotyping for Analysis of Leprosy Distribution and Transmission. *Japan Journal of Infectious Diseases* 65: 52–56.

Pietrusewsky, M., and M. T. Douglas. 2012. History of Paleopathology in the Pacific. In *The Global History of Paleopathology: Pioneers and Prospects*, edited by J. Buikstra and C. Roberts, 594–615. Oxford: Oxford University Press.

Pinhasi, R., R. Foley, and H. D. Donoghue. 2005. Reconsidering the Antiquity of Leprosy. *Science* 312: 846.

Pinto, P., S. Salgado, N. P. Carneiro Santos, S. Santos, and A. Ribeiro-dos-Santos. 2015. Influence of Genetic Ancestry on INDEL Markers of NFKβ1, CASP8, PAR1, IL4 and CYP19A1 Genes in Leprosy Patients. *PLoS Neglected Tropical Diseases* 9(9): e0004050.

Pollard, T. T. M., and S. B. Hyatt, eds. 1999. *Sex, Gender and Health*. Cambridge: Cambridge University Press.

Popescu, E. 2009. *Norwich Castle: Excavations and Historical Survey, 1987–98*. Part I, *Anglo-Saxon to c. 1345*. East Anglian Archaeology 132. Dereham, Norfolk: Historic Environment, Norfolk Museums and Archaeology Service.

Porter, R. 1997. *The Greatest Benefit to Mankind: A Medical History of Humanity from Antiquity to the Present*. London: Fontana Press.

Potekhina, I. 2011. Ukraine/УКРАЇНА. In *The Routledge Handbook of Archaeological Human Remains and Legislation*, edited by N. Márquez-Grant and L. Fibiger, 469–477. London: Routledge.

Powell, F. 1996. The Human Remains. In *Raunds Furnells: The Anglo-Saxon Church and Churchyard*, edited by A. Boddington, 113–124. English Heritage Archaeological Report 7. London: English Heritage.
Price, C. R. 2003. *Late Pleistocene and Early Holocene Small Mammals in South West Britain*. British Archaeological Report British Series 347. Oxford: Archaeopress.
Price, E. W. 1959. Studies of Plantar Ulcers in Leprosy. *Leprosy Review* 30: 98–107.
———. 1960. Studies on Plantar Ulceration in Leprosy. 6. The Management of Plantar Ulcers. *Leprosy Review* 31: 159–171.
———. 1961. Studies of Plantar Ulcer in Leprosy. *Leprosy Review* 32: 97–103.
———. 1964a. The Etiology and Natural History of Plantar Ulcer. *Leprosy Review* 35: 259–266.
———. 1964b. The Problem of Plantar Ulcer. *Leprosy Review* 35: 267–272.
Pridham, C. 1846. *England's Colonial Empire*. London: Smith, Elder and Company.
Pring, D. J., and N. Casiebanca. 1982. Simple Plantar Ulcers Treated by Below-Knee Plaster and Moulded Double-Rocker Plaster Shoe: A Comparative Study. *Leprosy Review* 53: 261–264.
Purtilo, D. T., G. P. Walsh, E. E. Storrs, and I. S. Banks. 1974. Impact of Cool Temperatures on the Transformation of Human and Armadillo Lymphocytes (*Dasypus novemcinctus*, Linn.) as Related to Leprosy. *Nature* 248: 450–452.
Queiroz, J. W., G. H. Dias, M. L. Nobre, M. C. De Sousa Dias, S. F. Araújo, J. D. Barbosa, P. B. da Trinidade-Neto, J. M. Blackwell, and S. M. Jeronimo. 2010. Geographic Information Systems and Applied Spatial Statistics Are Efficient Tools to Study Hansen's Disease (Leprosy) and to Determine Areas of Greater Risk of Disease. *American Journal of Tropical Medicine and Hygiene* 82: 306–314.
Rafferty, J. 2005. Curing the Stigma of Leprosy. *Leprosy Review* 76: 119–126.
Rafi, A., and F. Feval. 1995. PCR to Detect *Mycobacterium tuberculosis* DNA in Sputum Samples from Treated Leprosy Patients with Putative Tuberculosis. *Southeast Asian Journal of Tropical Medicine and Public Health* 26: 253–257.
Rafi, A., M. Spigelman, J. Stanford, E. Lemma, H. Donoghue, and J. Zias. 1994a. DNA of *Mycobacterium leprae* Detected by PCR in Ancient Bone. *International Journal of Osteoarchaeology* 4: 287–290.
———. 1994b. *Mycobacterium leprae* DNA from Ancient Bone Detected by PCR. *Lancet* 343: 1360–1361.
Rajalingam, R., D. P. Singal, and N. K. Mehra. 1997. Transporter Associated with Antigen Processing (TAP) Genes and Susceptibility to Tuberculoid Leprosy and Pulmonary Tuberculosis. *Tissue Antigens* 49: 168–172.
Raju, M. S., and P. S. Rao. 2011. Medical and Social Concerns of Leprosy Cured after Integration in India. *Indian Journal of Leprosy* 83: 145–155.
Ramachandran, A., and P. N. Neeelan. 1987. Autonomic Neuropathy in Leprosy. *Indian Journal of Leprosy* 59: 405–413 and 277–285.
Ramirez, J. 2008. A Day at Carville: My Home—Mi Casa. *Public Health Reports* 123: 122–124.
Ramos, J. M., B. Alonso-Castañeda, D. Eshetu, D. Lemma, F. Reyes, I. Belinchón, and M. Górgolas. 2014. Prevalence and Characteristics of Neuropathic Pain in Leprosy Patients Treated Years Ago. *Pathogens and Global Health* 108: 186–190.

Ramos, J. M., D. Romero, and I. Belinchón. 2016. Epidemiology of Leprosy in Spain: The Role of the International Migration. *PLoS Neglected Tropical Diseases* 10(3): e0004321.

Ranjit, J. H., and A. Verghese. 1980. Psychiatric Disturbances among Leprosy Patients: An Epidemiological Study. *International Journal of Leprosy and Other Mycobacterial Diseases* 48(4): 431–434.

Rao, P. S. 2008. A Study on Non-Adherence to MDT among Leprosy Patients. *Indian Journal of Leprosy* 80: 149–154.

Rao, P. S., and A. S. John. 2012. Nutritional Status of Leprosy Patients in India. *Indian Journal of Leprosy* 84: 17–22.

Rao, P. S., A. B. A. Karat, V. G. Kaliaperumal, and S. Karat. 1972. Prevalence of Leprosy in Gudiyatham Taluk, South India. Part 1. Specific Rates with Reference to Age, Sex and Type. *International Journal of Leprosy and Other Mycobacterial Diseases* 40: 157–163.

Rao, S., V. Garole, S. Walawalkar, S. Khot, and N. Karandikar. 1996. Gender Differentials in the Social Impact of Leprosy. *Leprosy Review* 67: 190–199.

Rao, S., S. Khot, S. Walawalkar, V. Garole, and N. Karandikar. 1996. *Differences in Detection Patterns between Male and Female Leprosy Patients in Maharashtra*. Pune, India: Agharkar Research Institute.

Rasmussen, K. L., J. L. Boldsen, H. K. Christensen, L. Skytte, K. L. Hansen, L. Mølholm, P. M. Grootes, M. J. Nadeau, and K. M. F. Eriksen. 2008. Mercury Levels in Danish Medieval Human Bones. *Journal of Archaeological Science* 35: 2295–2306.

Rastogi, N., and R. C. Rastogi. 1984. Leprosy in Ancient India. *International Journal of Leprosy and Other Mycobacterial Diseases* 52: 541–543.

Rawcliffe, C. 2006. *Leprosy in Medieval England*. Woodbridge, Suffolk: Boydell Press.

———. 2013. *Urban Bodies: Communal Health in Late Medieval English Towns and Cities*. Woodbridge, Suffolk: Boydell Press.

Rawlani, S. M., S. Rawlani, S. Degwekar, R. R. Bhowte, and M. Motwani. 2011. Oral Health Status and Alveolar Bone Loss in Treated Leprosy Patients of Central India. *Indian Journal of Leprosy* 83(4): 215–224.

Rawson, T. M., V. Anjum, J. Hodgson, A. K. Rao, K. Murthy, P. S. Rao, J. Subbanna, and P. V. Rao. 2014. Leprosy and Tuberculosis Concomitant Infection: A Poorly Understood, Age-Old Relationship. *Leprosy Review* 85: 288–295.

Reader, R. 1974. New Evidence for the Antiquity of Leprosy in Early Britain. *Journal of Archaeological Science* 1: 205–207.

Redman J. E., M. J. Shaw, A. I. Mallet, A. L. Santos, C. A. Roberts, A. M. Gernaey, and D. E. Minnikin. 2009. Mycocerosic Acid Biomarkers for the Diagnosis of Tuberculosis in the Coimbra Skeletal Collection. *Tuberculosis* 89: 267–277.

Reenstierna, J. 1941. Leprosy in Sweden. *Acta Dermato-Venereologica* 22: 257–261.

Rees, R. J. W., M. F. R. Waters, A. G. M. Weddell, and E. Palmer. 1967. Experimental Lepromatous Leprosy. *Nature* 215: 599–602.

Reichart, P. 1974. Pathologic Changes in the Soft Palate in Lepromatous Leprosy: An Evaluation of Ten Patients. *Oral Surgery* 38: 898–904.

———. 1976. Facial and Oral Manifestations in Leprosy: An Evaluation of Seventy Cases. *Oral Surgery* 41: 385–399.

Reichart, P., T. Ananatasan, and G. Reznik. 1976. Gingiva and Periodontium in Leprosy: A Clinical, Radiological and Microscopical Study. *Journal of Periodontology* 47: 455–460.

Reis, F. J. J., D. Lopes, J. Rodrigues, A. P. Gosling, and M. K. Gomes. 2014. Psychological Distress and Quality of Life in Leprosy Patients with Neuropathic Pain. *Leprosy Review* 85: 186–193.

Rendall, J. R., A. C. McDougall, and L. A. Wilks. 1976. Intra-Oral Temperatures in Man with Special Reference to Involvement of the Central Incisors and Premaxillary Alveolar Process in Lepromatous Leprosy. *International Journal of Leprosy and Other Mycobacterial Diseases* 44: 462–468.

Resnick, D. 1995. Disorders of Other Endocrine Glands and of Pregnancy. In *Diagnosis of Bone and Joint Disorders*, 3rd ed., edited by D. Resnick, 2076–2104. Edinburgh: W. B. Saunders.

Resnick, D., T. G. Goergen, and G. Niwayama. 1995. Physical Injury: Concepts and Terminology. In *Diagnosis of Bone and Joint Disorders*, 3rd ed., edited by D Resnick, 2561–2692. Edinburgh: W. B. Saunders.

Resnick, D., and D. Niwayama. 1995a. Osteomyelitis, Septic Arthritis, and Soft Tissue Infection: Mechanisms and Situations. In *Diagnosis of Bone and Joint Disorders*, 3rd ed., edited by D. Resnick, 2325–2418. Edinburgh: W. B. Saunders.

———. 1995b. Osteomyelitis, Septic Arthritis, and Soft Tissue Infection: Organisms. In *Diagnosis of Bone and Joint Disorders*, 3rd ed., edited by D. Resnick, 2448–2558. Edinburgh: W. B. Saunders.

Reveiz, L., J. A. Buendía, and D. Téllez. 2009. Chemoprophylaxis in Contacts of Patients with Leprosy: A Systematic Review and Meta-Analysis. *Revista Panamericana de Salud Pública* 26: 341–349.

Richards, P. 1960. Leprosy in Scandinavia: A Discussion of Its Origins, Its Survival, and Its Effect on Scandinavian Life over the Course of Nine Centuries. *Centaurus* 7: 101–133.

———. 1977. *The Medieval Leper and His Northern Heirs*. Cambridge: D. S. Brewer.

———. 1990. Leprosy: Myth, Melodrama and Medievalism. The Fitzpatrick Lecture 1989. *Journal of the Royal College of Physicians of London* 24: 55–62.

Ridley, D. S., and W. H. Jopling. 1966. Classification of Leprosy According to Immunity: A Five-Group System. *International Journal of Leprosy and Other Mycobacterial Diseases* 34: 255–273.

Ridley, M., W. H. Jopling, and D. S. Ridley. 1976. Acid-Fast Bacilli in the Fingers of Long-Treated Lepromatous Patients. *Leprosy Review* 47: 93–96.

Roa, R., and M. Morris. 2006. Leprosy in Mexico. *Japanese Journal of Leprosy* 75: 51–58.

Robbins, G., V. Mushrif Tripathy, V. N. Misra, R. K. Mohanty, V. S. Shinde, K. M. Gray, and M. D. Schug. 2009. Ancient Skeletal Evidence for Leprosy in India (2000 BC). *PLoS ONE* 4(5): 1–8.

Roberts, C. A. 1986. Leprogenic Odontodysplasia. In *Teeth and Anthropology*, edited by E. Cruwys and R. A. Foley, 137–147. British Archaeological Reports International Series 291. Oxford: BAR.

———. 1987. Leprosy and Tuberculosis in Britain: Diagnosis and Treatment in Antiquity. *Museum of Applied Science Center for Archaeology Journal* 4: 166–171.

———. 1993. A Note on the Skeletons C and E from Burial 62/63/64. In A. G. Kinsley, *Broughton Lodge: Excavations on the Romano-British Settlement and Anglo-Saxon Cemetery at Broughton Lodge, Willoughby-on-the-Wolds, Nottinghamshire 1964–8*, 58.

Nottingham Archaeological Monographs 4. Nottingham: University of Nottingham, Department of Classical and Archaeological Studies.

———. 2002. The Antiquity of Leprosy in Britain: The Skeletal Evidence. In *The Past and Present of Leprosy: Archaeological, Historical, Palaeopathological and Clinical Approaches,* edited by C. A. Roberts, M. E. Lewis, and K. Manchester, 213–222. British Archaeological Reports International Series 1054. Oxford: Archaeopress.

———. 2007. A Bioarchaeological Study of Maxillary Sinusitis. *American Journal of Physical Anthropology* 133: 792–807.

———. 2012. Keith Manchester (1938–). In *The Global History of Paleopathology: Pioneers and Prospects,* edited by J. Buikstra and C. Roberts, 56–59. Oxford: Oxford University Press.

———. 2015. What Did Agriculture Do for Us? The Bioarchaeology of Health and Diet. In *The Cambridge World History,* vol. 2, *A World with Agriculture, 12,000 BCE–500 CE,* edited by G. Barker and C. Goucher, 93–123. Cambridge: Cambridge University Press.

———. 2017. Applying the "Index of Care" to a Person Who Experienced Leprosy in Late Medieval Chichester, England. In *New Developments in the Bioarchaeology of Care,* edited by L. Tilley and A. A. Schrenk, 101–124. Cham, Switzerland: Springer International.

———. 2018. *Human Remains in Archaeology: A Handbook.* 2nd ed. York: Council for British Archaeology.

———. 2019. Ethical Challenges of Working with Archaeological Human Remains, with a Focus on the UK. In *Ethical Approaches to Human Remains: A Global Challenge for Bioarchaeology and Forensic Anthropology,* edited by K. Squires, D. Errickson, and N. Márquez-Grant, 133–155. Cham, Switzerland: Springer International.

Roberts, C. A., and J. E. Buikstra. 2003. *The Bioarchaeology of Tuberculosis: A Global View on a Reemerging Disease.* Gainesville: University Press of Florida.

Roberts, C. A., and M. Cox. 2003. *Health and Disease in Britain: From Prehistory to the Present Day.* Stroud, Gloucestershire: Sutton Publishing.

Roberts, C. A., and R. Dixon. 1995. Reply (2). *International Journal of Osteoarchaeology* 5: 299.

Roberts, C. A., and S. Ingham. 2008. Using Ancient DNA Analysis in Palaeopathology: A Critical Analysis of Published Papers, with Recommendations for Future Work. *International Journal of Osteoarchaeology* 18: 600–613.

Roberts, C. A., A. Lagia, S. Triantaphyllou, C. Bourbou, and A. Tsaliki. 2005. Health and Disease in Greece: Past, Present and Future. In *Health in Antiquity,* edited by H. King, 32–58. London: Routledge.

Roberts, C. A., M. E. Lewis, and K. Manchester, eds. 2002. *The Past and Present of Leprosy: Archaeological, Historical, Palaeopathological and Clinical Approaches.* British Archaeological Reports International Series 1054. Oxford: Archaeopress.

Roberts, C. A., D. Lucy, and K. Manchester. 1994. Inflammatory Lesions of Ribs: An Analysis of the Terry Collection. *American Journal of Physical Anthropology* 95: 169–182.

Roberts, C. A., and K. Manchester. 2010. *The Archaeology of Disease.* Stroud: History Press.

Roberts, J. 2007. Human Remains from the Cists. In A. Dunwell, *Cist Burials and an Iron Age Settlement at Dryburn Bridge, Innerwick, East Lothian,* 18–25. Scottish Archaeological Internet Report 24. http://archaeologydataservice.ac.uk/archiveDS/archiveDownload?t=arch-310-1/dissemination/pdf/sair24.pdf.

———. 2002. The International Leprosy Association (ILA) Global Project on the History of Leprosy. In *The Past and Present of Leprosy: Archaeological, Historical, Palaeopathological and Clinical Approaches,* edited by C. A. Roberts, M. E. Lewis, and K. Manchester, 3–5. British Archaeological Reports International Series 1054. Oxford: Archaeopress.

Robertson, L. M., P. G. Nicholls, and R. Butlin. 2000. Delay in Presentation and Start of Treatment in Leprosy: Experience in an Out-Patient Clinic in Nepal. *Leprosy Review* 71: 511–516.

Rocha-Leite, C. I., R. Borges-Oliveira, L. Araújo-de-Freitas, P. R. Machado, and L. C. Quarantini. 2014. Mental Disorders in Leprosy: An Underdiagnosed and Untreated Population. *Journal of Psychomosomatic Research* 76: 422–425.

Rodrigues, G. A., N. P. Qualio, L. D. de Macedo, L. Innocentini, A. Ribeiro-Silva, N. T. Foss, M. Frade, and A. Motta. 2017. The Oral Cavity in Leprosy: What Clinicians Need to Know. *Oral Diseases* 23: 749–756.

Rodrigues, L. C., and D. N. Lockwood. 2011. Leprosy Now: Epidemiology, Progress, Challenges, and Research Gaps. *Lancet Infectious Diseases* 11: 464–470.

Rodriguez, J. N. 1926. Studies of Early Leprosy in Children of Lepers. *Philippine Journal of Science* 31: 115–145.

Rodríguez-Martín, C. 2012. A History of Paleopathology in Spain. In *The Global History of Paleopathology: Pioneers and Prospects,* edited by J. Buikstra and C. Roberts, 541–548. Oxford: Oxford University Press.

Roffey, S., and K. Tucker. 2012. A Contextual Study of the Medieval Hospital and Cemetery of St Mary Magdalen, Winchester, England. *International Journal of Paleopathology* 2: 170–180.

Roffey, S., K. Tucker, K. Filipek-Ogden, J. Montgomery, J. Cameron, T. O'Connell, J. Evans, P. Marter, and G. M. Taylor. 2017. Investigation of a Medieval Pilgrim Burial Excavated from the *Leprosarium* of St Mary Magdalen Winchester, UK. *PLoS Neglected Tropical Diseases* 11(1): e0005186. https://journals.plos.org/plosntds/article?id=10.1371/journal.pntd.0005186.

Rogers, L., and E. Muir.1946. *Leprosy.* 3rd ed. Bristol: John Wright and Sons Ltd.

Rohtagi, S., S. Naveen, S. Salunke, S. Someshwar, H. R. Jerajani, and R. Joshi. 2016. The Study of the Deformed Leprous Foot. *Leprosy Review* 87: 104–108.

Rojas-Espinosa, O. 1994. Active Humoral Immunity in the Absence of Cell-Mediated Immunity in Murine Leprosy: Lastly an Explanation. *International Journal of Leprosy and Other Mycobacterial Diseases* 62: 143–147.

Rojas-Espinosa, O., and M. Løvik. 2001. *Mycobacterium leprae* and *Mycobacterium lepraemurium* in Domestic and Wild Animals. *Scientific and Technical Review of the Office International des Épizooties (Paris)* 20: 219–251.

Rongioletti, F., R. Gallo, E. Cozzani, and A. Parodi. 2009. Leprosy: A Diagnostic Trap for Dermatopathologists in Nonendemic Area. *American Journal of Dermatopathology* 31: 607–610.

Roosta, N., D. S. Black, and T. H. Rea. 2013. A Comparison of Stigma among Patients with Leprosy in Rural Tanzania and Urban United States: A Role for Public Health in Dermatology. *International Journal of Dermatology* 52: 432–440.

Rose, P., and C. MacDougall. 1975. Adverse Reactions following Pregnancy in Patients with Borderline (Dimorphous) Leprosy. *Leprosy Review* 46: 109–114.

Ross, W. F. 1960. Etiology and Treatment of Plantar Ulcers. *Leprosy Review* 31: 25–40.

Rothschild, B. M., and C. Rothschild. 2001. Skeletal Manifestations of Leprosy: An Analysis of 137 Patients from Different Clinical Settings in the Pre- and Postmodern Treatment Eras. *Journal of Clinical Rheumatology* 7: 228–237.

Royal College of Physicians. 1867. *Report on Leprosy*. London: Her Majesty's Stationery Office.

Rubini, M., V. Dell'Anno, R. Giuliani, P. Favia, and P. Zaio. 2012. The First Probable Case of Leprosy in Southeast Italy (13th–14th centuries AD, Montecorvino, Puglia). *Journal of Anthropology* 2012: Article 262790. https://www.hindawi.com/journals/janthro/2012/262790/.

Rubini, M., Y. S. Erdal, M. Spigelman, P. Zaio, and H. D. Donoghue. 2012. Palaeopathological and Molecular Study on Two Cases of Ancient Childhood Leprosy from the Roman and Byzantine Empires. *International Journal of Osteoarchaeology* 24(5): 570–582.

Rubini, M., and P. Zaio. 2009. Lepromatous Leprosy in an Early Medieval Cemetery in Central Italy (Morrione, Campochiaro, Molise, 6th–8th century AD). *Journal of Archaeological Science* 36: 2771–2779.

———. 2011. Warriors from the East: Skeletal Evidence of Warfare from a Lombard-Avar Cemetery in Central Italy (Campochiaro, Molise, 6th–8th century AD). *Journal of Archaeological Science* 38: 1–9.

Rubini, M., P. Zaio, and C. A. Roberts. 2014. Tuberculosis and Leprosy in Italy: New Skeletal Evidence. *HOMO: Journal of Comparative Human Biology* 65: 13–32.

Rubini, M., P. Zaio, M. Spigelman, and H. D. Donoghue. 2017. Leprosy in a Lombard-Avar Cemetery in Central Italy (Campochiaro, Molise, 6th–8th century AD): Ancient DNA Evidence and Demography. *Annals of Human Biology* 44: 510–521.

Ruffin, L., R. Palich, X. Demondion, J. Blondiaux, Th. Colard, A. Alduc-Le Bagousse, C. Niel, and R.-M. Flipo. 2010. X-Rays as a Sensitive Method for the Study of Periosteal Lesions of Leg Bones in Leprosy. Poster presented at the 18th European meeting of the Paleopathology Association, Vienna, Austria.

Rühli, F. J., and M. Henneberg. 2013. New Perspectives on Evolutionary Medicine: The Relevance of Microevolution for Human Health and Disease. *BMC Medicine* 11: Article 115.

Ruis González, J., and C. Serrano Sánchez. 2015. *A Case of Leprosy in the Colonial Funerary Complex of Tlatelolco, México*. Paper presented at the 6th meeting of the Paleopathology Association Meeting in South America.

Ryrie, G. A. 1948. Some Impressions of Sungei Bulok Leper Hospital under Japanese Occupation. *Leprosy Review* 18: 10.

———. 1951. The Psychology of Leprosy. *Leprosy Review* 22: 13–24.

Saarinen, K., J. Jantunen, and T. Haahtela. 2011. Birch Pollen Honey for Birch Pollen Allergy—A Randomized Controlled Pilot Study. *International Archives of Allergy and Immunology* 155: 160–166.

Sabin, T. D. 1981. The Penikese Hospital: A Massachusetts Hospital for the Treatment of Hansen's Disease. *New England Journal of Medicine* 304: 1610–1612.

Sachdeva, S., S. S. Amin, Z. Khan, P. K. Sharma, and S. Bansal. 2011. Childhood Leprosy: Lest We Forget. *Tropical Doctor* 41: 163–165.

Saha, K., V. Sharma, and M. A. Siddiqui. 1982. Decreased Cellular and Humoral Anti-Infective Factors in the Breast Secretions of Lactating Mothers with Lepromatous Leprosy. *Leprosy Review* 53: 35–44.

Sahoo, M. R., S. P. Dhanabal, A. N. Jadhav, V. Reddy, G. Muguli, U. V. Babu, and P. Rangesh. 2014. Hydnocarpus: An Ethnopharmacological, Phytochemical and Pharmacological Review. *Journal of Ethnopharmacology* 154: 17–25.

Saint-Martin, P., N. Telmon, N. Dabernat, C. Theureau, P. O'Byrne, and C. Crubezy. 2006. Étude Paléopathologique d'une Nécropole La Léproserie de la Chapelle Saint-Lazare (Tours, Indre-et-Loire). Paper presented at the 7th Colloque d'Anthropologie Médico-Légale, Nice, France.

Salipante, S. J., and B. G. Hall. 2011. Towards an Epidemiology of *Mycobacterium leprae*: Strategies, Successes, and Shortcomings. *Infection, Genetics and Evolution* 11: 1505–1513.

Sampaio, P. B., A. I. Bertolde, E. Leonor, N. Maciel, and E. Zandonade. 2013. Correlation between the Spatial Distribution of Leprosy and Socioeconomic Indicators in the City of Vitória, State of ES, Brazil. *Leprosy Review* 84: 256–265.

Sane, S. B., V. N. Kulkarni, R. C. Sharangpani, and J. M. Mehta. 1985. "A Study on Disintegration of Carpal Bones in Leprosy." In *Proceedings of the International Conference on Biomechanics and Clinical Kinesiology of Hand and Foot, I. T. T., Madras, India, December 16–18, 1985*, edited by K. M. Patil and H. Srinivasan, 61–64. Madras: Indian Institute of Technology.

Sano, M. A. 1958. Leprous Pink Spots of the Tooth. *La Lepro* 27: 398–340.

Santos, A. L., and E. Cunha. 2012. Portuguese Developments in Paleopathology: An Outline History. In *The Global History of Paleopathology: Pioneers and Prospects*, edited by J. Buikstra and C. Roberts, 503–518. Oxford: Oxford University Press.

Santow, G. 1995. Social Roles and Physical Health: The Case of Female Disadvantage in Poor Countries. *Social Science and Medicine* 40: 147–161.

Sapolsky, R. 2004. *Why Zebras Don't Get Ulcers: The Acclaimed Guide to Stress, Stress-Related Diseases, and Coping*. 3rd ed. New York: Henry Holt.

Saporta, L., and A. Yuksel. 1994. Androgenic Status in Patients with Lepromatous Leprosy. *British Journal of Urology* 74: 221–224.

Saraya, M. A., M. A. Al-Fadhli, and J. A. Qasem. 2012. Diabetic Status of Patients with Leprosy in Kuwait. *Journal of Infection and Public Health* 5: 360–365.

Sarno, E. N., X. Illarramendi, J. A. Costa Nery, A. M. Sales, M. C. Gutierrez-Galhardo, M. L. Fernandes Penna, E. Pereiera Sampaio, and G. Kaplan. 2008. HIV-*M. leprae* Interaction: Can HAART Modify the Course of Leprosy? *Public Health Reports* 123: 206–212.

Satchell, M. 1998. The Emergence of Leper-Houses in Medieval England, 1100–1250. DPhil thesis, University of Oxford.

Sathiaraj, Y., G. Norman, and J. Richard. 2010. Long Term Sustainability and Efficacy of Self-Care Education on Knowledge and Practice of Wound Prevention and Management among Leprosy Patients. *Indian Journal of Leprosy* 82: 79–83.

Sauer, M. E., H. Salomão, G. B. Ramos, H. R. D'Espindula, R. S. Rodrigues, W. C. Macedo, R. H. Sindeaux, and M. T. Mira. 2016. Genetics of Leprosy: Expected and Unexpected Developments and Perspectives. *Clinical Dermatology* 34: 96–104.

Saunders, S. R., and D. L. Rainey. 2008. Nonmetric Trait Variation in the Skeleton: Abnormalities, Anomalies and Atavisms. In *Biological Anthropology of the Human Skeleton*, edited by M. A. Katzenberg and S. R. Saunders, 533–559. Chichester: Wiley-Liss.

Saunderson, P. R. 2008. Leprosy Elimination: Not as Straightforward as It Seemed. *Public Health Reports* 123: 213–216.

Scarre, G. 2006. Can Archaeology Harm the Dead? In *The Ethics of Archaeology: Philosophical Perspectives on Archaeological Practice*, edited by C. Scarre and G. Scarre, 181–198. Cambridge: Cambridge University Press.

Schäfer, J. 1998. Leprosy and Disability Control in the Guévra Prefecture of Chard, Africa: Do Women Have Access to Leprosy Control Services? *Leprosy Review* 69: 267–278.

Scheepers, A. 1998. Correlation of Oral Surface Temperatures and the Lesions of Leprosy. *International Journal of Leprosy and Other Mycobacterial Diseases* 66: 214–217.

Schepartz, L. A., S. C. Fox, and C. Bourbou. 2009. *New Directions in the Skeletal Biology of Greece*. Hesperia Supplement 43. Princeton, NJ: American School of Classical Studies at Athens.

Schilling, A. K., C. Avanzi, R. G. Rainer, P. Busso, B. Pisanu, N. Ferrari, C. Romeo, M. V. Mazzamuto, J. McLuckie, C. M. Shuttleworth, J. Del-Pozo, P. W. W. Lurz, W. Escalante-Fuentes, J. Orcampo-Candian, L. Vera-Cabrera, K. Stevenson, J-L Chapuis, A. L. Meredith, and S. T. Cole. 2019. British Red Squirrels Remain the Only Known Wild Rodent Host for Leprosy Bacilli. *Frontiers of Veterinary Science* 6: 8.

Schlagel, C. J., T. L. Hadfield, and W. M. Meyers. 1985. Leprosy in the Armed Forces of the United States: Newly Reported Cases from 1970–1983. *Military Medicine* 150: 427–430.

Schmitz-Cliever, E. 1971. Zur Osteo-archaeologie der Mittel-alterlichen Lepra: Ergebnis einer Probegrabung im Melaten bei Aachen. *Medizinhistorisches Journal* 6: 249–263.

———. 1972. Das Mittelalterliche Leprosarium Melatan bei Aachen. *Clio Medica* 7: 13–34.

———. 1973a. St Jakobspilger-Muscheln in Einem Mittelalterlichen Leprosengrab. *Aachener Kunstblätter* 42: 317–342.

———. 1973b. Zur Osteoarchäologie der mittelalterlichen Lepra II. *Medizinhistoriches Journal* 8: 182–200.

Schoonbaert, D., and V. Demedts. 2008. Analysis of the Leprosy Literature Indexed in Medline (1950–2007). *Leprosy Review* 79: 387–400.

Schuenemann, V. J., C. Avanzi, B. Krause-Kyora, A. Seitz, A. Herbig, S. Inskip, M. Bonazzi, E. Reiter, C. Urban, D. D. Pedersen, et al. 2018. Ancient Genomes Reveal a High Diversity of *Mycobacterium leprae* in Medieval Europe. *PLoS Pathogens* 14(5): e1006997.

Schuenemann, V. J., P. Singh, T. A. Mendum, B. Krause-Kyora, G. Jager, K. I. Bos, A. Herbig, C. Economou, A. Benjak, P. Busso, et al. 2013. Genome-Wide Comparison of Medieval and Modern *Mycobacterium leprae*. *Science* 341: 179–183.

Schug, G. R., K. E. Blevins, B. Cox, K. Gray, and V. Mushrif-Tripathy. 2013. Infection, Disease, and Biosocial Processes at the End of the Indus Civilization. *PLoS ONE* 8(12): e84814. https://journals.plos.org/plosone/article?id=10.1371/journal.pone.0084814.

Schultz, M. 2001. Paleohistology of Bone: A New Approach to the Study of Ancient Diseases. *Yearbook of Physical Anthropology* 44: 106–147.

Schultz, M., and C. A. Roberts. 2002. Diagnosis of Leprosy in Skeletons from an English Later Medieval Hospital Using Histological Analysis. In *The Past and Present of Leprosy: Archaeological, Historical, Palaeopathological and Clinical Approaches*, edited by C. A. Roberts, M. E. Lewis, and K. Manchester, 89–104. British Archaeological Reports International Series 1054. Oxford: Archaeopress.

Schuring, R. P., L. Hamann, W. R. Faber, D. Paha, J. H. Richardu, R. R. Schuman, and L.

Oskam. 2009. Polymorphism N248S in the Human Toll-Like Receptor 1 Gene Is Related to Leprosy and Leprosy Reactions. *Journal of Infectious Diseases* 199: 1816–1819.

Schurr, E., A. Alcaïs, L. de Léséleuc, and L. Abel. 2006. Genetic Predisposition to Leprosy: A Major Gene Reveals Novel Pathways of Immunity to *Mycobacterium leprae*. *Seminars in Immunology* 18: 404–410.

Schutkowski, H., and M. Fernández-Gil. 2010. Histological Manifestations of Mycobacterial Bone Lesions. Paper Presented at the 18th European meeting of the Paleopathology Association, Vienna, Austria.

Schwarcz, H. P., T. Dupras, and S. Fairgrieve. 1999. ^{15}N Enrichment in the Sahara: In Search of a Global Relationship. *Journal of Archaeological Science* 26: 629–636.

Scollard, D. M., L. B. Adams, T. P. Gillis, J. L. Krahenbuhl, R. W. Truman, and D. L. Williams. 2006. The Continuing Challenge of Leprosy. *Clinical Microbiology Reviews* 19: 338–381.

Scollard, D. M., and O. K. Skinses. 1999. Oropharyngeal Leprosy in Art, History and Medicine. *Oral Surgery, Oral Medicine, Oral Pathology* 87: 463–471.

Scrimshaw, N. 2000. Infection and Nutrition: Synergistic Reactions. In *The Cambridge World History of Food*, vol. 2, edited by K. F. Kiple and K. C. Ornelas, 1397–1411. Cambridge: Cambridge University Press.

Seboka, G., and P. Saunderson. 1996. Cost-Effective Footwear for Leprosy Control Programmes: A Study in Rural Ethiopia. *Leprosy Review* 67: 208–216.

Segal, K. L. 2001. Bioarchaeological Analysis of St Jørgensgård, a Medieval Leprosy Hospital in Odense, Denmark. PhD diss., University of Chicago.

Sehgal, V. N., V. L. Rege, M. F. Mascarenhas, and M. Reys. 1977. The Prevalence and Pattern of Leprosy in a School Survey. *International Journal of Leprosy and Other Mycobacterial Diseases* 45: 360–363.

Selye, H. 1950. *The Physiology and Pathology of Exposure to Stress*. Oxford: Acta.

Sen, R., S. S. Yadav, V. Singh, P. Sehgal, and V. B. Dixit. 1991. Patterns of Erythropoiesis and Anaemia in Leprosy. *Leprosy Review* 62: 158–170.

Senior, K. 2009. Stigma, Prophylaxis and Leprosy Control. *Lancet Infectious Diseases* 9(1): 10.

Sermrittirong, S., W. H. Van Brakel, and J. F. G. Bunbers-Aelen. 2014. How to Reduce Stigma in Leprosy—A Systematic Literature Review. *Leprosy Review* 85: 149–157.

Setia, M. S., C. Steinmaus, C. S. Ho, G. W. Rutherford, and S. B. Lucia. 2006. The Role of BCG in Prevention of Leprosy: A Meta-Analysis. *Lancet Infectious Diseases* 6: 162–170.

Sharma, J., S. Gaorola, Y. P. Sharma, and R. D. Gaur. 2014. Ethnomedicinal Plants to Treat Skin Diseases by Tharu Community of District Udham Singh Nagar, Uttarakhand, India. *Journal of Ethnopharmacology* 158: 140–206.

Sharma, R., P. Singh, W. J. Loughry, J. M. Lockhart, B. W. Inman, M. S. Duthie, M. T. Pena, L. A. Marcos, D. M. Scollard, S. T. Cole, and R. W. Truman. 2015. Zoonotic Leprosy in the Southeastern United States. *Emerging Infectious Diseases* 21: 2127–2134.

Shen, J., M. Liu, M. Zhou, and W. Li. 2011. Causes of Death among Active Leprosy Patients in China. *International Journal of Dermatology* 50: 57–60.

Shennan, S. 1978. Report on the Skeletons from Bevis Grave, Bedhampton, Hampshire. Unpublished.

Shepard, C. C., and D. H. McRae. 1965. *Mycobacterium leprae* in Mice: Minimal Infectious

Dose, Relationship between Staining Quality and Infectivity, and Effect of Cortisone. *Journal of Bacteriology* 89: 365.

Shepard, C. C., and E. W. Saitz. 1967. Lepromin and Tuberculin Reactivity in Adults Not Exposed to Leprosy. *Journal of Immunology* 99: 637–642.

Shinde, V., H. Newton, R. M. Sakamuri, V. Reddy, S. Jain, A. Joseph, T. Gillis, I. Nath, G. Norman, and V. Vissa. 2009. VNTR Typing of *Mycobacterium leprae* in South Indian Leprosy Patients. *Leprosy Review* 80: 290–301.

Shumin, C., L. Diangchang, L. Bing, Z. Lin, and Y. Xioulu. 2003. Role of Leprosy Villages and Leprosaria in Shandong Province, People's Republic of China: Past, Present and Future. *Leprosy Review* 74: 222–228.

Siddiqui, M. R., S. Meisner, K. Tosh, K. Balakrishnan, S. Ghei, S. E. Fisher, M. Golding, N. P. Shanker Narayan, T. Sitaraman, U. Sengupta, R. Pitchappan, and A. V. Hill. 2001. A Major Susceptibility Locus for Leprosy in India Maps to Chromosome 10p13. *Nature Genetics* 27: 439–441.

Silla, E. 1998. *People Are Not the Same: Leprosy and Identity in Twentieth-Century Mali*. Oxford: James Currey.

Silva Nery, J. S., A. Ramond, J. M. Pescarini, A. Alves, A. Strina, M. Y. Ichihara, M. L. Fernandes Penna, L. Smeeth, L. C. Rodrigues, M. L. Barreto, E. B. Brickley, and G. O. Penna. 2019. Socioeconomic Determinants of Leprosy New Case Detection in the 100 Million Brazilian Cohort: A Population-Based Linkage Study. *Lancet Global Health* 7(9):e1226–e1236.

Singh, A. V., and D. S. Chauhan. 2018. *Mycobacterium lepromatosis* Lepromatous Leprosy in US Citizen Who Traveled to Disease-Endemic Areas. *Emerging Infectious Diseases* 24: 951–952.

Singh, G. P. 2012. Psychosocial Aspects of Hansen's Disease (Leprosy). *Indian Dermatology Online Journal* 3: 166–170.

Singh, P., A. Benjak, V. J. Schuenemann, A. Herbig, C. Avanzi, P. Busso, K. Nieselt, J. Krause, L. Vera-Cabrera, and S. T. Cole. 2015. Insight into the Evolution and Origin of Leprosy Bacilli from the Genome Sequence of *Mycobacterium lepromatosis*. *Proceedings of the National Academy of Sciences USA* 112: 4459–4464.

Singh, S., A. K. Sinha, B. G. Banerjee, and N. Jaswal. 2009. Participation Level of Leprosy Patients in Society. *Indian Journal of Leprosy* 81: 181–7.

Skamene, E., P. Gros, A. Forget, P. A. Kongshavn, C. St Charles, and B. A. Taylor. 1982. Genetic Regulation of Resistance to Intracellular Pathogens. *Nature* 297(5866): 506–509.

Skeates, R. 2010. *An Archaeology of the Senses: Prehistoric Malta*. Oxford: Oxford University Press.

Skinsnes, O. K. 1964. Leprosy in Society III. The Relationship of the Social to the Medical Pathology of Leprosy. *Leprosy Review* 34: 175–181.

———. 1977. Coughing, Sneezing and Mosquitoes in the Transmission of Leprosy. *International Journal of Leprosy and Other Mycobacterial Diseases* 43: 378–381.

———. 1980. Leprosy in Archeologically Recovered Bamboo Book in China. *International Journal of Leprosy and Other Mycobacterial Diseases* 48: 333.

Skinsnes, O. K., and P. H. C. Chang. 1985. Understanding of Leprosy in Ancient China. *International Journal of Leprosy and Other Mycobacterial Diseases* 53: 289–307.

Šlaus, M. 2006a. *Bioarheologija. Demografija, Zdravlie, Trauma I. Prehan Starohrvatskih Populacija.* Zagreb: Školska knjiga.

———. 2006b. Osteological and Dental Markers of Health in the Transition from the Late Antique to the Early Medieval Period in Croatia. *American Journal of Physical Anthropology* 136: 455–469.

———. 2010. Osteological Evidence of Leprosy in an 8th–9th Century Cemetery from Croatia. Paper presented at the 18th European meeting of the Paleopathology Association, Vienna, Austria.

Šlaus, M., N. Novak, and M. Vodanović. 2011. Croatia/Hrvatska. In *The Routledge Handbook of Archaeological Human Remains and Legislation,* edited by N. Márquez-Grant and L. Fibiger, 83–96. London: Routledge.

Smith, B. B., B. E. Crawford, G. Peterson, and J. Thomas. 2004. The Papar Project: Papa Stour (Walls & Sandness Parish). *Discovery and Excavation in Scotland* 5: 119. Unwin.

Smith, G. E. 1908. Anatomical Report. *Bulletin of the Archaeological Survey of Nubia* 1: 25–35.

Smith, G. E., and W. R. Dawson. 1924. *Egyptian Mummies.* London: George Allen and Smith, G. E., and D. E. Derry. 1910. Anatomical Report. *Bulletin of the Archaeological Survey of Nubia* 6: 9–30.

Smith, G. E., and F. W. Jones. 1910. *The Archaeological Survey of Nubia Report for 1907–1908.* Vol. 2, *Report on the Human Remains.* Cairo: National Print Department.

Smith, J. C., E. D. Brown, E. G. McDaniel, and W. Chan. 1976. Alterations in Vitamin A Metabolism during Zinc Deficiency and Food and Growth Restriction. *Journal of Nutrition* 106: 589–191.

Smith, W. C. 2004. What Is the Best Way to Use BCG to Protect against Leprosy: When, for Whom and How Often? *International Journal of Leprosy and Other Mycobacterial Diseases* 72: 8–15.

Sommerfelt, H., L. M. Irgens, and M. Christian. 1985. Geographical Variation in the Occurrence of Leprosy: Possible Roles Played by Nutrition and Some Other Environmental Factors. *International Journal of Leprosy and Other Mycobacterial Diseases* 53: 524–532.

Soni, N. K. 1988. Radiological Study of the Paranasal Sinuses in Lepromatous Leprosy. *Indian Journal of Leprosy* 60:285–289.

———. 1989. Antroscopic Study of the Maxillary Antrum in Lepromatous Leprosy. *Journal of Laryngology and Otology* 103: 502–503.

———. 1992. Leprosy of the Larynx. *Journal of Laryngology and Otology* 106: 518–520.

Spickett, S. G. 1962a. Genetics and the Epidemiology of Leprosy: I. The Incidence of Leprosy. *Leprosy Review* 33: 76–93.

———. 1962b. Genetics and the Epidemiology of Leprosy: II. The Form of Leprosy. *Leprosy Review* 33: 173–181.

Spigelman, M., and H. D. Donoghue. 2001. Brief Communication. Unusual Pathological Condition in the Lower Extremities of a Skeleton from Ancient Israel. *American Journal of Physical Anthropology* 114: 92–93.

———. 2002. The Study of Ancient DNA Answers a Palaeopathological Question. In *The Past and Present of Leprosy: Archaeological, Historical, Palaeopathological and Clinical*

Approaches, edited by C. A. Roberts, M. E. Lewis, and K. Manchester, 293–296. British Archaeological Reports International Series 1054. Oxford: Archaeopress.

Springer, K. W., O. Hankivsky, and L. M. Bates. 2012. Gender and Health: Relational, Intersectional, and Biosocial Approaches. *Social Science and Medicine* 74: 1661–1666.

Squires, K., T. Booth, and C. A. Roberts. 2019. The Ethics of Sampling Human Skeletal Remains for Destructive Analyses: A UK Perspective. In *Ethical Challenges in the Analysis of Human Remains,* edited by K. Squire, D. Errickso, and N. Márquez-Grant, 265–297. Amsterdam: Elsevier.

Sreevatasan, H. 1993. Leprosy and Arthropods. *Indian Journal of Leprosy* 65: 189–200.

Srinivasan, H. 1976. Heel Ulcers in Leprosy Patients. *Leprosy in India* 48: 355–361.

Srinivasan, H., and Dharmendra. 1978a. Deformities of Hands. In *Leprosy,* edited by Dharmendra, 205–217. Bombay: Kothari Medical Publishing House.

———. 1978b. Deformities of Feet. In *Leprosy,* edited by Dharmendra, 218–236. Bombay: Kothari Medical Publishing House.

———. 1978c. Neuropathic Ulceration. In *Leprosy,* edited by Dharmendra, 224–236. Bombay: Kothari Medical Publishing House.

———. 1978d. Deformities in Leprosy (General Considerations). In *Leprosy,* edited by Dharmendra, 197–204. Bombay: Kothari Medical Publishing House.

Stanford, J. L. 1995. Reply (1). *International Journal of Osteoarchaeology* 5: 300–301.

Stang, F. 1895. *St Jørgen's Hospital i Bergen.* Bergen: Kristiana.

Stearns, A. T. 2002. Leprosy: A Problem Solved by 2000? *Leprosy Review* 73: 215–224.

Stearns, S. C. 2012. Darwin Review. Evolutionary Medicine: Its Scope, Interest, and Potential. *Proceedings of the Royal Society B* 279: 4305–4312.

Stearns, S. C., and R. Medzhitov. 2016. *Evolutionary Medicine.* Sunderland, MA: Sinauer Associates.

Steckel, R. H., C. S. Larsen, C. A. Roberts, and J. Baten, eds. 2018. *The Backbone of Europe: Health, Diet, Work and Violence over Two Millennia.* Cambridge: Cambridge University Press.

Stefansky, V. K. 1902. Zabolevanija Ukrys, Vyzvannyja Kislotoupornoj Palotsjkoj. *Russkij Vratsj* 47: 1726–1727.

Stein, S., and L. G. Blochman. 1973. *Alone no Longer.* Carville, LA: The Star.

Steinbock, R. T. 1976. *Paleopathological Diagnosis and Interpretation: Bone Diseases in Ancient Human Populations.* Springfield, IL: Charles. C. Thomas.

Štěpán, M. 1963. Epidemie v Českych Zemich v Době Předhusitké (Epidemics in Czech Countries before the Time of the Hussite Wars). *Čs. Zdravotnictví* 13: 303.

Stini, W. A. 1985. Growth Rates and Sexual Dimorphism in Evolutionary Perspective. In *The Analysis of Prehistoric Diets,* edited by R. I. Gilbert and J. H. Mielke, 191–226. London: Academic Press.

Stinson, S. 1985. Sex Differences in Environmental Sensitivity during Growth and Development. *Yearbook of Physical Anthropology* 28: 123–147.

Stone, A. C. 2008. DNA Analysis of Archaeological Remains. In *Biological Anthropology of the Human Skeleton,* 2nd ed., edited by M. A. Katzenberg and S. R. Saunders, 461–483. New York: Wiley-Liss.

Storrs, E. E., G. P. Walsh, H. P. Burchfield, and C. H. Binford. 1974. Leprosy in the Armadillo: A New Model for Biomedical Research. *Science* 183: 851.

Strouhal, E., L. Horáčkova, J. Likovsky, L. Vargová, and J. Daneš. 2002. Traces of Leprosy from the Czech Kingdom. In *The Past and Present of Leprosy: Archaeological, Historical, Palaeopathological and Clinical Approaches*, edited by C. A. Roberts, M. E. Lewis, and K. Manchester, 223–232. British Archaeological Reports International Series 1054. Oxford: Archaeopress.

Stuart-Macadam, P. 1985. Porotic Hyperostosis: Representative of a Childhood Condition. *American Journal of Physical Anthropology* 66: 391–398.

———. 1992. Porotic Hyperostosis: A New Perspective. *American Journal of Physical Anthropology* 87: 39–47.

Subramaniam, K., S. C. Marks, and S. H. Nah. 1983. Rate of Bone Loss of Maxillary Anterior Alveolar Bone Height in Patients with Leprosy. *Leprosy Review* 54: 119–127.

Subramaniam, K., S. H. Nah, and S. C. Marks. 1994. A Longitudinal Study of Alveolar Bone Loss around Maxillary Central Incisors in Patients with Leprosy in Malaysia. *Leprosy Review* 65: 137–142.

Sundelin, A., and A. Sörman. 2004. *Skammens Hud: Om Spetälska i Sverige*. Stockholm: Bokförl, D. N.

Suttles, W., ed. 1990. *Handbook of North American Indians*. Vol. 7, *Northwest Coast*. Washington, DC: Smithsonian Institution.

Suzuki, J., T. Oshima, K. Watanabe, H. Suzuki, T. Kobayashi, and S. Hashimoto. 2013. Chronic Rhinosinusitis in Ex-Lepromatous Leprosy Patients with Atrophic Rhinitis. *Journal of Laryngology and Otology* 127: 265–270.

Suzuki, K., A. Saso, K. Hoshino, J. Sakurai, K. Tanigawa, Y. Luo, Y. Ishido, S. Mori, K. Hirata, and N. Ishii. 2014. Paleopathological Evidence and Detection of *Mycobacterium leprae* DNA from Archaeological Skeletal Remains of Nabe-kaburi (Head-Covered with Iron Pots) Burials in Japan. *PLoS ONE* 9(2): e88356.

Suzuki, K., W. Takigawa, K. Tanigawa, K. Nakamura, Y. Ishido, A. Kawashima, H. Wu, T. Akama, M. Sue, A. Yoshihara, S. Mori, and N. Ishii. 2010. Detection of *Mycobacterium leprae* DNA from Archaeological Skeletal Remains in Japan Using Whole Genome Amplification and Polymerase Chain Reaction. *PLoS ONE* 5(8): e12422.

Suzuki, K., T. Udono, M. Fujisawa, K. Tanigawa, G. Idani, and N. Ishii. 2010. Infection during Infancy and Long Incubation Period for Leprosy Suggested in the Case of a Chimpanzee Used for Medical Research. *Journal of Clinical Microbiology* 48: 3432–3434.

Suzuki, T. 1986. Paleopathological and Paleoepidemiological Study on Human Skeletal Remains from the Mariana Islands. In *Anthropological Studies on the Origin of Pacific Populations, with Special Reference to the Micronesians: A Preliminary Report*, edited by K. Hanihara, 15–57. Tokyo: Tokyo Ministry of Education, Science and Culture.

———. 2001. Leprosy. *The Bone* 15(5): 507–511. (In Japanese.)

Tadesse Argaw, A., E. J. Shannon, A. Assefa, F. Silassie Mikru, B. Kidane Mariam, and J. B. Malone. 2006. A Geospatial Risk Assessment Model for Leprosy in Ethiopia Based on Environmental Thermal-Hydrological Regime Analysis. *Geospatial Health* 1: 105–113.

Tagaya, A. 2014. Human Remains from Tappeh Sang-e Chakhmaq. In *The First Farming Village in Northeast Iran and Turan: Tappeh Sang-e Chakhmaq and Beyond*, edited by A. Tsuneki, 39–42. Tsukuba: Research Center for West Asian Civilization, University of Tsukuba.

Taheri, J. B., H. Mortazavi, M. Moshfeghi, S. Bakhtiari, S. Azari-Marhabi, and S. Alirezaei.

2012. Oro-Facial Manifestations of 100 Leprosy Patients. *Medicina Oral, Patologia Oral y Cirugia Bucal* 17: e728–732.

Tait, E. S. R. 1939. The Lepers of Papa Stour. *Hjaltland Miscellany* 3: 24–39.

Tajiri, I. 1936. Leprosy and Childbirth. *International Journal of Leprosy and Other Mycobacterial Diseases* 4: 189–195.

Talhari, C., M. T. Mira, C. Massone, A. Braga, A. Chruscuak-Talhari, M. Santos, A. T. Orsi, C. Matsuo, R. Rabelo, L. Nogueira, et al. 2010. Leprosy and HIV Coinfection: A Clinical, Pathological, Immunological, and Therapeutic Study of a Cohort from a Brazilian Referral Center for Infectious Diseases. *Journal of Infectious Diseases* 202: 345–354.

Tayles, N. 1996. Tooth Ablation in Prehistoric Southeast Asia. *International Journal of Osteoarchaeology* 6: 333–345.

Tayles, N., and H. R. Buckley. 2004. Leprosy and Tuberculosis in Iron Age Southeast Asia? *American Journal of Physical Anthropology* 125: 239–256.

Tayles, N., S. Halcrow, and N. Pureepatpong. 2012. Regional Developments: Southeast Asia. In *The Global History of Paleopathology: Pioneers and Prospects*, edited by J. Buikstra and C. Roberts, 528–540. Oxford: Oxford University Press.

Taylor, G. M., S. Blau, S. A. Mays, M. Monot, C. O.-Y. Lee, D. E. Minnikin, G. S. Besra, S. T. Cole, and P. C. Rutland. 2009. *Mycobacterium leprae* Genotype Amplified from an Archaeological Case of Lepromatous Leprosy in Central Asia. *Journal of Archaeological Science* 36: 2408–2414.

Taylor, G. M., and H. D. Donoghue. 2011. Multiple Loci Variable Number Tandem Repeat (VNTR) Analysis (MLVA) of *Mycobacterium leprae* Isolates Amplified from European Archaeological Human Remains with Lepromatous Leprosy. *Microbes and Infection* 13: 923–929.

Taylor, G. M., E. M. Murphy, T. A. Mendum, A. W. G. Pike, B. Linscott, H. Wu, J. O'Grady, H. Richardson, E. O'Donovan, C. Troy, and G. R. Stewart. 2018. Leprosy at the Edge of Europe—Biomolecular, Isotopic and Osteoarchaeological Findings from Medieval Ireland. *PLoS ONE* 13(December 26): e.0209495.

Taylor, G. M., K. Tucker, R. Butler, A. W. Pike, J. Lewis, S. Roffey, P. Marter, O. Y. Lee, H. H. Wu, D. E. Minnikin, G. S. Besra, et al. 2013. Detection and Strain Typing of Ancient *Mycobacterium leprae* from a Medieval Leprosy Hospital. *PLoS ONE* 8(4): e62406. https://journals.plos.org/plosone/article?id=10.1371/journal.pone.0062406.

Taylor, G. M., C. L. Watson, A. S. Bouwman, D. N. J. Lockwood, and S. A. Mays. 2006. Variable Nucleotide Tandem Repeat (VNTR) Typing of Two Palaeopathological Cases of Lepromatous Leprosy from Mediaeval England. *Journal of Archaeological Science* 33: 1569–1579.

Taylor, G. M., S. Widdison, I. N. Brown, and D. Young. 2000. A Mediaeval Case of Lepromatous Leprosy from 13th–14th Century Orkney, Scotland. *Journal of Archaeological Science* 27: 1133–1138.

Taylor, G. M., D. B. Young, and S. A. Mays. 2005. Genotypic Analysis of the Earliest Known Prehistoric Case of Tuberculosis in Britain. *Microbiology* 43: 2236–2240.

Taylor, J. 2015. *Body by Darwin: How Evolution Shapes Our Health and Transforms Medicine.* Chicago, IL: University of Chicago Press,

Taylor, J. P., I. Vitek, V. Enriquez, and J. W. Smedley. 1999. A Continuing Focus of Hansen's Disease in Texas. *American Journal of Tropical Medicine and Hygiene* 60: 449–452.

Tayman, J. 2006. *The Colony: The Harrowing True Story of the Exiles of Molokai.* London: Scribner.

Teixeira, C. S. S., D. S. Medeiros, C. H. Alencar, A. N. Ramos Júnior, and J. Heukelbach. 2019. Nutritional Aspects of People Affected by Leprosy, between 2001 and 2014, in Semi-Arid Brazilian Municipalities. *Ciência y Saúde Coletiva* 24: 2431–2441.

Teschler-Nicola, M., and G. Grupe. 2012. Paleopathology in Germanic Countries. In *The Global History of Paleopathology: Pioneers and Prospects,* edited by J. Buikstra and C. Roberts, 387–404. Oxford: Oxford University Press.

Thappa, D. M., V. K. Sharma, S. Kaur, and S. Suri. 1992. Radiological Changes in Hands and Feet in Disabled Leprosy Patients: A Clinicoradiological Correlation. *Indian Journal of Leprosy* 64: 58–66.

Thilakavathi, S., P. Manickam, and S. M. Mehendale. 2012. Awareness, Social Acceptance and Community Views on Leprosy and Its Relevance for Leprosy Control, Tamil Nadu. *Indian Journal of Leprosy* 84: 233–240.

Thomson, T. A. 1854. On the Peculiarities in the Disfigurations and Customs of New Zealanders; with Remarks on Their Diseases and Their Mode of Treatment. *British and Foreign Medico-Chirurgical Reviews* 13: 489–502.

Tiendrebeogo, A., I. Toyre, and P-J. Zerbo. 1996. A Survey of Leprosy Impairments and Disabilities among Patients Treated by MDT in Burkina Faso. *International Journal of Leprosy* 64: 15–25.

Tió-Coma, M., H. Sprong, M. Kik, J. T. van Dissel, X. Y. Han, T. Pieters, and A. Geluk. 2019. Lack of Evidence for the Presence of Leprosy Bacilli in Red Squirrels from North-West Europe. *Transboundary and Emerging Diseases:* doi 10.1111.

Tió-Coma, M., T. Wijnands, L. Pierneef, A. K. Schilling, K. Alam, J. Chandra Roy, W. R. Faber, H. Menke, T. Pieters, K. Stevenson, J. H. Richardus, and A. Geluk. 2019. Detection of *Mycobacterium leprae* DNA in Soil: Multiple Needles in the Haystack. *Scientific Reports* 9: Article 3165.

Tisseuil, J. 1975. La Régression de la Lèpre ne fut-elle pas Aussi Fonction de l'Évolution Éconimique au 14e siècle. *Bulletin de la Société de Pathologie Exotique* 68: 33–37.

Trapnell, D. H. 1965. Calcification of Nerves in Leprosy. *British Journal of Radiology* 38: 796–797.

Trembly, D. 1995. On the Antiquity of Leprosy in Western Micronesia. *International Journal of Osteoarchaeology* 5: 377–384.

———. 2002. Perspectives on the History of Leprosy in the Pacific. In *The Past and Present of Leprosy: Archaeological, Historical, Palaeopathological and Clinical Approaches,* edited by C. A. Roberts, M. E. Lewis, and K. Manchester, 233–238. British Archaeological Reports International Series 1054. Oxford: Archaeopress.

Trevathan, W. R., E. O. Smith, and J. J. McKenna, eds. 2008. Introduction and Overview of Evolutionary Medicine. In *Evolutionary Medicine and Health: New Perspectives,* edited by W. R. Trevathan, E. O. Smith, and J. J. McKenna, 1–11. Oxford: Oxford University Press.

Trimble, D., S. Unsworth, and T. Hurley. 1991. Excavation of a Medieval Hospital Cemetery in Grantham. *Lincolnshire Past and Present* 5: 10–11.

Troy, C. 2010. Final Report on the Human Remains from Ardreigh, Co. Kildare. Headland Archaeology (Ireland) Ltd. Unpublished.

Truman, R. W. 2005. Leprosy in Wild Armadillos. *Leprosy Review* 76: 198–208.

Truman, R. W., and P. E. Fine. 2010. Environmental Sources of *Mycobacterium leprae:* Issues and Evidence. *Leprosy Review* 81:2233–2239.

Truman, R. W., A. B. Fontes, A. B. de Miranda, P. Suffys, and T. Gillis. 2004. Genotypic Variation and Stability of Four Variable-Number Tandem Repeats and Their Suitability for Discriminating Strains of *Mycobacterium leprae. Journal of Clinical Microbiology* 42: 2558–2565.

Truman, R. W., and J. J. Krahenbuhl. 2001. Viable *M. leprae* as a Research Reagent. *International Journal of Leprosy* 69: 1–12.

Truman, R. W., P. Singh, R. Sharma, P. Busso, J. Rougemont, A. Paniz-Mondolfi, A. Kapopoulo, S. Brisse, D. M. Scollard, T. P. Gillis, and S. T. Cole. 2011. Probable Zoonotic Leprosy in the Southern United States. *New England Journal of Medicine* 364: 1626–1633.

Tukhbatova, R. I., V. J. Schuenemann, V. I. Selezneva, N. Berezina, and A. Buzhilova. 2016. Morphological and Genetic Approaches for Confirmation of *leprae* on the Skull from Rokhlin's Paleopathological Collection (St Petersburg). Poster presented at the 21st European meeting of the Paleopathology Association, Moscow, Russia.

Turankar, R. P., M. Lavania, J. Darlong, K. S. R. Siva Sai, U. Sengupta, and R. S. Jadhav. 2019. Survival of *Mycobacterium leprae* and Association with Acanthamoeba from Environmental Samples in the Inhabitant Areas of Active Leprosy Cases: A Cross-Sectional Study from Endemic Pockets of Purulia, West Bengal. *Infection, Genetics and Evolution* 72: 199–204.

Turankar, R. P., M. Lavania, M. Singh, U. Sengupta, K. Siva Sai, and R. S. Jadhav. 2016. Presence of Viable *Mycobacterium leprae* in Environmental Specimens around Houses of Leprosy Patients. *Indian Journal of Medical Microbiology* 34: 315–321.

Turankar, R. P., M. Lavania, M. Singh, K. S. Siva Sai, and R. S. Jadhav. 2012. Dynamics of *Mycobacterium leprae* Transmission in Environmental Context: Deciphering the Role of Environment as a Potential Reservoir. *Infection, Genetics and Evolution* 12: 121–126.

Turk, J. L., and R. J. W. Rees. 1988. Editorial: AIDS and Leprosy. *Leprosy Review* 59: 193–194.

Tze-Chun, L., and Q. Ju-Shi. 1984. Pathological Findings on Peripheral Nerves, Lymph Nodes, and Visceral Organs of Leprosy. *International Journal of Leprosy and Other Mycobacterial Diseases* 52: 377–383.

Ulrich, M., A. M. Zulueta, G. Cácares-Dittmar, C. Sampson, M. E. Pinardi, E. M. Rada, and N. Aranzazu. 1993. Leprosy in Women: Characteristics and Repercussions. *Social Science and Medicine* 37: 445–456.

Underhill, D. M., A. Ozinsku, K. D. Smith, and A. Aderem. 1999. Toll-Like Receptor-2 Mediates Mycobacteria-Induced Proinflammatory Signals in Macrophages. *Proceedings of the National Academy of Sciences USA* 96: 14459–14463.

Ustianowski, A. P., and D. N. Lockwood. 2003. Leprosy: Current Diagnostic and Treatment Approaches. *Current Opinion in Infectious Diseases* 16: 421–427.

Üstündağ, H. 2011. Turjey/Türkiye. In *The Routledge Handbook of Archaeological Human Remains and Legislation,* edited by N. Márquez-Grant and L. Fibiger, 455–467. London: Routledge.

Van Buynder, P., J. Eccleston, J. Leese, and D. N. J. Lockwood. 1999. Leprosy in England and Wales. *Communicable Disease and Public Health* 2: 119–121.

Van Schie, C. H., F. J. Slim, R. Keukenkamp, W. R. Faber, and F. Nollet. 2013. Plantar Pressure and Daily Cumulative Stress in Persons Affected by Leprosy with Current, Previous and No Previous Foot Ulceration. *Gait & Posture* 37: 326–330.

Varkevisser, C. M., P. Lever, O. Alubo, K. Burathoki, C. Idawani, T. M. Moreira, P. Patroba, and M. Yulizar. 2009. Gender and Leprosy: Case Studies in Indonesia, Nigeria, Nepal and Brazil. *Leprosy Review* 80: 65–76.

Venkatatraju, B., and S. Prasad. 2013. Psychosocial Trauma of Diagnosis: A Qualitative Study on Rural TB Patients' Experiences in Nalgonda District, Andra Pradesh. *Indian Journal of Tuberculosis* 60: 162–167.

Verma, C., P. S. Rao, and M. S. Raju. 2011. Public Awareness on Integration of Leprosy Services at Primary Health Centres in Uttar Pradesh, India. *Indian Journal of Leprosy* 83: 95–100.

Vlassoff, C., S. Khot, and S. Rao. 1996. Double Jeopardy: Women and Leprosy in India. *World Health Statistics Quarterly* 49: 120–126.

Vogelsang, T. M. 1965. Leprosy in Norway. *Medical History* 9: 29–35.

———. 1978. Gerhard Henrik Armauer Hansen 1841–1912: The Discoverer of the Leprosy Bacillus. His Life and His Work. *International Journal of Leprosy and Other Mycobacterial Diseases* 46: 257–332.

Von Hunnius, T., C. A. Roberts, A. Boylston, and S. R. Saunders. 2006. Histological Identification of Syphilis in Pre-Columbian England. *American Journal of Physical Anthropology* 129: 559–566.

Vuorinen, H. S. 2002. History of Leprosy in Finland. In *The Past and Present of Leprosy: Archaeological, Historical, Palaeopathological and Clinical Approaches*, edited by C. A. Roberts, M. E. Lewis, and K. Manchester, 239–246. British Archaeological Reports International Series 1054. Oxford: Archaeopress.

Waaler, E. 1953. Changes in the Maxillary Bone in Leprosy. *International Journal of Leprosy and Other Mycobacterial Diseases* 21: 617.

Wachholz, L. 1921. *Szpitale Krakowskie 1220–1920*. Kraków: WL Anczyc I. Sp.

Wade, H. W., and V. Ledowski. 1952. The Leprosy Epidemic at Nauru: A Review with Data on the Status since 1937. *International Journal of Leprosy* 20: 1–29.

Wagenaar, I., L. van Muiden, K. Alam, R. Bowers, M. A. Hossain, K. Kispotta, and J. H. Richardus. 2015. Diet-Related Risk Factors for Leprosy: A Case-Control Study. *PLoS Neglected Tropical Diseases* 9(5): e0003766. https://www.ncbi.nlm.nih.gov/pmc/articles/PMC4428634/.

Waldron, T. 1994a. *Counting the Dead: The Epidemiology of Skeletal Populations*. Chichester: John Wiley and Sons.

———. 1994b. The Human Remains. In V. I. Evison, *An Anglo-Saxon Cemetery at Great Chesterford, Essex*, 52–66. Council for British Archaeology Research Report 91. York: Council for British Archaeology.

Walker, D. 2009. The Treatment of Leprosy in 19th-Century London: A Case Study from St Marylebone Cemetery. *International Journal of Osteoarchaeology* 19: 364–374.

———. 2012. *Disease in London, 1st–19th Centuries: An Illustrated Guide to Diagnosis*. London: Museum of London Archaeology.

Walker, J. 1934. *Folk Medicine in Modern Egypt: Selections from the* Tibb al-Rukka *or Old Wives' Medicine of ʿAbd al-Rahmān Ismāʿīl*. London: Luzac.

Walker, P. L. 2008. Bioarchaeological Ethics: A Historical Perspective on the Value of Human Remains. In *Biological Anthropology of the Human Skeleton,* edited by M. A. Katzenberg and S. R. Saunders, 3–40. Chichester: Wiley-Liss.

Walker, P. L., R. R. Bathurst, R. Richman, T. Gjerdrum, and V. A. Andrushko. 2009. The Cause of Porotic Hyperostosis and Cribra Orbitalia: A Reappraisal of the Iron-Deficiency Anemia Hypothesis. *American Journal of Physical Anthropology* 139: 109–125.

Walker, P. L., and D. C. Cook. 1998. Brief Communication. Gender and Sex: Vive la Différence. *American Journal of Physical Anthropology* 106: 255–259.

Walsh, G. P., W. M. Meyers, P. J. Binford, P. J. Gerone, R. H. Wolf, and J. R. Leininger. 1981. Leprosy—A Zoonosis. *Leprosy Review* 52 supplement 1: 77–82.

Walsh, G. P., W. M. Meyers, P. J. Binford, B. J. Gormus, G. B. Baskin, R. H. Wold, and P. J. Gerone. 1988. Leprosy as a Zoonosis: An Update. *Acta Leprologica* 6: 51–60.

Walsh, G. P., E. E. Storrs, H. P. Burchfield, E. H. Cottrell, M. F. Vidrine, and C. H. Binford. 1975. Leprosy-Like Disease Occurring Naturally in Armadillos. *Journal of the Reticuloendothelial Society* 18: 347.

Walsh, G. P., E. E. Storrs, and W. M. Meyers. 1977. Naturally Acquired Leprosy-Like Disease in the Nine-Banded Armadillo (*Dasypus novemcinctus*): Recent Epizootic Findings. *Journal of the Reticuloendothelial Society* 22: 363–367.

Wang, C., N. Wang, Y. Yu, G. Yu, Z. Wang, X. Fu, H. Liu, and F. Zhang. 2015. Tuberculosis Risk-Associated Single Nucleotide Polymorphisms Do Not Show Association with Leprosy in Chinese Population. *International Journal of Infectious Diseases* 35: 1–2.

Wang, D., J. Q. Feng, Y. Y. Li, D. F. Zhang, X. A. Li, Q. W. Li, and Y. G. Yao. 2012. Genetic Variants of the MRC1 Gene and the IFNG Gene Are Associated with Leprosy in Han Chinese from Southwest China. *Human Genetics* 131: 1251–1260.

Wang, D., D. F. Zhang, G. D. Li, R. Bi, Y. Fan, Y. Wu, X. F. Yu, H. Long, Y. Y. Li, and Y. G. Yao. 2017. A Pleiotropic Effect of the APOE Gene: Association of APOE Polymorphisms with Multibacillary Leprosy in Han Chinese from Southwest China. *British Journal of Dermatology* 178: 931–939.

Warinner, C., J. F. M. Rodrigues, R. Vyas, N. Shved, J. Grossman, A. Radini, Y. Hancock, R. Y. Tito, S. Fiddyment, C. Speller, et al. 2014. Pathogens and Host Immunity in the Ancient Human Oral Cavity. *Nature Genetics* 46: 336–344.

Washburn, W. L. 1950. Leprosy among Scandinavian Settlers in the Upper Mississippi Valley 1932–1964. *Bulletin of the History of Medicine* 24: 123–248.

Waters, M. F. R. 1981. Leprosy. *British Medical Journal* 283: 1321–1324.

Waters, M. F. R., I. B. Bakri, H. J. Isa, R. J. Rees, and A. C. McDougall. 1978. Experimental Lepromatous Leprosy in the White-Handed Gibbon (*Hylobatus lar.*): Successful Inoculation with Leprosy Bacilli of Human Origin. *British Journal of Experimental Pathology* 59: 551–557.

Watson, C. L., E. Popescu, J. Boldsen, M. Slaus, and D. N. J. Lockwood. 2009. Single Nucleotide Polymorphism Analysis of European Archaeological *M. leprae* DNA. *PLoS ONE* 4(10): e7547. https://journals.plos.org/plosone/article?id=10.1371/journal.pone.0007547.

———. 2010. Correction: Single Nucleotide Polymorphism Analysis of European Archaeological *M. leprae* DNA. *PLoS ONE* 5(1). https://journals.plos.org/plosone/article?id=10.1371/journal.pone.0007547.

Webb, S. 1995. *Palaeopathology of Aboriginal Australians: Health and Disease across a Hunter-Gatherer Continent*. Cambridge: Cambridge University Press.

Weiss, D. L., and V. Møller-Christensen. 1971a. Leprosy, Echinococcosis and Amulets: A Study of a Medieval Danish Inhumation. *Medical History* 15: 260–267.

———. 1971b. An Unusual Case of Tuberculosis in a Medieval Leper. *Danish Medical Bulletin* 18: 11–14.

Wells, C. 1962. A Possible Case of Leprosy from a Saxon Cemetery at Beckford. *Medical History* 6: 383–387.

———. 1967. A Leper Cemetery at South Acre, Norfolk. *Medieval Archaeology* 11: 242–248.

———. 1996. Human Burials. In V. I. Evison and P. Hill, *Two Anglo-Saxon Cemeteries at Beckford, Hereford and Worcester*, 41–66. Council for British Archaeology Research Report 103. York: Council for British Archaeology.

Weng, X., J. Vander Heiden, Y. Xing, J. Liu, and V. Vissa. 2011. Transmission of Leprosy in Qiubei County, Yunnan, China: Insights from an 8-Year Molecular Epidemiology Investigation. *Infection, Genetics and Evolution* 11: 363–374.

Weston, D. A. 2008. Investigating the Specificity of Periosteal Reactions in Pathology Museum Specimens. *American Journal of Physical Anthropology* 137: 48–59.

———. 2009. Brief Communication. Paleohistological Analysis of Pathology Museum Specimens: Can Periosteal Reaction Microsctructure Explain Lesion Etiology? *American Journal of Physical Anthropology* 140: 186–193.

Weymouth, A. 1938. *Through the Leper-Squint: A Study of Leprosy from Pre-Christian Times to the Present Day*. London: Selwyn and Blount.

Wheat, W. H., A. L. Casali, V. Thomas, J. S. Spencer, R. Lahiri, D. L. Williams, G. E. McDonnell, M. Gonzalez-Juarrero, P. J. Brennan, and M. Jackson. 2014. Long-Term Survival and Virulence of *Mycobacterium leprae* in Amoebal Cysts. *PLoS Neglected Tropical Diseases* 8(12): e3405.

White, C. 2005. Explaining a Complex Disease Process: Talking to Patients about Hansen's Disease (Leprosy) in Brazil. *Medical Anthropology Quarterly* 19: 310–333.

———. 2010. Déjà Vu: Leprosy and Immigration Discourse in the Twenty-First Century United States. *Leprosy Review* 81: 17–26.

White, C., and C. Franco-Paredes. 2015. Leprosy in the 21st Century. *Clinical Microbiology Reviews* 28: 80–94.

Wiggins, R., A. Boylston, and C. A. Roberts. 1993. Report on Human Skeletal Remains from Blackfriars, Gloucester (19/91). University of Bradford. Unpublished.

Wiker, H. G., G. G. Tomazella, and G. A. de Souza. 2011. A Quantitative View on *Mycobacterium leprae* Antigens by Proteomics. *Journal of Proteomics* 74: 1711–1719.

Wilbur, A. K., J. E. Buikstra, and C. Stojanowski. 2002. Mycobacterial Disease in North America: An Epidemiological Test of Chaussinand's Cross-Immunity Hypothesis. In *The Past and Present of Leprosy: Archaeological, Historical, Palaeopathological and Clinical Approaches*, edited by C. A. Roberts, M. E. Lewis, and K. Manchester, 247–258. British Archaeological Reports International Series 1054. Oxford: Archaeopress.

Wilbur, A. K., and A. C. Stone. 2012. Using Ancient DNA Techniques to Study Human Disease. In *The Global History of Paleopathology: Pioneers and Prospects*, edited by J. Buikstra and C. Roberts, 703–717. Oxford: Oxford University Press.

Wilczak, C. A., and D. J. Ortner. 2013. A Case of Lepromatous Leprosy from Kodiak Island,

Alaska. Paper presented at the 40th annual meeting of the Paleopathology Association 40th, Knoxville, Tennessee.

Wiley, S., and J. S. Allen. 2013. *Medical Anthropology: A Biocultural Approach*. 2nd ed. New York: Oxford University Press.

Wilkinson, R., and K. Pickett. 2009. *The Spirit Level: Why Equality Is Better for Everyone*. Penguin: London.

———. 2019. *The Inner Level: How More Equal Societies Reduce Stress, Restore Sanity and Improve Everyone's Well-Being*. London: Penguin.

Williams, D. L., and T. P. Gillis. 2012. Drug-Resistant Leprosy: Monitoring and Current Status. *Leprosy Review* 83: 269–281.

Wilson, J. V. K. 1966. Leprosy in Ancient Mesopotamia. *Revue d'Assyriologie* 60: 47–58.

Wilson, K. J. W. 1990. *Ross & Wilson Anatomy and Physiology in Health and Illness*. 7th edition. London: Churchill Livingstone.

Witas, H. W., H. D. Donoghue, D. Kubiak, M. Lewandowska, and J. J. Gladykowska-Rzeczycka. 2015. Molecular Studies on Ancient *M. tuberculosis* and *M. leprae*: Methods of Pathogen and Host DNA Analysis. *European Journal of Clinical Microbiology and Infectious Diseases* 34: 1733–1749.

Wokaunn, M., I. Jurćc, and Z. Vrbica. 2006. Between Stigma and Dawn of Medicine: The Last Leprosarium in Croatia. *Croatian Medical Journal* 47: 759–766.

Wood, J. W., G. R. Milner, H. C. Harpending, and K. M. Weiss. 1992. The Osteological Paradox: Problems of Inferring Prehistoric Health from Skeletal Samples. *Current Anthropology* 33: 343–370.

Wood, S. R. 1991. A Contribution to the History of Tuberculosis and Leprosy in 19th Century Norway. *Journal of the Royal Society of Medicine* 84: 428–430.

Woolgar, C. M. 2006. *The Senses in Late Medieval England*. New Haven, CT: Yale University Press.

World Health Assembly. 1991. *Resolution to Eliminate Leprosy as a Public Health Problem by the Year 2000. (Resolution W. H. A 44.9)*. Geneva: World Health Assembly.

World Health Organization. 1982. *Chemotherapy of Leprosy for Control Programmes: Report of a WHO Study Group*. Technical Report Series 675. Geneva: World Health Organization.

———. 1985. *Epidemiology of Leprosy in Relation to Control: Report of a World Health Organization Study Group*. Technical Report Series 716. Geneva: World Health Organization.

———. 1988a. *Leprosy Prophylaxis: A Report of the WHO Study Group*. WHO Technical Report Series 6: 27–30. Geneva: World Health Organization.

———. 1988b. *A Guide to Leprosy Control*. 2nd ed. Geneva: World Health Organization.

———. 1998a. *Expert Committee on Leprosy*. Technical Report Series 874. Geneva: World Health Organization.

———. 1998b. Leprosy Elimination Campaigns (LECs): Progress during 1997/98. *Weekly Epidemiological Record* 73: 177–182.

———. 1998c. *Model Prescribing Information: Drugs Used in Leprosy*. Geneva: World Health Organization.

———. 2000a. *Guide to Eliminate Leprosy as a Public Health Problem*. Geneva: World Health Organization Leprosy Elimination Group.

———. 2000b. Leprosy—Global Situation. *Weekly Epidemiological Record* 75: 226–231.

———. 2000c. Progress towards Leprosy Elimination. *Weekly Epidemiological Record* 75: 361–368.

———. 2005. *Global Strategy for Further Reducing the Leprosy Burden and Sustaining Leprosy Control Activities (Plan Period 2006–2010)*. Geneva: World Health Organization.

———. 2009a. *Enhanced Global Strategy for Reducing the Disease Burden Due to Leprosy (Plan Period 2011–2015)*. New Delhi: World Health Organization Regional Office for South-East Asia.

———. 2009b. Global Leprosy Situation. *Weekly Epidemiological Record* 84: 333–340.

———. 2013. Global Leprosy: Update on the 2012 Situation. *Weekly Epidemiological Record* 88: 365–380.

———. 2015. Global Leprosy: Update 2014: Need for Early Case Detection. *Weekly Epidemiological Record* 90:461–476.

———. 2016. *Global Leprosy Strategy 2016–2020. Accelerating towards a Leprosy-Free World*. Geneva: World Health Organization.

———. 2017. Global Leprosy: Update 2016: Accelerating Reduction of Disease Burden. *Weekly Epidemiological Record* 92:501–520.

———. 2018. Global Leprosy Update, 2017: Reducing the Disease Burden Due to Leprosy. *Weekly Epidemiological Record* 93: 445–456.

———. 2019. Global Leprosy Update, 2018: Moving toward a Leprosy-Free World. *Weekly Epidemiological Record* 94 (35–36): 389–412.

Worth, R. M. 1963. The Disappearance of Leprosy in a Semi-Isolated Island (Niihao Island, Hawaii). *International Journal of Leprosy* 31: 34–45.

Xavier, M. B., M. G. B. do Nascimento, K. N. M. Batista, D. N. Somensi, F. O. M. Juca Neto, T. X. Carneiro, C. M. C. Gomes, and C. E. P. Corbett. 2018. Peripheral Nerve Abnormality in HIV Leprosy Patients. *PLoS Neglected Tropical Diseases* 12(7): e0006633.

Xing, Y., J. Liu, R. M. Sakamuri, Z. Wang, Y. Wen, V. Vissa, and X. Weng. 2009. VNTR Typing Studies of *Mycobacterium leprae* in China: Assessment of Methods and Stability of Markers during Treatment. *Leprosy Review* 80: 261–271.

Yadin, Y. 1983. *The Temple Scroll*. 3 vols. London: Weidenfeld and Nicolson.

Yamaguchi, N., K. C. Poudel, and M. Jimba. 2013. Health-Related Quality of Life, Depression, and Self-Esteem in Adolescents with Leprosy-Affected Parents: Results of a Cross-Sectional Study in Nepal. *BMC Public Health* 13: 22.

Yang, Q., H. Liu, H.-Q. Low, H. Wang, Y. Yu, X. Fu, G. Yu, M. Chen, X. Yan, S. Chen, et al. 2012. Chromosome 2p14 Is Linked to Susceptibility to Leprosy. *PLoS ONE* 7(1): e29747.

Yawalkar, S. J. 2009. *Leprosy for Medical Practitioners and Paramedical Workers*. 8th rev. ed. Basel: Novartis Foundation for Sustainable Development.

Young, D. B. 2001. Leprosy and the Genome—Not Yet a Burnt-Out Case. *Lancet* 357: 1639–1640.

Yu, R., P. Jarrett, D. Holland, J. Sherwood, and C. Pikholz. 2015. Leprosy in New Zealand: An Epidemiological Update. *New Zealand Medical Journal* 128: 9–14.

Zapletal, V. 1952. The Old Hospitals in Brno. *Lékařské Listy* 7: 347–351. (In Czech.)

Zhang, F., S. Chen, Y. Sun, and T. Chu. 2009. Healthcare Seeking Behaviour and Delay in Diagnosis of Leprosy in a Low Endemic Area of China. *Leprosy Review* 80: 416–423.

Zhang, F. R., W. Huang, S. M. Chen, L. D. Sun, H. Liu, Y. Li, Y. Cui, X. X. Yan, H. T. Yang, R.

D. Yang, et al. 2009. Genomewide Association Study of Leprosy. *New England Journal of Medicine* 361: 2609–2618.

Zhang, F., H. Liu, S. Chen, C. Wang, C. Zhu, L. Zhang, T. Chu, D. Liu, X. Yan, and J. Liu. 2009. Evidence for an Association of HLA-DRB1*15 and DRB1*09 with Leprosy and the Impact of DRB1*09 on Disease Onset in a Chinese Han Population. *BMC Medical Genetics* 10: 133.

Zhenbiao, Z. 1994. The Skeletal Evidence of Human Leprosy and Syphilis in Ancient China. *Acta Anthropologica Sinica* 13: 294–298.

Zias, J. 1985. Leprosy in the Byzantine Monasteries of the Judaean Desert. *Koroth* 9: 242–248.

———. 1991. Leprosy and Tuberculosis in the Byzantine Monasteries of the Judean Desert. In *Human Paleopathology: Current Syntheses and Future Options*, edited by D. J. Ortner and A. C. Aufderheide, 197–199. Washington, DC: Smithsonian Institution Press.

———. 1995. Reply (2). *International Journal of Osteoarchaeology* 5: 301.

———. 2002. New Evidence for the History of Leprosy in the Ancient Near East: An Overview. In *The Past and Present of Leprosy: Archaeological, Historical, Palaeopathological and Clinical Approaches*, edited by C. A. Roberts, M. E. Lewis, and K. Manchester, 259–268. British Archaeological Reports International Series 1054. Oxford: Archaeopress.

Zias, J., and P. D. Mitchell. 1996. Psoriatic Arthritis in a Fifth-Century Judean Desert Monastery. *American Journal of Physical Anthropology* 101: 491–502.

Zimmerman, C., L. Kiss, and M. Hossain. 2011. Migration and Health: A Framework for 21st Century Policy-Making. *PLoS Medicine* 8(5): e1001034. https://journals.plos.org/plosmedicine/article?id=10.1371/journal.pmed.1001034.

Zink, A. R., and A. G. Nerlich. 2005. Notes and Comments. Long-Term Survival of Ancient DNA in Egypt: Reply to Gilbert et al. *American Journal of Physical Anthropology* 128: 115–118.

Zodpey, S. P., N. N. Ambadekar, and A. Thakur. 2005. Effectiveness of Bacillus Calmette Guerin (BCG) Vaccination in the Prevention of Leprosy: A Population-Based Case-Control Study in Yavatmal District, India. *Public Health* 119: 209–216.

Zoëga, G., and H. Gestsdóttir. 2011. Iceland/Ísland. In *The Routledge Handbook of Archaeological Human Remains and Legislation*, edited by N. Márquez-Grant and L. Fibiger, 203–208. London: Routledge.

Zubriczky, A. 1924. *Az Európai Nagy Leprajárvány Törtenete*. Budapest: Magyar Tud., Társ., Sajtóvállalata, RT.

Zuckerman, M. K., B. L. Turner, and G. J. Armelagos. 2012. Evolutionary Thought in Paleopathology and the Rise of the Biocultural Approach. In *A Companion to Paleopathology*, edited by A. L. Grauer, 35–57. Oxford: Wiley-Blackwell.

Index

Page numbers in *italics* refer to illustrations.

Abel, L., 42
Aberastury Law, Argentina, 276
Abony-Turjányos, Hungary, 230
Accelerating Towards a Leprosy-Free World (WHO), 20
Acroosteolysis, 160
Actinomycetes, 250
Active detection, 48
Acute inflammation in leprosy, 68, 85–88
Æbelholt, Zealand, Denmark, 218, 222
Africa, sub-Saharan, bioarchaeological evidence, 257–259
Agarwal, S. C., 53
Age in relation to frequency rates, 44–47
Agnew, A. M., 242
Aitutaki Island, 268
Åkirkeby, Bornholm, Denmark, 218
ALERT (All Africa Leprosy, Tuberculosis and Rehab Training Centre), 102
Alexander the Great, 246, 282
Ali ibn-Rabban Tabari, 265
al-Majusi, Ali ibn al-Abbas, 114
Alveolar bone loss, 135
American Association of Physical Anthropologists, 193
American Leprosy Missions, 104
Amniotic fluid, role in transmission, 36, 47
Amoebae, role in transmission, 39
Amputation, 118, *119*
Andersen, J. G.: changes to alveolar process of maxilla, 136–137; changes to palatine process of maxilla, 138–139; early work on skeletal changes, 18, 128–129, 130; hand and foot bone alterations, 162, 176; leprosaria in Denmark, 224–225; osteoarthropathia centralis metatarsophalangea prima, 168; palmar grooving, 158; periosteal lesions of lower legs, 166–167; rhinomaxillary syndrome, 131, 132, 133, 218
Anderson, S., 203
Anemia, 53–54
Anesthesia, 75–77, 112, 146–147
Angel, J. L., 248, 265
Animal to human transmission, 56–62
Annan, Kofi, 306
Antibiotic resistance, 2, 283
Antibiotic treatment, 17, 67–68, 305
Antigenic shift, 293
Anti-immigration sentiment, 56
Archaeology. *See* Bioarchaeological evidence
Arches, in feet, loss of, 155
Archival material, 307
Arcini, C., 243, 245
Ardagna, Y., 248, 257
Arensberg, B., 250
Argaw, Tadesse, 39–40
Armadillos, 16, 58–61, *59*, 285, 287
Arnamagnæanic manuscript collection, 117, 120
Arsenic ointment, 117
Asypmtomatic leprosy. *See* Subclinical leprosy
Austria, bioarachaeological evidence, 200–201
Autonomic nerves, 85; autonomic nerve damage, 16–17; autonomic neuropathy, 159–162
Avanzi, C., 286
Aycock, W. L., 274
Ayurvedic medicine, 121

Bacillus Calmette-Guérin (BCG) vaccine, 98, 103–104, 292, 297–299
Baker, B. J., 142, 257
Bakija-Konsuo, A., 43

Barker's developmental origins hypothesis, 47, 54–55, 223
Barnetson, J., 162
Barriers to treatment, 99
Barton, R. P. E., 171
Bathing and skin treatments, 116–117
Bayliss, J., 293
BCG (Bacillus Calmette-Guérin) vaccine, 98, 103–104, 292, 297–299
Beckford, England, 203, *204*
Bedbugs, role in transmission, 34
Bed burial, 204, *205*
Bedić, Z., 215
Beguiristan, M. A., 254
Beit Guvrin, Israel, 250
Belcastro, M. G., 252
Bélmegyer-Csömöki, Hungary, 232
Bending (flexion deformity), 81–83, *83*
Ben Hur, 6
Bennike, P., 169
Benzene derivative, as treatment, 120
Bergel, M., 291
Bergen, Norway, 23–24
Bergen Syndrome, 131
Bernard de Gordon, 113
Berns, John, 213
Bessastaðir, Iceland, 234
Bhoyroo, J., 213–214
Bible, the, and mistranslation, 36, 123–124, 249, 274, 305
Biehler, R., 162
Bijelo Brdo, Croatia, 215
Bioarchaeological evidence: Austria, 200–201; British Isles, 201–215, 323–328; Canada, 272–273; Caribbean, 275; Central America, 274–275; China, 259–260; Croatia, 215–216; Cyprus, 247; Czech Republic, 216–217; Denmark, 217–225; Egypt, 255–256; ethical concerns, 192; Finland, 225–226; France, 247–248; Germany, 226–228; Greece, 248; Hungary, 228–234, *229*; Iceland, 234–235; India, 261–263; Iran, 263–265; Ireland, 201–215; Israel, 249–251; Italy, 251–253; Japan, 260–261; Norway, 235–241; Nubia, 256–257; Pakistan, 261–263; Poland, 242–243; Portugal, 253–254; role in interpretation of history, 2–3, 14–15, 18, 19, 307–308; Scotland, 212–213; South America, 276; southeast Asia, 266–269; Spain, 254–255; sub-Saharan Africa, 257–259; Sweden, 243–246; Turkey, 265–266; Ukraine, 246; United States, 273–274; Uzbekistan, 263–265
Biological sex. *See* Sex and frequency rates
Biomolecular analysis, 179–184, *181, 182,* 207
Biting flies (*Stomoxys*), role in transmission, 35
Bjerkedal, T., 240
Björned, Västerbotten, Sweden, 244
Bjune, G., 36
Blackfriars monastery site, Ipswich, England, 210, *211,* 223–224
Blau, S., 263
Blindness, 83–85
Blondiaux, J., 185, 247
Bloodletting, 118
Blood use in treatment, 120
Bluebottle flies (*Calliphora*), role in transmission, 35
Boeck, C. W., 40, 193
Boeckl, C. M., 4
Bog vegetation, 38–39
Boldsen, J., 135, 177–178, 222–223, 223–224, 228, 244, 300
Bolek, E. C., 139
Bolhofner, K. L., 142
Bone alterations: from autonomic nerve damage, 159–162; bone absorption, 148; bone lesions, 186–187; bone loss, *79;* bone remodeling, 159–161; facial bone damage, 224; facial lesions, 142–143; fractures, 173–175; frontal bone grooves, *174,* 175; hands and feet, 162–163; and joint defects, 167–169; osteopenia and osteoporosis, 163–164; periodontal disease, 173; periosteal lesions on tubular bones, 164–167; porosity of orbits, 169–170; respiratory involvement, 170–173
Bonnar, P. E., 61
Boocock, P., 171
Borbognoni, Theodoric, 121
Borderline leprosy, 66; and nerve damage, 75; and skin lesions, 69, 71; and Type 1 reactions, 87
Bouchez, I., 248, 257
Bourland, J., 47
Boyd, R., 271
Boylston, A., 164
Brand, P. W., 142
Brander, T., 222
Brandsma, J. W., 158

Bratschi, M. N., 31
Braveheart, 6
Brazil: frequency rates, 276; terminology used, 7
Breastmilk, role in transmission, 35, 47
Bridgford, S., 248
British Isles, bioarchaeological evidence, 201–215, 329–330
Brno, site of Czech leprosaria, 216
Brosch, R., 37
Brothwell, D. R., 201, 202, 213
Brough, St Giles, England, 210, *212*
Browne, S. G., 213, 249
Brownsea Island, 286
Brubaker, M. L., 47
Bruintjes, T. D., 172
Bubonic plague and leprosy, 292–293
Buikstra, J. E., 295, 300
Burial practices: Germany, 226–227; grave goods and bed burials, 203–205, *205*; Japan, 261; leprosaria contrasted with non-leprosy hospital cemeteries, 280; segregation in burial, 303–304; table of non-standard practices, *278*; Thailand, 267

Cadiz, Spain, 254
Calculus (dental), 142, 173
Cambridgeshire, Midland Road cemetery, England, 210
Campochiaro, Italy, 252
Canada, bioarchaeological evidence, 272–273
Cancer and leprosy misdiagnosis, 110
Capilla de San George, Spain, 254
Carayan, A., 292
Carbolic acid, as treatment, 120
Caribbean, bioarchaeological evidence, 275
Carpal disintegration, 148–150
Carpal dislocation, 156
Carville, Louisiana, United States, leprosy hospital, 273, 274
Casalecchio di Reno, Italy, 252
Cats, leprosy in, 57
Cautery, 118, *119*
Cell mediated immunity (CMI), 64, 66
Cementochronological evidence, 247
Cemeteries: hospital, 212, 329–330; leprosaria cemeteries contrasted with non-leprosy hospital cemeteries, 280. *See also* Burial practices

Central America, bioarchaeological evidence, 274–275
Chakrabarty, A. N., 38–39
Chapelle Saint-Lazare Hospital, France, 247
Charcot joints, 155
Chauliac, Guy de, 113, 115–116
Chaulmoogra oil, 121–123
Chaussinand, R., 297
Chemotherapy. *See* Multidrug therapy
Children: childhood onset of leprosy, 45–46; children's books and portrayals of leprosy, 6; skeletal remains of children, 193–195, 203
China: ancient treatments, 118, 121, 123; bioarchaeological evidence for leprosy, 259–260
Cholesterol levels, as risk factor, 54
Chu Chen-heng, 106
Clarithromycin, 98
"Cleansing," in *Bible,* 123–124
Climate conditions and decrease in leprosy, 293–294
Climatic Worsening, 294
Clofazimine, 68, 96, 97, 100
Close-Brooks, J., 202
Clotting, blood, 110
Clubbing of fingers, 155
CMI (cell mediated immunity), 64, 66
Cockburn, Aidan, 191
Coimbra Identified Skeletal Collection, 183, 253–254
Community integration, 50, 206, 210. *See also* Segregation of leprosy patients
Comorbidities, 104–106
Comox Indian skull, 269–271, *270*
Compendium of Medicine (1510), 110
Confidential Text Regarding Malign Ulcers (Hsueh Li-Chai), 118
Congenital leprosy, 36, 47
Contact dermatitis, 69
Contact reporting, 48
Contractures, 146; in fingers, 158, 159; in toes, 81, 158–159
Contreras, Barionuevo, 254
Cook, D. C., 142
Cook Islands, 268
Cool areas of body and leprosy, 64, 142, 144–145
Corneal ulcers, 140
Corticosteroids, 86
Corvaro, Italy, 251
Cosenza Cathedral, Italy, 252

Cottle, W., 122
Coutelier, L., 185, 247
Crespo, F., 296
Cribra orbitalia, 169–170, 223
Croatia, bioarchaeological evidence, 215–216
Croc shoes, 100, *101*
Cross, A. B., 151
Cross-immunity, leprosy and tuberculosis, 240, 295–297, 300–301
Crowding, and risk of leprosy transmission, 33, 51, 52. *See also* High-density living
Crusades, and spread of leprosy, 293
Csengele-Bogárhát, Hungary, 232
Culion Island, 266–267
Culpeper's Herbal (1653), 120
Cup-and-peg deformity, 147, 154, *156,* 160
Curability of leprosy, 1, 86, 99
Curieuse Island, 262
Cybulski, Jerry, 271
Cyprus, bioarchaeological evidence, 247
Czech Republic, bioarchaeological evidence, 216–217

Dactylitis, 148, 154
Dafeng, 259
Daffe, M., 37
Dakhleh Oasis, Egypt, 255–256
Dalby, G., 173
Danielsen, K., 145
Danielssen, D.-C., 40, 193
Dapsone, 67–68, 96, 97, 100
D'Arcy Island, 273
Dastidar, S. G., 38–39
Dayal, R., 34, 281
Deacon, H., 259
Dead Sea Scrolls, 249
Decentralization of treatment, 100
Decrease in leprosy, 291–301; changes in climate, 293–294; changes in living conditions, 291–293; cross-immunity with tuberculosis, 295–297, 300–301; and leprosaria, 224; Norway, 238, 240–241; and third epidemiological transition, 283. *See also* Elimination of leprosy
Dehumanization of leprosy patients, 307–308
De las Aguas, J. T., 276
Delayed diagnosis, 89–90, 94–95
Demaitre, Luke, 4, 110, 112, 115
Demedts, V., 2

De Miguel Ibáñez, M. P., 255
Demyelination, 75
Denmark, bioarchaeological evidence, 217–225
Dental caries as co-morbidity, 105
Depression, 86, 93, 106
Destructive analysis, guidance on, 184
Developmental origins hypothesis, 47, 55, 223
Devkesken 6, Uzbekistan, 263
De Witte, S. N., 189, 223
Dey, N. C., 140
Dharmendra, 121, 140
Diabetes, 154; as co-morbidity, 105; as confounding diagnosis, 264
Diagnosis: accuracy of, 112–116; in ancient China, 106, 109–110, 112; confounding factors, 186–189; delayed, 89–90, 94–95; diagnostic criteria, 90–91; diagnostic tests, 93–95; differential diagnoses, 185–189; in medieval Europe, 106–107, 110–112, 113
Diaphyseal remodeling, 160, *161*
Diet: effect on immunity, 53, 65; and frequency rates, 291; Iceland, 235; and poverty, 52–54; in relation to transmission, 52–54; Sweden, 245; as treatment, 117–118
Differential diagnoses, 185–189
Ditch sign, 156
DNA: DNA evidence and migration, 284–291; role in diagnosis, 179–185
Donoghue, H. D., 180, 206, 232, 233, 242, 289, 296
Dor, Israel, 251
Dormant infection. *See* Subclinical infection
Dorsal tarsal "bars" or exostosis, *157,* 158–159
Downgrading of disease process, 68
Downsizing of genome, 284
Draper, P., 37
Droplet infection, 31–33
Drug resistance, 97
Drug therapy. *See* Multidrug treatment
Drutz, D. J., 36
Dryburn Bridge, Scotland, 202
Dudar, Chris, 271
Duncan, K., 293
Duncan, M. E., 35, 36
Duvette, J.-F., 247
Dwidevi, V. P., 53
Dyer, I., 158, 166
Dzierzykray-Rogalski, Tadeusz, 191, 255

Ear bones, *172*, 172–173
Earhart, T. W., 146
Ear infections, 170
Easter Island, 268
Economou, C., 289
Edema of lower extremities, 91
Edix Hill, Cambridgeshire, England, 204, *205*
Education for self-care, 100–102
Egypt, bioarchaeological evidence, 255–256
Ehlers, E., 234–235
Eight-banded armadillos, 59
El Bigha, Sudan, 256
Elephantine, Egyptian Nubia, 257
Elimination of leprosy, 11, 13–14, 20. *See also* Decrease in leprosy
Ell, S. R., 54, 114–115, 291
Endocrine dysfunction, 46–47; as co-morbidity, 105
Environment: climate conditions and decrease in leprosy, 293–294; environment-to-person transmission, 27–28; factors in transmission, 36–40; in relation to frequency rates, 51–52
Epidemiological transitions, 283–284
Erdal, Y. S., 252
Ermida de Santo André, Portugal, 253
Erythema nodosum leprosum, 85, 87
Eshiet, A. L., 49
Ethical concerns about human remains, 192
Evolution, reductive, 25
Evolutionary decay, 25
Evolutionary medicine, 308–309
Examen des éléphaniques on lépreux, 108
Exteberria, F., 254
Eye care, 83–85, 101

Facial involvement: facial bone damage, 224; facial bone lesions, 142–143; facial deformity, *109*; facial features and diagnosis, 108–109; facial nerve damage, 81, 83, 84, 155; facial palsy, 140
Facies leprosa, 131, 141–143
Faget, G., 97, 176
Faking leprosy, 116
Farkas, G., 232
Faye, L., 236
Fear of leprosy, 8, 56
Feet: arches, 152, 155; bone alterations, 146–147, 162–163; effects of *M. leprae*, 150–154; ligament damage, 152; protection from injury, 100–101; toe contractures, 158–159; ulcers, 79–80, 150–151, 155–156
Feldman, R. A., 274
Feline leprosy, 57
Fernandez, J. M. M., 297
Fernández-Gill, M., 184, 185
Fernandez reaction, 93–94
Ferreira, M. T., 253
Field Book of Wound Treatment, 110–111, *111*
Fiji, 268
Film, portrayals of leprosy, 6
Fine, P. E. M., 9–10, 22, 31, 34, 35, 41, 64, 293, 297
Finland, bioarchaeological evidence, 225–226
Fish consumption, 53, 291
Five Deaths classification, 106
Flexion deformity, 81–83, *83*, *157*
Flies, role in transmission, 35
Flores, López, 254
The Fog, 6
Foley, R., 206
Food, role in transmission, 35
Foot drop, 155, 156, 158, 159
Foot ulcers, 79–80, 150–151, 155–156
Forson, R. B., 6
Forster, R., 146
Fóthi, E., 232, 233
Fractures, 173–175
France, bioarchaeological evidence, 247–248
Frequency of leprosy: and age, 44–47; calculation of rates, 188–189; current cases, table and map, *12*; current cases in Britain, 214–215; current rates, 11–12; and environmental influences, 277; family clusters, 40–41; female contrasted with male, 48–50; historical, 9–11; manipulation of figures, 13; Norway, 19th and 20th centuries, 237, *238*, *239*; and occupation, 49; rates, underreported, 10; in relation to environment, 51–52; and sex, 47–50
Frontal bone grooves, *174*, 175
Frota, C. C., 60

Galen, 226
Gambier-Parry Lodge cemetery, England, 202, *203*
Gastrointestinal route and transmission, 35
Gausterer, C., 289
Geater, J. G., 35

Gender. *See* Sex
Genesse, J., 1, 269, 274
Genetic markers and diagnosis, 94
Genetic predisposition, 40–44
Genome analysis, 287
Gerla monastery, Hungary, 232
Germany, bioarchaeological evidence, 226–228
Gersdorff, Hans von, 110–111, *111*
Ghosh, K. K., 140
Gibson, J., 271
Gilbertus Anglicus, 110, 112
Gill, A. L., 214
Gillis, T. P., 28
Gilmore, J. K., 171
Gingivitis as co-morbidity, 105
GIS (geographical information systems), 26, 39–40
Gladykowska-Rzeczycka, J., 242
Gloskär, Finland, 226
Gomacin, Spain, 254
Goode, J. L., 175
Graciano-Machuca, O., 4
Grange, J. M., 7, 37
Granulomatous lesions, 154, *174,* 175
The Grateful Leper: Tales of Two Birds (Forson), 6
Grave goods, 203–204; Japan, 261; Thailand, 267
Great Chesterford, England, 204, 206
Greece, bioarchaeological evidence, 248
Green, M., 307
Greyfriars monastery, 223–224
Grmek, M., 255, 282
Grossetete, G., 134
Gruczno, Poland, 242
Guam, 268
Guillaume des Innocens, 108
Gurjon oil, 117

Haas, C. J., 180, 228, 232
Hall, B. G., 14, 27
Han, X. Y., 22
Handbook of North American Indians, 271
Hands: bone absorption, 148; bone alterations, 146–147, 162–163; clubbing of fingers, 155; effects of *M. leprae,* 148–150; finger contractures, 158, *159;* protection from injury, 100–102
Hansen, Gerhard Armauer, 7, 23–24, *24,* 38, 56, 105, 235, 237

Hanseniasis, 276
Hansen's Disease, 7, 276
Hansieníase, 7
Harbitz, F., 146
Harboe, M., 36
Hatani, Japan, 261
Healing springs, 117
Hearing issues, 172–173
Heel ulcers, 151
Heiberg, H., 146
Helsinki, Finland, 226
Herbs, 120–123
Hereditary transmission: hereditary theory of transmission, 40; South Africa, 258–259
Herrast, L., 254
Herrick, A. B., 146
Hershkovitz, I., 180, 250
High-density living, 206. *See also* Crowding, and risk of leprosy transmission
High-resistance leprosy. *See* Tuberculoid leprosy
Hinkle, C. A., 6
Hinnom Valley, Israel, 249
Hippocrates, 308
Hirata, K., 260
Hirschberg, M., 162
HIV (human immunodeficiency virus) as co-morbidity, 104–105
HLA (human leucocyte antigen), 41
Hohmann, N., 296, 300
Holistic treatment, 96
Holton, J., 232, 233, 242
Hopkins, R., 158
Hospital cemeteries, 212, 329–330. *See also* Leprosaria
Hospital-Colónia Rovisco Pais, Portugal, 254
Houseflies (*Musca*), role in transmission, 35
Hoxne, Suffolk, England, 206
Hrdlička, Aleš, 271–272
Hsueh Li-Chai, 118, 123
Huang, C. L., 32, 34
Hudson, T., 1, 269, 274
Hughes, M. S., 176, 256
Humanistic view of patients, 307–308
Human remains, ethical concerns about excavation, 309
Humidity, Norway, 240, *246*
Humlegården, Uppland, Sweden, 244–245
Hungary, bioarchaeological evidence, 228–234

Hunter, J. M., 277, 293, 299
Hutchinson, D. L., 269
Hutchinson, J., 291
Hutter, H. J., 161
Hydnocarpus, 121–123
Hyoid bone, 171
Hyperextension, 155, 158
Hyperflexion, 158

Ibrány-Esbohalom, Hungary, 233
Iceland, bioarchaeological evidence, 234–235
ILEP (International Federation of Anti-Leprosy Associations), 8
Images of Leprosy, 4
Imaging (radiography), 185
Immigration and spread of leprosy, 272, 273–274, 281–282
Immunity: cell mediated, 64, 66; cross-immunity with tuberculosis, 240, 295–297, 300–301; immune function and *M. leprae* resistance, 127–128; immune strength as factor in transmission, 32; immune system, 1, 66; immunization, 292; immunosuppression, 46, 53; susceptibility to leprosy, 187–188
Impetigo and leprosy misdiagnosis, 110
Incidence of leprosy. *See* Frequency of leprosy
Incubation period, 65
India: bioarchaeological evidence, 261–263; leprosy treatment in ancient India, 121
Indian Journal of Leprosy, 2
Indo-Gangetic basin, 246
Indonesia, 266
Infants: congenital leprosy, 47; infant onset of leprosy, 45–46. *See also* Breastmilk, role in transmission
Infectious Disease Research Institute, 104
Injury, from anesthesia, 77–79, 78, 173–175
Inkster, R. G., 213, 255, 269
Insects and transmission, 34–35
Inskip, S., 290
Integration. *See* Segregation of leprosy patients
Integration of leprosy patients in community, 50; evidence for, 206, 210
International Federation of Anti-Leprosy Associations (ILEP), 8
International Journal of Leprosy, 2
International Leprosy Association global history project, 306–307
International Leprosy Congress, 6–7, 97

Inventarium sive Collectorium Partis Chirurgicalis Medicinae, 113, 115–116
Iodine, as treatment, 120
Iran, bioarchaeological evidence, 263–265
Ireland, bioarchaeological evidence, 201–215
Irgens, L., 38, 237–238, 240
Iron: iron deficiency, 53–54; as treatment, 120
Ishikawa, S., 164
Isolation. *See* Segregation of leprosy patients
Israel, bioarchaeological evidence, 249–251
Italy, bioarchaeological evidence, 251–253

Japan, 260–261
Jean de Todon Chapel, Gard, France, 248
Job, C. K., 131, 148, 154, 155, 161, 166
John of Gaddesden, 108, 112
Joint involvement: Charcot joints, 155; joint defects, 168–169; joint degeneration, 154; joint disease as co-morbidity, 105
Jones, F. W., 256
Jopling, W. H., 34, 37, 66, 257, 281
Judd, M., 173
Juniper, 117
Justus, H. M., 242

Kalau papa settlement, Hawaii, 274
Kaldus, Poland, 242
Kalisch, P. A., 272, 273
Kamakura, Japan, 260–261
Karat, A. B., 146
Karat, S., 146
Karataş-Semayük, Turkey, 265
Kaufman, M. H., 213
Kazda, J., 38, 39
Kellersberger, E. R., 96
Kelmelis, K. S., 223
Kerr-Pontes, L. R. S., 54
Kerudin, A., 290
Kidney disease as co-morbidity, 105
King Baldwin IV, 124, 193, 251
King Henry VII (Germany), 252
King Philip V (Spain), 275
Kingsmead Quarry, England, 202
Kirchheimer, W. F., 59
Kiskundorozsma-Daruhalom, Hungary, *209,* 210 230
Kiskundorozsma-Kettőshatár, Hungary, 230
Kjellstrom, A., 191, 245
Köhler, K., 230

Ko Hung, 112
Kolk, A. H., 38
Koran, 250
Kovuklukaya, Turkey, 265
Kozak, A. D., 257
Krahenbuhl, J. L., 39
Krause-Kyora, B., 42
Kulkarni, V. N., 148, 153
Kuzlowski, T., 242
Kvarteret St. Jørgen, Malmö, Sweden, 244
Kwiatkowska, B., 242

Lactation, 35, 47
Lagophthalmos, 83–84, *84,* 140, 155
Lahiri, R., 39
Lake of the lepers, 117
Language, choice of words, 308
Lászlófalva-Szentkirály, Hungary, 232
Latent leprosy. *See* Subclinical leprosy
Lauchheim, Germany, 228
Lavania, M., 32, 37
Law of signatures, 121–122
Laxatives, 118
Lechat, M. F., 281
Lee, C., 206
Lee, F., 138, 141, 153, 158, 159, 162, 164, 168, 177
Lee, O. Y.-C., 230
Lei Yang Chi Yao, 118, 123
Leloir, H., 146
"Leper": definitions, 322; as stigmatizing term, 8; as unacceptable term, 6–7; and word choice, 4–5, 308
"Leper" as outcast. *See* Segregation
Leper Festival (Devon, England), 6
Leper villages, China, 260
Lepra mutilans, 146
Leprogenic odontodysplasia, 143–145, *144*
Lepromas, 154, *174,* 175
Lepromatous leprosy, 1, 10, 23, 296; age of onset, 45–46; alveolar bone changes, 135; anesthesia, 147; and bone changes, 128, 129–130, 132; classification, 66–67; and co-morbidities, 105; contrasted with tuberculoid leprosy, 190; cure rates, 99; development of disease, 64; diagnosis, 93; diagnosis based on skeletal remains, 176; drug treatment regimens, 67–68, 96, 98; and immunology, 43–44; incubation period, 65; leprosy reactions, 85; and Lucio's phenomenon, 87–88; MLFT (*M. leprae* Lateral Flow Test), 94; nasal changes, 170–171; osteomyelitis, 143; and periodontal disease, 173; prognosis, 104; rhinomaxillary syndrome and bone change, 143; signs and symptoms, 91–93; and skin lesions, 71, *72;* symmetrical nerve damage, 75; transmission, 31; and Type 2 reactions, 87
Lepromin conversion, 297, 299
Lepromin test, 66, 93
Leprophobia, 8, 56
Leprosaria: Britain, 201, 206–207, 212–213; Canada, 272–273; Caribbean, 275; cemeteries, 115; Central America, 275; Czech Republic, 216–217; and decline in leprosy rates, 224; Denmark, 220, 224–225; Finland, 226; France, 247–248; Germany, 226–227; Hungary, 228–229, 233; Iceland, 234–235; India, 262–263; Iran, 265; late Medieval Britain, 210, 212; Malaysia, 266; medieval Europe, 206, 304; medieval hospitals where skeletal remains have been excavated, 329–330; Mexico, 274–275; near healing springs, 117; Norway, 235–237, *236,* 241; overview, 124–125; Philippine Islands, 266–267; Portugal, 253; Roman Empire, 249–251; South America, 276; Sweden, 243–244; United States, 273, 274
Leprosy: accuracy of diagnosis, 112–116; acute inflammation, 68, 85–88; age as factor in frequency rates, 44–47; and anesthesia, 112; associated with sin, 36; bathing and skin treatments, 116–117; BCG vaccine, 98, 103–104, 292, 297–299; biomolecular analysis contrasted with tuberculosis, *208, 209;* and blood, 110–112, 120; bone alterations, 185–187; childhood onset, 45–46; classification of *M. leprae*, 66–68; clinical features, 91–93; clustering of rates, 10; comorbidities, 104–106; congenital, 36; as contemporary problem, 305; contrasted with tuberculosis, 2, 154, 183; and cross-immunity with tuberculosis, 240, 269–270, 300–301; and curability, 1, 86, 99; current rates, 11–12, 214–215; declining rates, 2, 13–14, 212–214, 217–218, 238, 240–241; and depression, 86, 93; diagnosis, 17; diagnostic tests, 93–95; diet as treatment, 117–118; as disability, 8; droplet transmission, 32–33; and edema of lower extremities, 91; elimination of, 11, 13–14, 20; environmental factors in

transmission, 36–40; and facial features, 108–109; faking leprosy, 116; and fractures, 173–175; frequency rates, 10, 13, 237–239; gastrointestinal route and transmission, 35; genetic predisposition, 40–44; historical rates, 9–11; history of, 97–99, 306–307; holistic treatment, 96; incubation period, 65; inheritance, 305; insects and transmission, 34–35; and lactation, 47; as lived experience, 8, 307–308, 310; mathematical modeling of rates, 13–14; and mental health, 92–93; migration and transmission, 29–30, 214–215, 286; misconceptions concerning, 1, 304–305; and misdiagnosis, 110; multidrug therapy, 17, 21, 67–68, 96, 97–99, 100; nerve damage, 65; new cases v. prevalence rate, 13; noncompliance with treatment, 99–100; nondrug therapies, 100–103; overview, 22–23; patient education, 100; in popular culture, 6; and poverty, 33, 49, 52–55, 283; prognosis, 17; public perceptions of, 4–5; relapse after treatment, 96; research focus, 2; signs and symptoms, 91–93; stigmatization of, 1, 4–8; stress as predisposing factor, 223; subclinical infections, 31, 33, 65, 89, 180, 183; surgery, 118–120; survey of public knowledge base, 3–4; susceptibility to, 187–188; and trauma, 77–79, 173–175; as tropical disease, 305; Type 1 reactions, 87; Type 2 reactions, 87; types 1–4 (Monot), 29–30; urban/rural contrast, 10, 51, 206, 224, 283, 293, 297–300; and urine, 110–112
Leprosy Act (Canada), 273
Leprosy Acts (Norway), 240
Leprosy clinics, South Pacific, 268
Leprosy colonies, South Africa, 259
Leprosy hospitals. *See* Leprosaria
Leprosy in India, 2
Leprosy Mission Referral Hospital (Champa, India), 99
Leprosy reactions, 68, 85–86
Leprosy Registry (Bethesda, MD), 46
Leprosy Repression/Suppression Act (South Africa), 259
Leprosy Review, 2
Leprosy Study Group, 191–192
Leprosy Susceptible Human Gene Database, 26
Leprous-related osteomyelitis, 143
Lethaby, J. I., 7

Leung, A. K. C., 259–260
Lewis, M. E., 169, 193
Lice, role in transmission, 34
Licked candy stick appearance in bones, 160
The Life of Brian, 6
Ligament damage in feet, 152
Likovsky, J., 217
Limpopo, South Africa, 257–258
Linderholm, A., 245
Lisieux-Michelet, France, 247
Little Ice Age, 294
Lived experience of leprosy, 8, 307–308, 310
Liverpool School of Hygiene and Tropical Medicine, 214
Living conditions, as environmental risk factor, 36–37, 39–40, 51–52, 292. *See also* Poverty
Looft, C., 38, 105
Lopez, Pedro, 275
Loss of sensation. *See* Anesthesia
Losting, Johan Ludvig, *109*, 193
Løvik, M., 57
Lower temperature areas of body, 64, 142, 144–145
Low-resistance leprosy. *See* Lepromatous leprosy
Lucio's phenomenon, 25, 87–88
Lucy, D., 171
Luis Lopes Identified Skeletal Collection, 253–254
Lund, Sweden, 243–244
Luờng, K. V. Q., 51
Lynnerup, N., 222, 300

MacArthur, W. P., 201
MacLennan, W. J., 213
Madrid system, 66
Madura foot, 250
Mafeng, 259
Magilton, J. R., 107, 115, 279
Makogai Island, 268
Malaysia, 266
Malnutrition, 52–54
Manchester, Keith: on Asian skeletal remains, 192; bone alterations, Chichester, 171; changes to alveolar process of maxilla, 136–137; changes to oral surface of palatine process, 138–139; on Comox Indian skull, 269; on diagnosis based on skeletal remains, 176, 177; disuse atrophy, Chichester remains,

Manchester, Keith—*continued*
  162; early work on skeletal changes, 295–296; on leprosy decline in medieval Britain, 210, 212; leprosy in Poundbury skeletal remains, 202; Leprosy Study Group, 191; osteochondritis dissecans, 168–169; palmar grooves, 158; rhinomaxillary syndrome, 131, 133, 141; on skeletal evidence from Canada, 271; on skeletal evidence from Palestine, 250; tarsal bars, 159; tarsal disintegration, 153
The Man with Leprosy (McDonough), 6
Marchoux, N. E., 57
Marcsik, A., 232–233
Mark, S., 281, 282
Marks, S. C., 134
Marlow, S., 228
Martellona, Italy, 252
Martillous monastery, 250
Martin, A., 111
Massone, C., 20
Matos, V. M. J., 177, 190, 253, 264, 271
Matsuoka, M., 39
Maxillar loss, 135–137
Mayoral, H., 176
McDonough, A., 6
McDougall, A. C., 132
McKinley, J. I., 202
"Meddygon Myddfai" (1861), 120–121
Medical anthropology, 8
Medical historians, 2–3
Medieval Warm Period, 294
Medley, G. F., 303
Mehta, J. M., 148, 153
Melaten, Aachen, Germany, 226–227
Melsom, R., 36
Mendum, T. A., 289
Mental health, 86, 92–93, 106
Mercury: mercury ointment, 117; as treatment, 120
Meyers, W. M., 47, 56–57, 62
MHC (major histocompatibility complex), 41
Miao His-Yung, 123
Michailovsky Gold Cathedral, Kiev, Ukraine, 246
Michman, J., 134, 141–142
Microscopy, 184
Migrant groups and disease, 55–56
Migration, 290; and DNA evidence, 284–291; and transmission, 29, 29–30, 214–215, 286

Miles, J., 267
Minocycline, 68, 98
Miscarriage, 46
Misconceptions concerning leprosy, 1, 304–305; questionnaire used in survey in England, 3, 319–322
Mis Island, Sudan, 257
Mitchell, P. D., 124, 251
Mitsuda, K., 105
Mitsuda reaction, 93–94
M. leprae, 1, 22–23; classification, 66–68; discovery of bacillus, 23–24; gene sequencing, 24–30; map of ancient strains, *291*; multiplication rate, 65; reproductive evolution, 246; in skeletal remains, 183–184; in soil, 37–38; in sphagnum moss, 38–39; strains, differences between, 25–30, 245, 263–264, 269, 284–285, 297–299; in surface water, 39; Type 2, 245; Type 3, 206–207, 245, 253
M. leprae, direct effects: bone changes, grading, 140–141; maxillar loss, 135–137; nasal loss, 133–135; palatine process, nasal surface, 137–138; palatine process, oral surface, 138–140; rhinomaxillary syndrome, 141–145
M. leprae, indirect effects: autonomic nerve bone changes, 162; autonomic neuropathy, 159–162; bone changes due to motor nerve damage, 159; bone changes due to sensory nerve damage, 154–155; feet, 150–153; hands, 148–150; motor nerve involvement, 155–159; sensory nerve involvement, 146–148
M. leprae Lateral Flow Test (MLFT), 94
M. lepraemurium, 57
M. lepromatosis, 1, 22–23, 290; and bone changes, 131; gene sequencing, 25; and Lucio's phenomenon, 87–88
MLFT (*M. leprae* Lateral Flow Test), 94
Mobility: and health risks, 55–56; mobility studies, 245; and transmission, 55–56
Mogren, M., 243
Molecular tests and diagnosis, 94
Molecular typing, 26
Møller-Christensen, Vilhelm: on bone changes, 113, 140; on Comox Indian skull, 269; on diagnosis based on skeletal remains, 176, 177; early work on skeletal remains, 18, 225; on maxillary changes, 136–137, 138–139; on migration and spread of leprosy, 281–282; osteoarthropathia centralis

metatarsophalangea prima, 168–169; on periosteal lesions, 167; rhinomaxillary syndrome, 141; on rhinomaxillary syndrome, 131, 135, 142, 143; on Robert the Bruce's skull, 213; on skeletal changes, 128–129; on skeletal evidence from Africa, 255, 256; on skeletal remains in Australia, 269; on skeletal remains in Naestved, Denmark, 218; sketches of remains in Naestved, Denmark, *219, 220, 221;* ursura orbitae, 169–170
Mollerup, L., 223–224
Molnár, E., 232, 233
Molto, J. E., 255, 256
Monastery of St. John the Baptist, 250
Monot, M., 19; DNA analysis of human remains, 289; origins of *M. leprae* based on DNA evidence, 253, 255; strains of *M. leprae,* 27, 28–30, 179, 246, 279, 284, 285–286; theories of transmission of *M. leprae,* 206, 269, 290
Montecorvino, Italy, 252
Montgomerie, J. Z., 267
Morimato, I., 260
Morphea and leprosy misdiagnosis, 110
Mosquitoes, role in transmission, 34–35
Motor nerves, 81–85, *82,* 155–159
Mouat, F. J., 122
Mouse leprosy, 57
Mouth breathing, 142, 145, 173
Mouwies, Egypt, 257
Mud baths, volcanic, 117
Muir, E., 297
Mukherjee, R., 103
Mulić, R., 43
Muller, K., 38
Multibacillary leprosy. *See* Lepromatous leprosy
Multidrug therapy, 17, 21, 67–68, 96, 97–99, 100
Murdock, J. R., 161
Murine leprosy, 57
Muscle atrophy, 81
Musculoskeletal complaints as co-morbidity, 105
Mycobacterium leprae. See M. leprae
Mycobacterium lepromatosis. See M. lepromatosis
Mycocerosates, *209*
Mycocerosic acids, 183

Mycolic acids, 179, 183, *208,* 210
Myths concerning leprosy, 1, 304–305; questionnaire used in survey in England, 3, 319–322

Næstved, Denmark, 218–222, *219, 220, 221*
Nasal discharge, 32
Nasal guttering, 142
Nasal loss, 133–135
Nasal septum perforation, 139–140
National Hansen's Disease Program (Louisiana, US), 58
National Leprosy Registry of Norway, 237, 240
Natural selection, 294
Navicular dislocation, 158
Navicular joint stress, 158–159
Neglected tropical diseases, 2
Nephritis as co-morbidity, 105
Nerlich, A., 228
Nerve damage, 16–17, 65, 72, 100; prevention of, 75
Nesse, R. M., 54
Neuritis, chronic leprous, 146
Neuville-sur-Escaut, France, 247
Neves, M. J., 253
Newell, K. W., 34
Newman, G., 201
New Zealand, bioarchaeological evidence, 267
Nguyên, L. T. H., 51
Nidaros Cathedral, Trondheim, Norway, 241
Nine-banded armadillos, 58–61, *59*
Nippon Foundation, 96
NOD2 gene, 43
Noen U-Loke, Thailand, 267
Nomadic people, leprosy rates, 263
Nonadult skeletal remains, 193–195, 203
Noncompliance with treatment, 99–100
Nondrug therapies, 101–103
Nonhuman primates, leprosy in, 57, 61–62
Norway: bioarchaeological evidence, 235–241; and bog vegetation, 38; fish consumption, 53; fishermen, 23
Novartis Foundation for Sustainable Development, 96
NRAMP proteins, 42
Nshaga, D. S., 21
Nubia, bioarchaeological evidence, 256–257
Nunn, C. L., 308
Nutrient deficiencies, 52–54

Nutrient foramina, enlarged, 160–161, *161*
Nutrition, 291–292
Nykjöping, Sweden, 245

Ober, W. B., 112
Objectification of leprosy patients, 307–308
Odense, Denmark, cemeteries, 222, 223
Ofloxacin, 68, 98
Ogawa, M. A., 105
Oku, C., 260
Old Testament, 36, 249, 274
Øm, Denmark, 222
Øm Kloster, Jutland, 223
Oommen, S. T., 121
Ópusztaszer monastery, Hungary, 233
Oral disease as co-morbidity, 105
Oral histories, 307
Orbits, porosity of, 169–170
Order of the Knights of St. Lazarus, 124, 251
Orivesi, Finland, 226
Orosháza, Hungary, 233
Orosháza-Bléke Tsz, Hungary, 230
Ortner, D., 141, 142–143, 170, 177, 185, 271–272
Osler, William, 308
Osteitis, 147, 154, 155–156
Osteoarthropathia centralis metatarsophalangea prima, 168, 169
Osteochondritis dissecans, 168–169
Osteological paradox, 187–189
Osteomyelitis, 143, 147, 154, 155–156
Osteopenia and osteoporosis, 105, 163–164
Ősz, B., 233

Pakistan, bioarchaeological evidence, 261–263
Palatine process: nasal surface, 137–138; oral surface, 138–140
Paleopathology. *See* Bioarchaeological evidence
Paleopathology Association, 191, 193, 308–309
Pálfi, G., 229, 232, 234
Palmar grooving, 156, *157*, 158, 159
Palombara, Italy, 252
Palsy, facial, 140
Pamplona, Spain, 254
Pap, I., 232
Papa Stour Island, 212–213
Papillon, 6
Parahidrosis, 85
Paralysis, 81
Paris Catacombs, 248

PARK2 gene, 43
Passive detection, 48
Pastazote sandals, 102
Paterson, D. E., 131, 148, 154, 155, 161, 166
Patient education, 100
Patient registries, 240
Paucibacillary leprosy. *See* Tuberculoid leprosy
Pediatric cases of leprosy, 45–46
Penikese Island, 273–274
Pen Ts'ao Ching Shu, 123
Periodontal disease, 105, 173
Periosteal lesions on tubular bones, 164–167
Peripheral nerve thickening, 72
Peripheral neuropathy, 21
Person-to-person transmission, 27–28
Perspiration, 85
Peters, E. S., 49
Pfrengle, S., 241, 253
Phenol, as treatment, 120
Philippine Islands, 266–267
Pickett, K., 55
Pinglu County, Shanxi Province, China, 259
Pinhasi, R., 206
Pirates!, 6
Placenta, role in transmission, 36, 47
Plague, bubonic, and leprosy, 292–293
Plaque (dental), 142, 173
Plaster casts, as treatment, 151–152
Platelets, blood, 110
Poland, bioarchaeological evidence, 242–243
Polis Chrysochous, Cyprus, 247
Pompey the Great, 282
Popular culture, portrayals of leprosy, 6
Portugal, bioarchaeological evidence, 253–254
Pottenbrunn, Austria, 201
Poundbury, England, 202
Poverty, 33, 49, 52–55, 283. *See also* Living conditions, as environmental risk factor; Socioeconomic status
Poznan-Srudka, Poland, 243
Pregnancy, 46–47
Prescriptions for Emergencies, 112
Preservation bias, 187
Prevalence of leprosy. *See* Frequency of leprosy
Price, E. W., 79, 151
Price, M. H., 223
Primates, leprosy in, 57, 61–62
Priniatikos Pyrgos, Greece, 248
Promin, 97

Proprioception, 147
Protein calorie malnutrition (PCM), 52–53
Prusanky, Prague, Czech Republic, 217
Psiorasis and leprosy misdiagnosis, 110, 250
Psoriatic arthritis, 250
Psychological complications, 86, 92–93, 106
Public perceptions of leprosy, 4–5
PubMed database, 2
Purgatives, 118
Püspökladóny-Esperjes, Hungary, 232, 233
Pyogenic infection, 154, 155–156

Questionnaire used in survey in England, 3, 319–323; results, 323–322

Racism, 258–259
Radašinovci, Croatia, 215
Radiography, 185
Rafferty, J., 96
Rafi, A., 180
Rainfall and frequency rates of leprosy, 277
Rao, P. S., 100
Raunds, Northamptonshire, England, 206
Rawson, T. M., 296
Reader, R., 202
Red squirrels. *See* Squirrels
Reductive evolution, 25, 284
Reichart, P., 132
Relapse after treatment, 96
Rendall, S. R., 132
Respiratory disease as co-morbidity, 105
Respiratory involvement, 170–173
Restless leg syndrome, 91–92
Reversal of disease process, 68
Rhinomaxillary syndrome, 18, 131–133, 141–145; diagnosis based on skeletal remains, 180; differential diagnosis, 141–143
Rib lesions, 171
Richards, P., 6, 113, 115, 235
Ridley, D. S., 34, 66
Ridley, M., 34
Ridley-Jopling classification system, 66, 68, 93
Rifampicin, 67–68, 96, 97, 98, 100
Robben Island, 258–259
Robbins, G., 261, 262
Roberts, C. A., 171, 173, 176, 184, 185, 251, 252, 295, 296
Robert the Bruce, 213, *214*
Rodriguez, J. N., 35

Roffey, S., 141, 142, 143, 207, 290
Rohatgi, S., 5
Rojas-Espinosa, O., 57
Rubin, Stanley, 191
Rubini, M., 251, 252
Ruffin, L., 185
Ruis González, J., 274
Rural/urban contrast, 10, 51, 206, 224, 283, 293, 297–300

Sager, F., 134, 141–142
Saint-Martin, P., 247
Saitz, E. W., 299
Salipante, S. J., 14, 27
Sane, S. B., 149–150
Santiago de Compostela pilgrims, 227
Santos, A. L., 264
Sárrétudvari-Hízóföld, Hungary, 232
Satchell, M., 113
Saunderson, P., 102–103
Scabies mites, role in transmission, 34
Schmitz-Cliever, E., 115, 226–227
Schoonbaert, D., 2
Schuenemann, V. J., 286–287, 288, 289, 290
Schug, G. R., 262
Schultz, M., 184, 185
Schuring, R. P., 43
Schurr, E., 42
Schutkowski, H., 184, 185
Schwann cells, 24, 32, 65, 74
Scleroderma and leprosy misdiagnosis, 110
Scollard, D., 21, 191, 306
Scotland, bioarchaeological evidence, 212–213
Seboka, G., 102–103
Segal, K. L., 220, 222, 224
Segregation of leprosy patients: Argentina, 276; Britain, 213; in burial practices, *279, 280*, 303–304; Canada, 273; China, 259–260; Cuba, 275; and decline in frequency rates, 291; Denmark, 224; within health services, 95; historical contrasted with bioarchaeological evidence for, 123–125, 303–305; Iceland, 235; India, 50, 262–263; Iran, 265; South Africa, 258–259; United States, 273
Self-care education, 100–102
Sensation, loss of. *See* Anesthesia
Sensory involvement, 175–176
Sensory nerve damage, 16–17
Sensory nerves, *76*, 76–81, *77*, 146–148, 154–155

Septic arthritis, 147
Septicemia as co-morbidity, 105
Serrano Sánchez, C., 274
SES. *See* Socioeconomic status
Seven-banded armadillos, 59
Sex and frequency rates, 47–50
Sexual intercourse and transmission, 36, 114, 291, 305
Seychelles Islands, 262
Shepard, C. C., 299
Shetland Islands, 40, 212–213
Shoes, 100–103, *101;* canvas, 102–103
Signs and symptoms of leprosy, 91–93
Sigtuna, Sweden, 245
Silk Road and transmission, 30, 286
Silla, E., 4–5
Silva, F. J., 22
Sin, association with, 36
Sinus involvement, 170–171; sinusitis as co-morbidity, 105
Six-banded armadillo, 60
Skamene, E., 42
Skeletal effects of leprosy, 17–18
Skeletal remains: children, 193–195, 203; diagnosis of, 176–178
Skin: skin diseases, 69, 109–110; skin lesions, 67, 68–72, *70, 71,* 109–110; skin-to-skin contact and transmission, 34
Skinsnes, O. K., 35
Šlaus, M., 215–216
Slave trade and transmission, *29,* 29–30, 269, 275–276, 281–282
Smith, Grafton Elliot, 256
SNPs (single nucleotide polymorphisms), 26–27, 285–286; analysis of rare strains, 28–30
Socioeconomic status, 33, 36, 49, 54–55. *See also* Poverty
Soft tissue changes, 187; soft tissue damage, 175; soft tissue ulcers, 131–132
South Africa, bioarchaeological evidence, 257–259
South America, bioarchaeological evidence, 276
Southeast Asia, bioarchaeological evidence, 266–269
Spain, bioarchaeological evidence, 254–255
Spejlsby, Denmark, 218
Spermaceti, as treatment, 120
Sphagnum vegetation, 238

Spickett, S. G., 40
Spiers, M., 250
Spigelman, M., 180, 232, 233, 242, 250
Spina bifida, 155
Spinalonga, Crete, Greece, 248
Springs, healing, 117
Squirrels, 16, 58, 286, 287–288, 306
Sreevatasan, H., 34–35
St. Albani parish cemetery, 223–224
Stefansky, V. K., 57
Stein, C., 289
Stewart, G., 290
St. George, Turku, Finland, 225
St. Giles, Brompton Bridge, North Yorkshire, England, 210, *212*
Stigmatization of leprosy, 1, 4–8, 7–8, 9, 89, 93
St. James and St. Mary Magdalene, Chichester, Sussex, England, 206–207, 222
St. John, England, 203
St. Jørgen, Naestved, Denmark, 218
St. Jørgensgård, Odense, Denmark, 220–222, 224
St. Jørgens Leprosy Hospital, Bergen, Norway, 125, 235–237, *236*
St. Knud Cathedral, 223–224
St. Laurentius Church, Oslo, Norway, 241
St. Leonard's, Grantham, England, 210
St. Linhartus, Prague, Czech Republic, 216
St. Mary Magdalen leprosarium, Winchester, England, 207–210
St. Mary Magdalene cemetery, Chichester, England, *301*
St. Mary's Church, Oslo, Norway, 241
Stojanowski, C., 189, 300
Storrs, E. E., 59
St. Peter's Church, Tønsberg, Norway, 241
Stress as predisposing factor, 54–55, 223
St. Thomas d'Aizier Hospital, France, 247
Stuart-Macadam, P., 170
Stubbekøbing, Denmark, 222
Sturdivant, M., 274
Subclinical leprosy, 31, 33, 65, 89, 180, 183
Subramaniam, K., 135
Suicide, 106
Sulfurous water, as treatment, 117
Suraż, Poland, 242
Surface water pollution, 39
Surgery, 118–120
Sushsruta Samhita, 121

Suttles, W., 271
Suzuki, K., 65, 261, 289
Suzuki, T., 260, 268
Svenborg, Denmark, 218, 222
Sweating, 85
Sweden, bioarchaeological evidence, 243–246
Syphilis, 114, 120, 144, 155, 184
Szalai, F., 232
Szarvas-Grexa, Hungary, 230, *231*
Székesfehérvár-Basilica, Hungary, 233

Takigawa, W., 261, 289
TAP amino acid chains, 43
Tapeworm infections, 218
Tappeh Sang-e Chakhmaq, Iran, 264
Tarsal disintegration, 152–153
Taylor, G. M., 264, 289, 290
Temperature and bacilli growth, 132–133; *See also* Lower temperature areas of body
Temperature changes (climatic), 293–294
Teschler-Nicola, M., 289
Thailand, grave goods, 267
Thalidomide, 86
The Thankful Leper (Hinkle), 6
Theodoric of Cervia, 110
Theory of hereditary transmission, 40, 258–259
Thinning of bones, *81*
Third epidemiological transition, 2, 283, 309
Thomas, M. O., 277, 299
Ticks, role in transmission, 34
Tió-Coma, M., 38
Tirup, Jutland, 222, 224
Tissue necrosis, 79–80
TLRs (toll-like receptors), 43
Tracadie leprosarium, Canada, 272–273
Transmission, 15–16, 237, 304–305; and amniotic fluid, 36, 47; amoebae, 39; animal to human, 59–62; and atmospheric conditions, 240, *246*; in breastmilk, 35, 47; and diet, 52–54; droplets, 32–33; environmental factors, 36–40; environment-to-person, 27–28; gastrointestinal route, 35; GIS (geographical information systems), 26; hereditary theory, 40, 258–259; and immune strength, 32; insects, 34–35; lepromatous leprosy contrasted with tuberculoid leprosy, 31; and living conditions, 36–37, 39–40; and migration, 214–215; and mobility, 55–56; person-to-person, 27–28; and placenta, 36, 47; rates, 13; regional patterns, 28–30, *29*; respiratory role, 32; skin-to-skin contact, 34; and socioeconomic status, poverty and crowding, 33, 54–55; transmission routes, 290; in utero, 36
Trauma, resulting from anesthesia, 77–79, *78*, 173–175
Travel. *See* Mobility
Travel restrictions, 89
Treatment, 95–96; barriers to treatment, 99; chaulmoogra oil, 121–123; decentralization of, 100; herbal, 120–121; holistic, 96; during medieval period, 117, 120–121; noncompliance with treatment, 99–100; plaster casts, 151–152; treatment goals, 95. *See also* Multidrug therapy
Trembly, D., 268
Treponemal disease. *See* Syphilis
Tropical disease, leprosy as, 2
Truman, R. W., 285
Tsar'ath, 123, 249
Tuberculoid leprosy, 1, 22–23, 296–297; age of onset, 45–46; alveolar bone changes, 135; anesthesia, 147; asymmetrical nerve damage, 75; and bone changes, 128; classification, 66–67; contrasted with lepromatous leprosy, 190; cure rates, 99; diagnosis, 93; diagnosis based on skeletal remains, 177–178; drug treatment regimens, 67–68, 96; and immunology, 43–44; incubation period, 65; MLFT, 94; multidrug treatment, 98; rhinomaxillary syndrome and bone change, 143; and skin lesions, 69; transmission, 31
Tuberculosis, 65, 89, 295–297, *298*; and BCG vaccine, 103–104; biomolecular analysis contrasted with leprosy, *208, 209*; as co-infection with leprosy, 297; as co-morbidity, 105; contrasted with leprosy, 2, 154, 183, 299–301; and cross-immunity with leprosy, 240, 295–297, 300–301; dactylitis in, 148
Tucker, K., 141, 142, 143
Tumor necrosis factor, 43
Turankar, R. P., 37–38
Turkey, bioarchaeological evidence, 265–266
Turnbull, R. N., 269
Type 1 reactions, 85, 87
Type 2 reactions, 83–84, 85, 87
Type 3 reactions, 263–264

Type 4 reactions, 269
Types: *M. leprae* strains, 285, 287, 288–290, *289*, *290*, *291*

Udono, T., 65
Ukraine, bioarchaeological evidence, 246
Ulcers: and activity patterns, 151–152; corneal, 140; foot, 79–80, 150–151, 155–156; heel, 151; soft tissue, 131–132
Ulnar nerve paralysis, 158
Undernutrition, 54–55
United Nations Millennium Development Goals, 21
United States, bioarchaeological evidence, 273–274
Upgrading of disease process, 68
Urban/rural contrast, 10, 51, 206, 224, 283, 293, 297–300
Urethritis, leprous, 36
Ursura orbitae, 169–170
Uzbekistan, bioarchaeological evidence, 263–265

Vaccines, 98, 103–104, 292, 297–299
Vaison-la-Romaine, France, 247
Valle da Gafaria, Portugal, 253
Van Hunnius, T., 184, 185
Västerhus, Sweden, 222
Vendeuil-Caply, Oise, France, 247
Venereal disease. *See* Syphilis
Vidal, L., 248, 257
Viipuri, Finland, 226
Vincenne-Campochiaro, Italy, 252
Vision, 83–85
Vitamin A deficiency, 53
Vitamin D deficiency, 51–52
Vitiligo and leprosy misdiagnosis, 69, 110
VNTRs (variable number tandem repeats), 26–27
Voice changes, 91, 172
Volcanic mud baths, 117
Voluntary reporting, 48
Voss-Böhme, A., 296, 300

Waaler, E., 141–142
Wachholz, L., 242
Wasterlain, S. N., 253
Wasting, muscle, 81
Watson, C. L., 203, 215, 289

Weaver, D. S., 269
Webb, S., 269
Weiss, D. L., 218
Welhaven, Pastor of Bergen, 113
Welsh Leech Book, 120–121
Weston, D. A., 166–167
Weymouth, A., 4
Wheat, W. H., 39
White, C., 7, 86
White, J., 296
Wilbur, A. K., 300
Wilczak, Cynthia, 271–272
Wilkinson, R., 55
Wilks, L. A., 132
Williams, D. L., 54
Willoughby-on-the-Wolds, Nottinghamshire, England, 203
Wilson, J. V. K., 282
Women and noncompliance with treatment, 99–100
Wood, J. W., 187, 188, 189, 223
Wood, S. R., 240–241
Woolgar, C. M., 176
World Health Assembly, 11, 21, 96; resolution on neglected tropical diseases, 2
World Health Organization, 9; and antibiotic treatment, 67–68; and BCG vaccine, 103; and classification, 66–67, *67*, *68*; on integration of leprosy service into health services, 95; plan for elimination of leprosy, 13, 20; treatment protocol, 97–98; on use of word "leper," 6–7
Wrist trauma, 148–150
Wroclaw-Olbin, Poland, 242

Yagodin, V., 263
Yawalkar, S. J., 89
Yi Men Fa Lu, 121, 123
Young, D. B., 25
Yu Chang, 121, 123
Yuigahama-minami, Kamakura, Japan, 260

Zaio, P., 251, 252
Zhenbiao, Z., 259
Zias, J., 249, 250
Zinc deficiency, 53
Zink, A., 228
Zwölfaxing, Austria, 201

CHARLOTTE A. ROBERTS, professor in the Department of Archaeology at Durham University, is a fellow of the Wolfson Research Institute for Health and Wellbeing.

BIOARCHAEOLOGICAL INTERPRETATIONS OF THE HUMAN PAST: LOCAL, REGIONAL, AND GLOBAL PERSPECTIVES

Edited by Clark Spencer Larsen

Ancient Health: Skeletal Indicators of Agricultural and Economic Intensification, edited by Mark Nathan Cohen and Gillian M. M. Crane-Kramer (2007; first paperback edition, 2012)

Bioarchaeology and Identity in the Americas, edited by Kelly J. Knudson and Christopher M. Stojanowski (2009; first paperback edition, 2010)

Island Shores, Distant Pasts: Archaeological and Biological Approaches to the Pre-Columbian Settlement of the Caribbean, edited by Scott M. Fitzpatrick and Ann H. Ross (2010; first paperback edition, 2017)

The Bioarchaeology of the Human Head: Decapitation, Decoration, and Deformation, edited by Michelle Bonogofsky (2011; first paperback edition, 2015)

Bioarchaeology and Climate Change: A View from South Asian Prehistory, by Gwen Robbins Schug (2011; first paperback edition, 2017)

Violence, Ritual, and the Wari Empire: A Social Bioarchaeology of Imperialism in the Ancient Andes, by Tiffiny A. Tung (2012; first paperback edition, 2013)

The Bioarchaeology of Individuals, edited by Ann L. W. Stodder and Ann M. Palkovich (2012; first paperback edition, 2014)

The Bioarchaeology of Violence, edited by Debra L. Martin, Ryan P. Harrod, and Ventura R. Pérez (2012; first paperback edition, 2013)

Bioarchaeology and Behavior: The People of the Ancient Near East, edited by Megan A. Perry (2012; first paperback edition, 2018)

Paleopathology at the Origins of Agriculture, edited by Mark Nathan Cohen and George J. Armelagos (2013)

Bioarchaeology of East Asia: Movement, Contact, Health, edited by Kate Pechenkina and Marc Oxenham (2013)

Mission Cemeteries, Mission Peoples: Historical and Evolutionary Dimensions of Intracemetery Bioarchaeology in Spanish Florida, by Christopher M. Stojanowski (2013)

Tracing Childhood: Bioarchaeological Investigations of Early Lives in Antiquity, edited by Jennifer L. Thompson, Marta P. Alfonso-Durruty, and John J. Crandall (2014)

The Bioarchaeology of Classical Kamarina: Life and Death in Greek Sicily, by Carrie L. Sulosky Weaver (2015)

Victims of Ireland's Great Famine: The Bioarchaeology of Mass Burials at Kilkenny Union Workhouse, by Jonny Geber (2015; first paperback edition, 2018)

Colonized Bodies, Worlds Transformed: Toward a Global Bioarchaeology of Contact and Colonialism, edited by Melissa S. Murphy and Haagen D. Klaus (2017)

Bones of Complexity: Bioarchaeological Case Studies of Social Organization and Skeletal Biology, edited by Haagen D. Klaus, Amanda R. Harvey, and Mark N. Cohen (2017)

A World View of Bioculturally Modified Teeth, edited by Scott E. Burnett and Joel D. Irish (2017)

Children and Childhood in Bioarchaeology, edited by Patrick Beauchesne and Sabrina C. Agarwal (2018)

Bioarchaeology of Pre-Columbian Mesoamerica: An Interdisciplinary Approach, edited by Cathy Willermet and Andrea Cucina (2018)

Massacres: Bioarchaeology and Forensic Anthropology Approaches, edited by Cheryl P. Anderson and Debra L. Martin (2019)

Mortuary and Bioarchaeological Perspectives on Bronze Age Arabia, edited by Kimberly D. Williams and Lesley A. Gregoricka (2019)

Bioarchaeology of Frontiers and Borderlands, edited by Cristina I. Tica and Debra L. Martin (2019)

The Odd, the Unusual, and the Strange: Bioarchaeological Explorations of Atypical Burials, edited by Tracy K. Betsinger, Amy B. Scott, and Anastasia Tsaliki (2020)

Bioarchaeology and Identity Revisited, edited by Kelly J. Knudson and Christopher M. Stojanowski (2020)

Leprosy: Past and Present, by Charlotte A. Roberts (2020)

CPSIA information can be obtained
at www.ICGtesting.com
Printed in the USA
BVHW040913270221
600884BV00006B/92